EXTREME WEATHER
a guide & record book

CHRISTOPHER C. BURT

with cartography by Mark Stroud

W. W. NORTON & COMPANY
New York, New York

ROCKY RIVER PUBLIC LIBRARY
1600 Hampton Road • Rocky River, OH 44116

To my father, Nathaniel Burt; for my mother, Margaret Burt; and for my wife Jeernen, for putting up with a weather fanatic all of these years; and also to the late David M. Ludlum, who nurtured my early interest in weather's countless manifestations.

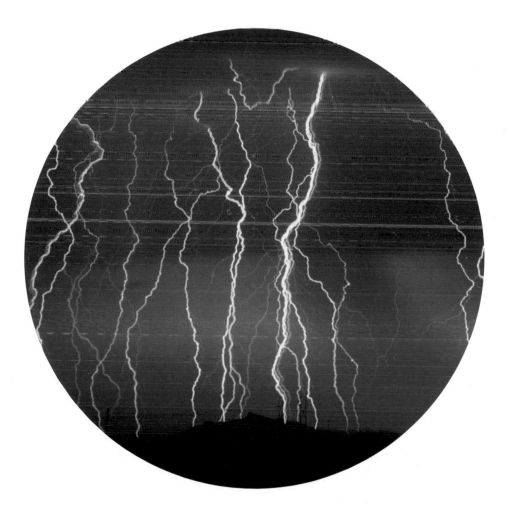

CONTENTS

extreme weather

MAPS, TABLES & GRAPHICS

extreme weather

INTRODUCTION

In the United States, weather records have been maintained by the official weather services since about 1870. In the 50 preceding years, records were kept intermittently by individuals and by some institutions, including the Smithsonian Institution. The figures accumulated in these records represent only a fraction of human experience with weather. Nevertheless, the weather records we do have are the yardstick used in determining climatic trends and in measuring the relative severity of weather events.

Many people wonder if the earth's climate has become more extreme in recent years. People ask themselves: "Was last summer really the hottest ever in the West? Were the floods in the spring unprecedented?" The answer to these kinds of questions would be an unqualified "no." After all, at one time a glacier sat on Chicago's doorstep; at another time palm trees grew in northern Canada. A more relevant question to ask ourselves is whether recent extreme weather was the worst in *recorded* history. It's the recorded history of the past century or two that provides a benchmark against which we measure what happens today.

Scientists, of course, have many ways of estimating the earth's temperature and climate in past millennia, including dendrology and the study of ice cores and ocean sediments. What they can't do is provide details of specific extreme weather events in eons past.

This guide contains a compilation of weather statistics from more than 300 weather stations across the United States that represent areas in which more than 90% of the U.S. population resides. These figures reflect both official figures, accepted by the U.S. Weather Bureau going back as far as 1870, and records from sources that predate the Weather Bureau, including Smithsonian Institution observations from as early as the 1850s.

In most cities the weather instruments have been moved over time from one location to another, and most weather statistics currently available are readings from the latest location. In this book, however, I have included statistics gathered from any official site used in a city at one time or another. This more fairly represents what the extreme records are for a community and provides a more complete historical record.

Weather is confined to the earth's troposphere, the five-mile thick shell which encases the globe. Occasionally, severe thunderstorms will project their tops into the stratosphere some 60,000 feet or more into the atmosphere. This photograph, taken during a space shuttle mission, shows giant thunderstorms forming over northern Australia. (Johnson Space Center)

Galileo invented the thermometer in 1593. Gabriel Fahrenheit introduced his temperature scale in 1714, and, in 1742, Anders Celsius presented his.

Has Weather Become More Extreme?

Has the weather really become more extreme over the period of official record? Since we now follow the weather ever more assiduously, we all know immediately when a weather record has been broken. The Weather Channel television station broadcasts new record temperatures almost daily. Are records being broken at an exponentially greater rate, indicating drastic climate change in the U.S. over the past hundred years?

Record-breaking events are by definition anomalies but nonetheless may help indicate whether or not the climate is becoming more extreme. In fact, records indicate that there has never been a U.S. heat wave to rival the July 1936 scorcher, when temperatures peaked at 121° in North Dakota, nor a cold wave as widespread and extreme as that of February 1899, when the temperature bottomed out at -2° in northern Florida. Climatologist Max Beran has concluded that

> *T*wo fundamental points are emphasized [about breaking weather records]. The first is that records will always be broken. The second is that the occurrence of a record breaker is not of itself an indication of any change in causative mechanisms. In other words, unprecedented events do not presage extraordinary explanations—an admonition that could usefully be pasted above the desk of every headline writer.
>
> —*Weather* (the bulletin of Britain's Royal Meteorological Society), August 2002

Recent confirmation of a global warming trend over the past 25 years has highlighted the significance of trends in extreme weather. Some short-term climate forecasting scenarios suggest dramatic changes within the lifetime of a single generation. However, in spite of global warming, extreme weather so far does not appear to be occurring more or less frequently than it has in the past. Tropical-storm activity waxes and wanes from one decade to the next. Tornadoes are not becoming more frequent or severe. Droughts and floods continue to occur in one place or another somewhere in the world every year. Extreme weather events are elusive phenomena, and monitoring them accurately is made difficult by numerous factors. Nevertheless, for climate scientists the statistics are the only way to know if extreme weather is increasing or decreasing.

(opposite) A supercell thunderstorm descends upon Turkey, Texas, unleashing hailstones the size of softballs. In this case, the orange-yellow area indicates the region of intense hail-fall. (Alan Moller)

A GUIDEBOOK?

This book collates extreme climate data in a concise and accessible form, and proposes to act as a "guidebook" as well; a guide to inform you about the places in the United States and around the world that experience the most extreme weather.

Do you fantasize experiencing a temperature of more than 120°? Probably not, but Death Valley is one of the three hottest places on earth, and the other two, the Sahara Desert and the Tigris-Euphrates River Valley, are a bit more of a challenge to visit these days. If you love snow, what better place to enjoy it than the Sierra Nevada of California, where ten feet of the white stuff are likely to have accumulated by March. If you're lucky, a big storm will increase the snow depth by feet rather than inches.

For aficionados of severe storms there are, of course, storm-chaser companies that can offer you a whirlwind tour of the High Plains, where you'll have the best chance in the world of seeing a tornado. For lightning displays, spend an evening on Tucson's Mt. Lemon during the month of July, when monsoon thunderstorms build over the desert landscape and provide a daily electrical display that can hardly be matched.

In short, there are few thrills as exciting as weather at its worst. And no show is ever the same.

chapter 1
HEAT & DROUGHT

Judging by the explosive population growth of Phoenix and Las Vegas, the two hottest big cities in the United States, one might well conclude that Americans are not fazed by extreme heat (especially if they have air-conditioning). Residents, of course, will tell you that the humidity stays low, making the heat less burdensome than it is during the humid summers common east of the Rockies. A popular cartoon in Arizona depicts two skeletons in the desert under a cactus with one telling the other, "But it's a dry heat!" No matter what the humidity level, a temperature over 110° in the shade is uncomfortably hot, and both Phoenix and Las Vegas regularly record such temperatures from June through August.

HOTTEST PLACES IN THE UNITED STATES

The usual method of comparing places in terms of warmth is to measure their average annual temperatures. A second method compares the average temperatures of the hottest month (usually July). A third method compares not the average temperature of the hottest month, but the average *maximum* temperature of the hottest month.

15 Hottest Major Cities in U.S.
Average Annual Temperatures
(period of record 1970–2000)

Location	Temperature
Key West, FL	78.0°
Honolulu, HI	77.5°
Miami, FL	76.6°
Fort Lauderdale, FL	75.7°
West Palm Beach, FL	75.3°
Fort Myers, FL	74.9°
Yuma, AZ	74.6°
Hilo, HI	74.1°
St Petersburg, FL	74.1°
Brownsville, TX	74.0°
Phoenix, AZ	73.9°
Palm Springs, CA	73.8°
Laredo, TX	73.7°
Orlando, FL	72.7°
Corpus Christi, TX	72.1°

source: NCDC

15 Hottest Major Cities in U.S.
Average July Maximum Temperatures
(period of record 1970–2000)

Location	Temperature
Palm Springs, CA	108.3°
Yuma, AZ	107.0°
Phoenix, AZ	106.0°
Las Vegas, NV	104.1°
Tucson, AZ	101.0°
Presidio, TX	100.6°
Laredo, TX	100.5°
Redding, CA	99.5°
Bakersfield, CA	98.2°
Fresno, CA	98.1°
Wichita Falls, TX	97.6°
Waco, TX	96.7°
Dallas-Ft Worth, TX	96.3°
Del Rio, TX	96.2°
El Paso, TX	95.5°

source: NCDC

Hottest Summertime Regions in the United States

1	Lower Colorado River Valley of Arizona, California, and Nevada
2	Rio Grande River Valley of Texas and southeastern region of Texas
3	Central valleys of California
4	North-central Texas and southern Oklahoma
5	Snake River Valley from Idaho Falls to Ontario, Idaho
6	Southeastern New Mexico
7	Lower inland region of the Deep South from Mississippi to South Carolina
8	Lower Mississippi River Valley from Southern Missouri to Louisiana
9	Southeast Colorado to south-central Nebraska, west-central Kansas
10	Lower basins of Wyoming

The desert Southwest is Ameria's hottest region in spite of frequent summer afternoon rainstorms. (Jake Dykinga)

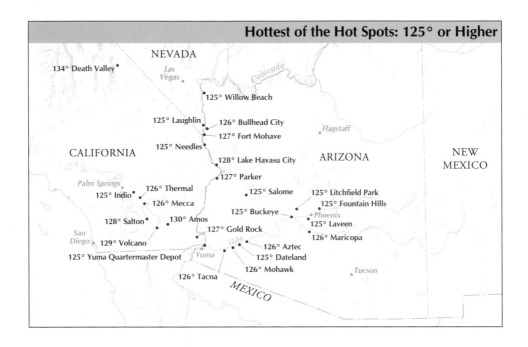

Hottest of the Hot Spots: 125° or Higher

NEVADA

134° Death Valley

Las Vegas

Colorado

125° Willow Beach

125° Laughlin · 126° Bullhead City

127° Fort Mohave · Flagstaff

CALIFORNIA · 125° Needles

128° Lake Havasu City · ARIZONA · NEW MEXICO

127° Parker

Palm Springs · 126° Thermal · 125° Salome · 125° Litchfield Park

125° Indio · 125° Fountain Hills

126° Mecca · 125° Buckeye · Phoenix

128° Salton · 130° Amos · 125° Laveen

San Diego · 127° Gold Rock · 126° Maricopa

129° Volcano · Gila

125° Yuma Quartermaster Depot · Yuma · 126° Aztec

125° Dateland

126° Tacna · 126° Mohawk · Tucson

MEXICO

Compared by the first method, average annual temperature, the cities of Florida and Hawaii rank in the top positions, with Key West topping the list with an annual average temperature of 78.1°. However, anyone who has spent time in Key West knows that it is a relatively comfortable place to live with little in the way of extreme temperatures. In fact, the temperature has reached 100° only once, and that was almost 120 years ago, in August of 1886.

The second method, average temperature in the warmest month, shows just how hot the summers usually are at any given location and is more useful in determining what are really the hottest cities in the country. The daily July maximum is even more relevant since people are actually awake to experience the heat during the daylight hours. So for the short analysis following, I will use daily July maximum temperatures to determine which of America's cities are the hottest.

Although Phoenix and Las Vegas both reach 104° on a typical July afternoon, making them the hottest major cities in the U.S., smaller towns suffer even more. Most notorious of all is Yuma, Arizona, which tops out at 107° most July days and peaked at an incredible 124° during the torrid July of 1995. Worse still, is the booming town of Lake Havasu City, Arizona, hottest inhabited place in America, where on an ordinary July day the temperature peaks at about 111°. It is safe to say that from Laughlin, Nevada (located just across the Colorado River from Lake Havasu City), to Yuma 150 miles south, comparable temperatures are attained on

July afternoons. Those regions of Arizona, Nevada, and California that border the Colorado River and the depression of Death Valley are unquestionably the hottest areas of the United States and among the hottest regions in the world.

Only a handful of other places in the United States register average July daytime temperatures over 100°. The region surrounding the Salton Sea depression of southern California is the second hottest area, with July afternoon temperatures averaging 108.6° at Mecca.

Palm Springs in the Mohave Desert registers 108.2° most July days. The next hottest city of any size in the United States is St. George in southwestern Utah, with 102.8° registered on most July afternoons. It is followed by Presidio, Texas, with 102.1°, although a place called Castolon, which is hard to find on any map of Texas, is the state's hottest spot, recording a blistering June-afternoon average temperature of 103.4°(June is usually the hottest month in western Texas). *For the hottest towns in each state, see the table on pp. 20–21.*

The hottest of the hot places, not just in the United States, but perhaps in the world, is Death Valley, the unusual depression lying 282 feet below sea level in interior California.

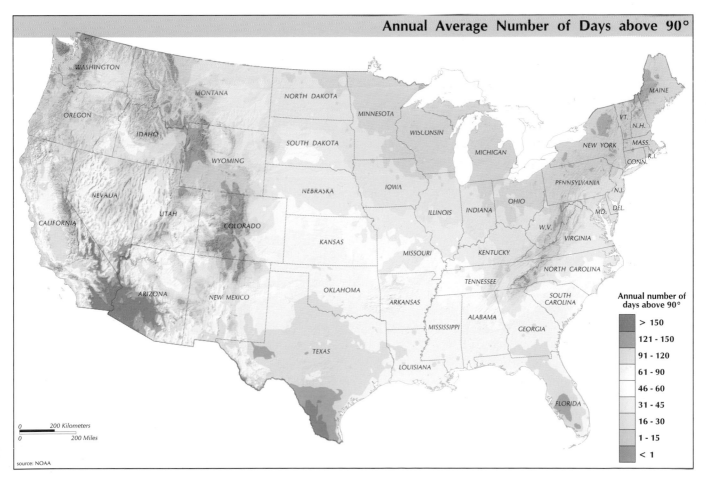

Annual Average Number of Days above 90°

Annual number of days above 90°

Color	Range
	> 150
	121 - 150
	91 - 120
	61 - 90
	46 - 60
	31 - 45
	16 - 30
	1 - 15
	< 1

0   200 Kilometers
0   200 Miles

source: NOAA

DEATH VALLEY: HOTTEST PLACE IN THE WORLD?

Temperatures in Death Valley, located 282 feet below sea level in interior California, have been maintained since 1911 at the Greenland Ranch near Furnace Creek. With an average daily high of 115° and low of 87° during the month of July, Death Valley is far and away the hottest location in North America and perhaps the hottest place in the world.

The Valley's absolute maximum temperature of 134°, recorded on July 10, 1913, stands as the hottest ever observed in the Western Hemisphere and has been surpassed globally only by a reading of 136° (57.8°C) measured in Al Aziziyah, Libya, located 20 miles south of Tripoli (not in the Sahara Desert, by the way). A 135° reading claimed by Tindouf, Algeria is of questionable veracity. The Greenland Ranch figure of 134° has been the center of a small controversy itself because there is no documentation of the accuracy of the thermometer and condition of its shelter, and no other official reading has ever since come close to this reading. A sandstorm was raging at the time of the observation, and some speculate hot sand or dust was driven into the thermometer casing, inflating the actual temperature.

A 130° temperature recorded at Amos (Mammoth Tank) in the Mohave Desert in 1887 is also suspect for the same reasons. So, Death Valley's second hottest readings of 129° recorded in July 1960, and July 1998 may, in fact, be the highest true maximum temperatures ever recorded in the United States. As weather historian David Ludlum once put it, "Apparently, what this country needs is a good 135° reading made under standard conditions, so that the figure, like Caesar's wife, may be beyond question."

The hottest summer of record in the Valley was that of 1917, when 43 consecutive days above 120° were recorded between July 6 and August 17. The average temperature for the entire month of July that year was 107.2°, just shy of yet another questionable national record of 107.4° recorded at Salton, California, in August 1897.

In July of 2002, Death Valley averaged 106.0°, its hottest month in modern records. On July 13 of 2002, the temperature ranged from a low of 100° to a high of 127°, a daily average of 113.5° and perhaps the hottest day (average temperature) ever recorded anywhere in the world.

Overnight lows above 100° seem to be unique to Death Valley (although the airport at Muscat, Oman, registered a low of 100° on the night of July 30, 1989). On the night of July 31, 2003, the temperature failed to drop below 104°.

The longest stretch of consecutive days with a maximum of 100° degrees or longer was 154 days in 2001. This compares favorably to the world record of such days recorded at Marble Bar, West Australia, with 161 straight days registering highs of 100° or more.

Death Valley is the site of an unsolved weather mystery. Large rocks, some weighing as much as 700 pounds, have evidently slid across the desert playa (a lake that forms and evaporates frequently) leaving trails an inch or so deep in their wake. Strong winds and the playa's slippery surface are not enough to account for their travels. One possible explanation is that water may freeze just below the surface, forming a sheet of barely submerged ice across which the rocks glide. (Kerrick James)

WARMEST LOCATION BY STATE

State	Temp	Location
ALABAMA	93.8°	Cordova
ALASKA	73.9°	Central
ARIZONA	112.5°	Willow Beach
ARKANSAS	94.7°	Blue Mountain Dam
CALIFORNIA	114.9°	Death Valley
COLORADO	94.9°	Uravan
CONNECTICUT	85.4°	Stamford
DELAWARE	87.6°	Newark
FLORIDA	93.7°	Archbold Bio Station
GEORGIA	94.7°	Waycross
HAWAII	88.9°	Honolulu (Airport)
IDAHO	94.5°	Swan Falls
ILLINOIS	91.7°	Kankakee
INDIANA	90.5°	Evansville (Museum)
IOWA	88.3°	Keosauqua
KANSAS	95.1°	Aetna
KENTUCKY	93.6°	Gilbertsville Dam
LOUISIANA	94.5°	Calhoun Research Station
MAINE	82.5°	Sanford
MARYLAND	90.6°	Baltimore
MASSACHUSETTS	85.5°	Chester
MICHIGAN	85.7°	Dearborn
MINNESOTA	85.8°	Chaska
MISSISSIPPI	92.9°	Meridian Key
MISSOURI	92.8°	Kennett
MONTANA	91.5°	Brandenberg
NEBRASKA	93.3°	Beaver City
NEVADA	108.4°	Laughlin
NEW HAMPSHIRE	83.2°	Durham
NEW JERSEY	88.1°	Woodstown
NEW MEXICO	97.7°	Waste Isolation Plant
NEW YORK	86.0°	Scarsdale
NORTH CAROLINA	92.6°	Whiteville 7 NW
NORTH DAKOTA	86.2°	Medora
OHIO	88.1°	Fairfield
OKLAHOMA	98.6°	Chattanooga
OREGON	95.4°	Pelton Dam
PENNSYLVANIA	87.5°	Marcus Hook
RHODE ISLAND	82.6°	Providence
SOUTH CAROLINA	95.2°	Columbia (Univ. of SC)
SOUTH DAKOTA	91.9°	Fort Pierre
TENNESSEE	92.1°	Memphis
TEXAS	103.4°	Castolon
UTAH	102.8°	St. George
VERMONT	83.8°	Vernon
VIRGINIA	91.1°	Stony Creek
WASHINGTON	91.0°	Smyrna
WEST VIRGINIA	89.4°	Williamson
WISCONSIN	85.6°	Lake Geneva
WYOMING	93.1°	Lingle

source: NOAA

AVERAGE DAILY MAXIMUM JULY TEMPERATURE

NORTH DAKOTA
▲ 86.2°

91.5°

MINNESOTA

SOUTH DAKOTA
▲ 91.9°

WISCONSIN

85.8° ▲

MICHIGAN

MAINE

VT.
N.H. ▲ 82.5°
▲ 83.2°

83.8° ▲
85.5° ▲ MASS.

NEW YORK

82.6° ▲
R.I.

86.0° ▲ 85.4° ▲ CONN.

3.1° ▲

NEBRASKA

IOWA
▲ 88.3°

91.7° ▲

ILLINOIS

INDIANA

OHIO

PENNSYLVANIA

85.6° ▲

85.7° ▲

87.5° ▲

88.1° ▲
90.6° ▲
87.6° ▲

N.J.

DEL.

MD.

COLORADO

KANSAS

MISSOURI

93.3° ▲

▲ 95.1°

90.5° ▲

KENTUCKY

93.6° ▲

W.V.
89.4° ▲

VIRGINIA
91.1° ▲

92.8° ▲

TENNESSEE

NORTH CAROLINA
92.6° ▲

OKLAHOMA

98.6° ▲

94.7° ▲

ARKANSAS

92.1° ▲

95.2° ▲
SOUTH
CAROLINA

97.7° ▲

03.4° ▲

TEXAS

94.5° ▲

LOUISIANA

MISSISSIPPI

92.9° ▲

93.8° ▲
ALABAMA

GEORGIA

94.7° ▲

Lake Superior

Lake Michigan

Lake Huron

Lake Ontario

Lake Erie

Atlantic Ocean

Gulf   of   Mexico

FLORIDA
93.7° ▲

105°
100°
95°
90°
85°
80°
75°
70° F

HAWAII

Kaua'i

Niihau
O'ahu
Moloka'i
88.9° ▲ Lanai Maui

HAWAIIAN ISLANDS

Pacific Ocean

Hawaii

0   200 km
0   200 mi

▲ 85.3° - Warmest location in each state

0   200 Kilometers
0   200 Miles

Scale for all areas except, Alaska and Hawaii.
Scale in those areas vary.

HIGHEST RECORDED TEMPERATURE BY STATE

State	Temp	Location	Date
ALABAMA	112°	Centerville	9/5/1925
ALASKA	100°	Fort Yukon	6/27/1915
ARIZONA	128°	Lake Havasu City	6/29/1994
ARKANSAS	120°	Ozark	8/10/1936
CALIFORNIA	134°	Greenland Ranch	7/10/1913
COLORADO	118°	Bennett	7/11/1888
CONNECTICUT	106°	Danbury	7/15/1995
DELAWARE	110°	Millsboro	7/21/1930
FLORIDA	109°	Monticello	6/29/1931
GEORGIA	112°	Greenville	8/20/1983
HAWAII	100°	Pahala	4/27/1931
IDAHO	118°	Orofino	7/28/1934
ILLINOIS	117°	East St. Louis	7/14/1954
INDIANA	116°	Collegeville	7/14/1936
IOWA	118°	Keokuk	7/20/1934
KANSAS	121°	Alton	7/24/1936
KENTUCKY	114°	Greensburg	7/28/1930
LOUISIANA	114°	Plain Dealing	8/10/1936
MAINE	105°	North Bridgeton	7/10/1911
MARYLAND	109°	Cumberland	7/10/1936
MASSACHUSETTS	107°	New Bedford	8/2/1975
MICHIGAN	112°	Mio	7/13/1936
MINNESOTA	114°	Moorhead	7/6/1936
MISSISSIPPI	115°	Holly Springs	7/29/1930
MISSOURI	118°	Warsaw	7/14/1954
MONTANA	117°	Medicine Lake	7/5/1937
NEBRASKA	118°	Minden	7/24/1936
NEVADA	125°	Laughlin	6/29/1994
NEW HAMPSHIRE	106°	Nashua	7/4/1911
NEW JERSEY	110°	Runyon	7/10/1936
NEW MEXICO	122°	Waste ISO Plant	6/27/1994
NEW YORK	108°	Troy	7/22/1926
NORTH CAROLINA	110°	Fayetteville	8/21/1983
NORTH DAKOTA	121°	Steele	7/6/1936
OHIO	113°	Gallipolis	7/21/1934
OKLAHOMA	120°	Tipton	6/27/1994
OREGON	119°	Pendleton	8/10/1898
PENNSYLVANIA	111°	Phoenixville	7/10/1936
RHODE ISLAND	104°	Providence	8/2/1975
SOUTH CAROLINA	111°	Camden	6/28/1954
SOUTH DAKOTA	120°	Gannvalley	7/5/1936
TENNESSEE	113°	Perryville	8/9/1930
TEXAS	120°	Seymore	8/12/1936
UTAH	117°	Saint George	7/5/1985
VERMONT	105°	Vernon	7/4/1911
VIRGINIA	110°	Balcony Falls	7/15/1954
WASHINGTON	118°	Ice Harbor Dam	8/5/1961
WEST VIRGINIA	112°	Martinsburg	7/10/1936
WISCONSIN	114°	Wisconsin Dells	7/13/1936
WYOMING	115°	Basin	8/8/1983

ALASKA

source: contours estimated based on 500 city plots

ABSOLUTE MAXIMUM TEMPERATURE ON RECORD

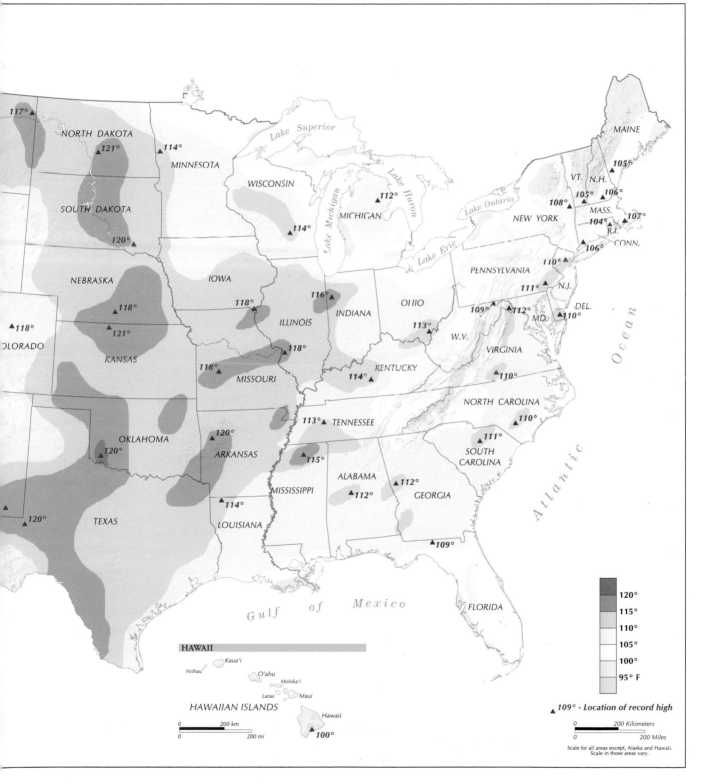

Scale	
	120°
	115°
	110°
	105°
	100°
	95° F

▲ 109° - Location of record high

0   200 Kilometers
0   200 Miles

Scale for all areas except, Alaska and Hawaii.
Scale in those areas vary.

HAWAII

Niihau · Kaua'i · O'ahu · Moloka'i · Lanai · Maui · Hawaii

HAWAIIAN ISLANDS

0   200 km
0   200 mi

100°

24 World's Hottest Places
Maximum Temperatures Recorded

Country	Temp	Location
AFRICA		
Libya	136°	Al Aziziyah*
hottest official temperature recorded in the world, Sep 13, 1922		
Algeria	135°	Tindouf
Tunisia	131°	Ben Gardene, Kebili
Mali	130°	Araouane, Timbuktu
Sudan	127°	Wadi Halfa
Egypt	124°	Aswan
Morocco	122°	Safi, Ouarzazate
Somalia	122°	Lugh Ferrandi
Mozambique	122°	Zumbo
Niger	122°	Agadez
Ethiopia	121°	Dallol
West Sahara	121°	Semara
Mauritania	121°	Bir Moghrein
Chad	121°	Faya
Senegal	120°	Matam
Burkina Faso	120°	Po
South Africa	118°	Komatipoort
Togo	118°	Lama-Kara
Zimbabwe	117°	Beitbridge
Nigeria	117°	Sokoto, Hadejia
Somalia	117°	Berbera
Madagascar	116°	Tranoroa
Kenya	115°	Garissa
Liberia	115°	Monrovia
Eritrea	115°	Massawa
Djibouti	115°	Djibouti
Guinea	115°	Labe
Cameroon	114°	Maroua
Gambia	113°	Banjul
Guinea-Bissau	113°	Bissau
Ivory Coast	113°	Bouake
Sierra Leone	111°	Kabala
Angola	111°	Huambo
Botswana	111°	Tshabong
Ghana	111°	Wa
Central African Rep.	109°	Birao
Ivory Coast	109°	Abidjan
Benin	109°	Ajatitingou
Canary Islands	109°	Santa Cruz
Zambia	109°	Mongu
Namibia	108°	Keetmanshoop
Rep. Of Congo	108°	Pointe-Noir
Gabon	104°	Port Gentil
Tanzania	103°	Moshi
Dem. Rep. of Congo	102°	Kinshasa
Cape Verde	100°	Santa Maria
Liberia	97°	Nowren

source: various and weatherbase.com

The Tidikelt Depression of Algeria may be the hottest place in the world, but because no official meteorological observations have ever been made here, this cannot be confirmed. (Frans Lemmens, Linair, Peter Arnold photo)

WORLD'S HOTTEST PLACES

The hottest places in the the world (aside from Death Valley) include oases in the Sahara Desert regions of Algeria and Mali. Bou-Bernous, Algeria, has an average daily temperature range in July of 87° to 116°, similar to Death Valley's, but has recorded an absolute maximum of only 123° compared to Death Valley's 134°. The annual rainfall at this remote station averages just .20" per year. Bou-Bernous's altitude of 1,509 feet indicates that temperatures are almost certainly hotter in the Tidikelt depression some 200 miles east and 1,000 feet lower than Bou-Bernous in the heart of the Sahara Desert. There are no permanent settlements or oases of note in this region.

The oasis of Araouane, 150 miles north of Timbuktu in Mali, has a July average temperature range of 80° to 116° and an extreme

Country	Temp	Location
NORTH AMERICA		
United States	134°	Death Valley, CA
Mexico	120°	Mexacali, BC
Canada	113°	Yellow Grass, SK
Cuba	109°	Santiago
Greenland	86°	Ivigtut
ASIA		
Israel	129°	Tirat Tsvi
Iran	128°	Ahwaz
Pakistan	127°	Jacobabdad
Iraq	126°	Ash Shu'aybah
United Arab Emirates	124°	Al Buraymi
Kuwait	124°	Kuwait City
Oman	124°	Al Ayn
India	123°	Alwar
Saudi Arabia	123°	Al Qaysumah
U.S.S.R. *former*	122°	Uch Adzhi, TKM
China	121°	Turfan
Syria	120°	Abu Kamal
Burma	120°	Pagan
Jordan	120°	Mahattat al Jufur
Afghanistan	119°	Jalalabad
Qatar	118°	Doha
Turkey	117°	Diyar Bakir
Indonesia	115°	Palembang, Sumatra
Nepal	115°	Bhairahawa
Bahrain	115°	Manama
Laos	113°	Luang Prabang
Vietnam	113°	(unknown)
Thailand	113°	Roi Et
Bangladesh	111°	Borga Thana
Yeman	110°	Al Qassasin
Malaysia	109°	Sibu, Sarawak
Lebanon	108°	Riyaq
Philippines	108°	Tuguegarao
Sri Lanka	108°	Batticaloa
Taiwan	108°	Paisha Tao
South Korea	108°	Chain
Japan	106°	Yamagata
Cambodia	106°	Batambang
Mongolia	104°	Buyant-Uhaa
Singapore	100°	Paya Lebar
Hong Kong	99°	Royal Observatory
OCEANIA and POLES		
Australia	128°	Cloncurry, QLD
New Zealand	108°	Rangiora
Antarctica	58.3°	Vanda Station
	7.5°	South Pole
North Pole	39°	North Pole

maximum of 130°. Absolute maximum temperatures of 120° plus have been recorded here every month of the year from March through October. This is also one of only ten weather stations in the world to record a temperature of 130° or higher. The others are Al Aziziyah in Libya with the world record temperature of 136°, Tindouf, Algeria, with 135°, In Salah, Algeria, with 133°, Gadames, Libya, with 131°, Ben Gardene and Kebili, Tunisia, with 131°, Death Valley's 134°, and Amos (Mammoth Tank) in California measuring 130°. Unofficial readings above 130° have been reported in Mexico and Iraq as well. An extreme maximum temperature of 136° was recorded in San Luis, Baja, Mexico, in 1933, and an incredible 140° was once reported from Delta, also in Baja. Both these readings have been discounted, however, because the thermometers used were improperly exposed. A press report in July 1989, indicated a temperature of 135° reported at Al-Amarha, Iraq. Even higher temperatures have occurred during heat bursts or heat flashes, when temperatures soared to incredible heights for a very brief period. *See sidebar pp. 36–37.*

World's Hottest Places
Maximum Temperatures Recorded

Country	Temp	Location
EUROPE		
Portugal	123°	Riodades (Alto Doura)
Spain	122°	Seville
Greece	118°	Elesis
Cyprus	116°	Nicosia
Italy	113.4°	Catania, Sicily
Romania	112°	Ion Sion
Yugoslavia *former*	112°	Skopje, Croatia
France	111°	Toulouse
Malta	111°	Valetta
Hungry	108°	Pecs
Albania	108°	Gjinokaster
Bulgaria	108°	Pleven, Ruse
Czech Republic	107.2°	Napajedla
Switzerland	106.7°	Grono
Germany	105.5°	Perl-Nenning
Belgium	104°	Sint-Joost-ten-Node
Holy See	104°	Vatican City
Poland	104°	Proszkow
Austria	103.5°	Dellach im Drautal
Netherlands	101.5°	Warnsveld
Great Britain	101.3°	Brogdale
Slovakia	101°	Bratislava
Sweden	100.4°	Ultuna & Malilla
Luxembourg	100.2°	Luxembourg
Denmark	98°	Holstedro
Finland	97°	Turku
Norway	96.1°	Nesbyen
Ireland	92°	Dublin, Kilkenny Castle
Iceland	87°	Teigarhorn
SOUTH AND CENTRAL AMERICA		
Argentina	120.4°	Villa de Maria
Brazil	113°	Cuiaba
Paraguay	112°	Mision Inglesia
Honduras	111°	San Pedro Sula
Venezuela	111°	Maturin
Columbia	111°	Valledupar
Uruguay	109°	Montevideo
Bolivia	109°	Yacuiba
Peru	108°	Pucallpa
Costa Rica	108°	Puntarenas
French Guiana	108°	St Laurent du Maroni
Belize	108°	Belize City
El Salvador	106°	Acajutla
Ecuador	105°	Trinidad
Chile	105°	Traiguen
Guatemala	104°	San Jose
Panama	104°	David
Nicaragua	104°	Puerto Cabezas

Another notorious world hot spot is Jacobabad, Pakistan, which suffers through an average July temperature range of 85° to 114°. Its absolute maximum of 127° is just two degrees lower than the hottest temperature ever measured in Asia. This was measured in Israel (the Middle East belongs to the continent of Asia) in June of 1942, when 129° was recorded in Tirat Tsvi, 770 feet below sea level on the shores of the Dead Sea.

The Tigris and Euphrates River Delta on the border of Iraq and Iran also reports phenomenal summertime heat, with Abadan, Iran, averaging a July high of 112° and a low of 82°, and an August high of 113° and low of 81°. The desert interior of Australia also experiences some of the hottest weather anywhere. Australia's hottest temperature of 128° may have occurred in Cloncury, a good-sized town in the western section of Queensland (this temperature is now considered questionable by the Australian Bureau of Meteorology). Marble Bar, on the edge of the Great Sandy Desert in Western Australia, once had a streak of 161 consecutive days above 100° (Yuma, Arizona's longest such streak is just 101 days and Death Valley's 154). In spite of these impressive extremes, Australia's hottest location, Marble Bar, averages a high of only 104° during its hottest months of December and January, not really in the same league as Death Valley or the world's other hot spots.

The location with the world's overall hottest annual average temperature is in the Danakil Depression of Ethiopia, where a prospecting company measured an annual average temperature of 94° at their location near Dallol between 1960 and 1966 (compare this to Key West's modest 78°, the warmest average annual temperature in the United States). On average, the temperature in the Danakil Depression exceeds 100° every day of the year except for January and February when it cools down to a pleasant 97°.

Below is the temperature table for Dallol, Ethiopia. Note the average nighttime low of 90° in July and September.

Dallol, Ethiopia

	Jan	Feb	Mar	Apr	May	Jun	Jul	Aug	Sep	Oct	Nov	Dec
Absolute Maximum	102°	108°	118°	115°	120°	119°	121°	119°	119°	115°	111°	106°
Average Maximum	97°	97°	102°	106°	111°	117°	115°	117°	109°	108°	102°	100°
Average Minimum	79°	79°	79°	81°	82°	88°	90°	88°	90°	86°	81°	81°
Absolute Minimum	72°	72°	70°	70°	73°	77°	75°	75°	81°	79°	75°	75°

source: World Climate Survey, Africa

REAL HEAT: THE HUMIDITY FACTOR

The temperature-humidity index, or THI, is an important way of measuring heat. It gauges the feeling moisture adds to heat by calculating humidity in conjunction with temperature, to produce the "apparent temperature." Using this index, coastal and southern Texas are the hottest summer locations in the United States. Del Rio, Texas, tops the list with an apparent July and August afternoon average temperature of 113°. Even Death Valley's real July average afternoon temperature of 115° would feel cooler than Del Rio's thanks to the dryness of its air. Corpus Christi ranks as the hottest major city, with an apparent July-August temperature of around 110°.

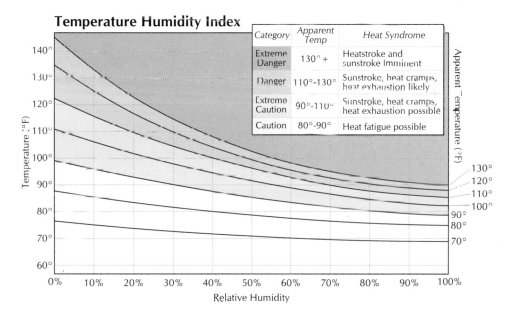

Temperature Humidity Index

Category	Apparent Temp	Heat Syndrome
Extreme Danger	130° +	Heatstroke and sunstroke Imminent
Danger	110°-130°	Sunstroke, heat cramps, heat exhaustion likely
Extreme Caution	90°-110°	Sunstroke, heat cramps, heat exhaustion possible
Caution	80°-90°	Heat fatigue possible

If hot nights are taken into consideration, Key West, Florida, ranks second to Corpus Christi with a day-long average apparent temperature of 100° during the summer months. The overnight-low, apparent summer temperature at Key West is a very uncomfortable 94°, even though the actual temperature drops down to 79°. Ocean breezes, however, tend to mitigate this effect.

During heat waves, incredible THI values have been reached. During the July 10 to 16, 1995, heat wave in the upper Midwest, THI values in excess of 120° were reported as far north as Green Bay, Wisconsin. Chicago's THI peaked at 119° causing 700 deaths. Waterloo, Iowa, recorded a 126° apparent temperature on July 13.

The heat wave moved east, and Philadelphia, not to be outdone by Waterloo, experienced an apparent temperature of 129° on July 15 (the actual temperature was 103°). On the afternoon of July 13, 2002, Winnemucca, Nevada—normally bone dry in the summer—experienced an apparent temperature of 130° when an approaching thunderstorm thrust a burst of moisture into the city, raising the humidity to 60% while the temperature stood at 104°.

WORST REAL HEAT IN THE WORLD

The worst heat-and-humidity combination worldwide occurs along the Red Sea coast of Ethiopia, the Gulf of Aden coast of Somalia, and in the Persian Gulf, where summer temperatures of 100° to 110° coincide with dew points as high as 85° to 90°, creating apparent temperatures of 135° to 145° on a daily basis.

Assab, Ethiopia, on the Red Sea coast, has an average afternoon dew point of 84° during June. The dew point is the temperature at which the atmosphere becomes completely saturated (100% humidity) when the air temperature (dry-bulb temperature) falls to that dew-point temperature. Dhahran, Saudi Arabia, on the Persian Gulf, reported a dew point of 95°, while the air temperature stood at 108° at 3 p.m. on July 8, 2003. The apparent temperature at that time would have been 172°. Earlier that day the airport in Dhahran reported zero visibility in fog with a temperature of 86° and 100% humidity. This is about as close to a natural steam bath as can be found anywhere in the world.

The Red Sea and the Persian Gulf have the hottest sea water to be found on the earth's surface, with summer sea temperatures averaging in the upper 80°s and once reaching as high as 96° in the Red Sea and 98° in the Persian Gulf. Consequently, the sea has no cooling affect on coastal communities and provides an ample source of moisture in the form of humidity. The Somalian town of Bender Qaasim (also known as Boosaso), has an average summer daily maximum (real) temperature of 105° combined with an average afternoon humidity of 61% (indicating a dew point of 83°). This creates an apparent temperature of 149° degrees. And that is just a typical day. The absolute maximum temperature at this station is 113° and every June and July it experiences a few days around 108°. The THI on these days would be literally off the chart somewhere in the 150–160° range. The absolute minimum temperature ever recorded in the month of July here is an astonishingly warm 82°, quite possibly the warmest monthly absolute minimum measured anywhere in the world.

Probably the hottest major city in the world is Bangkok, Thailand, where the temperature reaches 90° or more virtually every day of the year and humidity often raises the THI Index well above 120°. The months of April and May are its hottest, and the THI has hit 130° on several occasions.

HISTORIC HEAT WAVES IN THE UNITED STATES

Worst Heat Wave: Summer of 1936

Heat waves occur when strong upper atmospheric high pressure systems become locked over a region for an extended period of time.

The most intense and widespread heat wave ever to occur in the United States took place during the summer of 1936. In July and August of that year temperatures soared to levels that have not been matched since.

Fifteen state absolute-maximum temperature records still stand today as a result of this heat wave. Readings of 120° or higher were recorded in Arizona, Arkansas, California, Kansas, Oklahoma, South Dakota, Texas, and, incredibly, North Dakota, where the temperature peaked

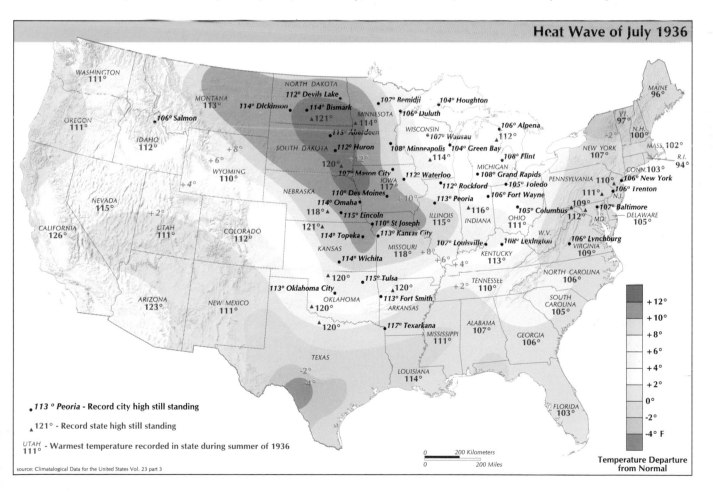

Heat Wave of July 1936

• 113° Peoria - Record city high still standing

▲ 121° - Record state high still standing

UTAH
111° - Warmest temperature recorded in state during summer of 1936

source: Climatological Data for the United States Vol. 23 part 3

0   200 Kilometers
0   200 Miles

Temperature Departure from Normal

+12°
+10°
+8°
+6°
+4°
+2°
0°
-2°
-4° F

at 121°. Even Duluth, Minnesota, on the shores of frigid Lake Superior, reached 106°. In the East, temperatures soared over 110° in New Jersey and Pennsylvania for the first and only time in their histories. New York City reached 106° in Central Park, Manhattan's hottest reading ever. Most cities in the Midwest from northern Louisiana to the Canadian border recorded their highest observed temperatures.

Nationwide, this was the hottest July ever. The apex of the heat wave occurred around July 15, when the average temperature of all 113 stations in Iowa was 108.7°. Lincoln, Nebraska, peaked at 115° after a nighttime low of 91°, perhaps the hottest nighttime temperature ever recorded outside the desert Southwest. The heat resulted in a desiccating drought, with some stations reporting only 15% of their normal summer precipitation. The entire state of Iowa averaged only 14% of normal rainfall for the month of July, eastern Nebraska only 8%. In Oklahoma the entire state averaged 7% normal precipitation for the month of August. In July the heat wave was centered over South Dakota, where the town of Kennebec had a monthly average high of 106.6° (normal is 90.8°). Twelve days of the month topped 110°, the peak being 119° on July 5.

Perhaps the hottest night ever recorded in the U.S. outside of the desert Southwest, occurred on the evening of July 15, 1936, at the unlikely location of Lincoln, Nebraska, when the minimum temperature fell to only 91°. The citizens of that city spent the night outdoors trying to sleep on the lawn of the state capitol. (Nebraska State Historical Society)

The temperature in Minneapolis peaked at an all-time record high of 108° on July 14, 1936. The want-ad staff of the St. Paul Daily News was provided with 400 pounds of ice and two electric fans to cool the air in the press room. (Minnesota Historical Society)

North Dakota was as hot, with Steele recording an astonishing 121° on July 6, tying with Alton, Kansas, for the hottest temperature ever measured in the United States outside the desert Southwest.

Bismarck, not usually associated with heat waves, recorded 100° or more every day from July 5 to July 12, with three days topping 110°. On July 11 the nighttime temperature fell only to 83°, the city's hottest night ever. On July 13, every weather station in Wisconsin, except Plum Island in Lake Michigan, exceeded 100°. Eau Claire hit 111° (the middle of a three-day stretch of 110°-plus temperatures), and Wisconsin Dells registered an unbelievable Death Valley–like 114°. Milwaukee suffered through five consecutive nights above 80° from July 8 to 13. Further east Corry, Pennsylvania, 20 miles from the usually cool shores of Lake Erie, had a temperature range of 85° to 110° on July 14.

In August, the center of heat drifted south over Oklahoma, where Altus averaged 109.8° for its monthly high, topping out at 120° on August 12, for the second time that summer. Ozark, Arkansas, exceeded 100° every day from August 3 to 23. Clarendon, Texas, was over 100° all but two days from July 12 to August 27. Needless to say, the country has never experienced before or since anything like this heat wave.

The summer of 1955 was one of the Northeast's hottest. At the time it was, in fact, Philadelphia's warmest on record, and sidewalk omelets were the rage. (Urban Archives, Temple University)

Heat Waves of the Northeast

The hottest months (as far as average temperatures are concerned) to be recorded in the Northeast were July and August 1955. Although few absolute-maximum temperature records were set, the persistence of the heat established many single-month and summer-heat temperature records. That July, Chicago, Cleveland, Detroit, Hartford, New York City, Philadelphia, Baltimore, and Washington, D.C. all recorded their hottest month ever.

Many of these records were broken in turn during the summer of 1999, when New York City had its hottest month on record with July averaging 81.4° (normal being 77.2°). Eighteen days exceeded 90° and a record-breaking 11 consecutive days above 90° occurred from July 23 to August 2. On July 5, 1999, temperatures soared past 100° at most locations from North only

Carolina to Massachusetts. At La Guardia airport in New York the overnight minimum fell only to 86° on the nights of July 4 and 5. At 7 a.m. on the 6th, the temperature stood at 88° in New York and 89° in Washington, D.C. A power outage affected much of New York City the night of the 5th, compounding the heat problem and making for a non-airconditioned nightmare. Later in the month, the heat extended southward where Charleston, South Carolina, recorded its hottest day ever on August 1, with a temperature of 105°.

New England's greatest heat wave happened during July of 1911. Record state maximum temperatures were established in Maine with 105°, New Hampshire with 106°, and Vermont 105°. Boston recorded its hottest day ever on July 4, when the temperature climbed to 104°. Portland, Maine, also achieved its all-time high of 103° that same day. Keene, New Hampshire, suffered through an unprecedented 12 consecutive days of 90°-plus heat July 1 to 12, including five days over 100°.

Heat Waves of the West

The American Southwest, already the hottest region of the United States, experienced its highest temperatures ever during late June and early July of 1994. From West Texas to Arizona most weather stations attained their all-time maximum temperatures, including Midland, Texas (116°); El Paso (114°); Albuquerque, New Mexico (107°); and Roswell, New Mexico (114°). Santa Fe,

Perhaps the worst drought ever to affect the West occurred during the 13th century, when the Anasazi, a group of Native Americans, vanished mysteriously from the remarkable dwellings they had built throughout the region. It is widely supposed that the prolonged drought was the major contributing factor to their disappearance.
(Kerrick James)

normally cool even in summer because of its 7,000-foot elevation, reached 100° for the first time in recorded history. Five state records were broken or tied in Nevada (125°), Arizona (128°), New Mexico (122°), Texas (120°), and Oklahoma (120°). El Paso experienced nine consecutive nights with temperatures at 80° or above, from June 26 to July 4, except one which cooled down to 79°. The night of July 1 was El Paso's hottest ever with a low of only 87°.

The Great Basin region had its worst heat wave during July of 2002. Records were broken at Reno, Nevada (108°), and Salt Lake City (107°). Boise, Idaho, came within one degree of their record with 110° on July 13. Virtually every weather station in Nevada experienced its hottest month ever in July 2002. Amazingly, most of these warmest-month-of-record temperatures were exceeded in July of 2003, when Phoenix, Las Vegas, and Salt Lake City all recorded their hottest month ever, along with most weather stations in Utah, Colorado, Wyoming, and northern New Mexico.

California rarely experiences a statewide heat wave because of its large size and varying topography. Its Pacific coastline alone is 1,200 miles long, and the state's eastern boundary is almost as long, embracing a variety of climatic zones, each one influenced by different weather systems.

Los Angeles, in the south, recorded its worst heat wave during late August and early September in 1955. For eight consecutive days (from August 31 to September 7) the temperature reached 100° or higher, peaking at 110° on September 1 at the Civic Center. The following night the temperature fell only to 83° at the Civic Center and 77° at the oceanside airport location, the city's warmest night ever. The heat was felt in Central California as well when Napa Valley's St. Helena hit 113° on September 2.

The San Francisco Bay Area had its hottest day on June 14, 2000, when San Francisco reached 103° at Mission Dolores and 106° at the airport. San Jose, Berkeley, and Oakland reached record temperatures respectively of 109°, 107°, and 106°. Vallejo and Martinez both hit 110°. Further north, Cloverdale reached an amazing 116°. This was a one-day heat wave, however, and fog rolled in the next day to cool things off. For all of coastal California, September of 1984 was the warmest month recorded, from Los Angeles, which averaged 81.3° (normal is 75.2°), to San Francisco with an average of 69.4° (normal 63.7°).

WHERE HEAT WAVES NEVER OCCUR

The most consistently cool places in the lower 48 states are either towns at very high altitudes or locations right on a coastline. The Pacific Coast never experiences prolonged heat waves. Even on San Francisco's hottest day ever, June 14, 2000, the temperature at Ocean Beach reached only a relatively modest 87° while a few miles inland the city center sizzled at 103°. The Farallon Islands, 25 miles off the coastline from San Francisco, have an all-time record high temperature of only 81°.

U.S. Locations with Least Extreme Absolute Range of Temperature (75°F or less between highest and lowest recorded temperatures)		U.S. Locations with Most Extreme Absolute Range of Temperature (170°F or more between highest and lowest recorded temperatures)	
Temp Range	*Location*	*Temp Range*	*Location*
43° (from 38° to 81°)	Farallon Island, CA	178° (from −78° to 100°)	Fort Yukon, AK
43° (from 52° to 95°)	Honolulu, HI*	175° (from −58° to 117°)	Medicine Lake, MT
56° (from 29° to 85°)	Point Piedras Blancas, CA	173° (from −63° to 110°)	Poplar, MT
59° (from 41° to 100°)	Key West, FL	172° (from −60° to 112°)	Parshall, ND
62° (from 37° to 99°)	Long Key, FL	172° (from −59° to 113°)	Glasgow, MT
67° (from 20° to 87°)	Eureka, CA	172° (from −58° to 114°)	McIntosh, SD
68° (from 32° to 100°)	Avalon Island, CA	172° (from −54° to 118°)	Pettibone, ND
73° (from 20° to 93°)	Fort Ross, CA	171° (from −57° to 114°)	Camp Crook, SD
73° (from 24° to 97°)	Belle Glade, FL	171° (from −50° to 121°)	Steele, ND
74° (from 19° to 93°)	Cresent City, CA	170° (from −58° to 112°)	Jordan, MT
74° (from 26° to 100°)	Miami, FL	170° (from −57° to 113°)	Culbertson, MT

*All stations in Hawaii have an extreme range of temperature less than 75°
The station with the least range is Kailua, Maui with a 36° range, (51° to 87°)

The coolest absolute-maximum temperatures in the United States, (except for the summit of Mt. Washington in New Hampshire) are along California's north coast where Eureka, for example, has never recorded a temperature higher than 87° in 130 years of record.

The coasts of Oregon and Washington are similarly moderate. Washington's Tattoosh Island has recorded a high of only 88°. The cold Humboldt Current is responsible for these equitable temperatures. Heat waves occur only when high pressure builds inland forcing hot air through the gaps along the coastal mountain ranges or river valleys.

Western mountain plateaus over 7,000 feet have rarely recorded temperatures above 100°. Such locations include Flagstaff in Arizona (97°) and Alamosa, Colorado (96°).

A few locations on the shores of the Great Lakes have also failed to reach 100° such as Muskegan and Sault Ste. Marie in Michigan. In the East, only extreme northern Maine and a few Atlantic coastal locations like Block Island, Rhode Island; Eastport, Maine; and Cape Hatteras, North Carolina, have never measured temperatures above 100°. The higher elevations of the Appalachians like Elkins, West Virginia, are also almost always comfortable during the summer, never recording temperatures above 100°.

In Florida, Tampa has never reached 100°; surprisingly, such temperatures are rare even in the interior of the Florida peninsula. The modifying influence is the surrounding ocean water, as is the case in Hawaii where only a single station has ever reached 100°.

The coolest location of all in the contiguous United States is the summit of Mt. Washington, New Hampshire, where the temperature has never exceeded 72°. In fact, even in Alaska, only St. Paul in the Pribilof Island group has stayed perpetually cooler than Mt. Washington, with a record high of 66°.

HEAT BURSTS

These rare and interesting events, sometimes called heat flashes, are characterized by a sudden increase of temperature, often lasting for no more than a few minutes, and usually occur in the vicinity of thunderstorms. Most happen at night, when relatively cool air contrasts strongly with a sudden burst of hot air. The cause of these heat bursts, at least in the cases reported in the High Plains, has been identified by researchers Ben Bernstein and Richard Johnson at Colorado State University, who recorded Doppler-radar measurements of heat bursts in Kansas. Their findings were reported in the *Monthly Weather Review* issue of February 1994. They explain,

> The heat burst, near the remnants of a thunderstorm cluster, is a downburst of air from the thunderstorm displacing a shallow pool of cool air hugging the ground. The phenomenon is akin to what would happen if one blew strongly down onto a shallow puddle of water; the puff from your lungs temporarily clears water from a spot on the surface.

HEAT BURSTS AROUND THE WORLD

Some extreme examples of heat bursts reported from around the world include temperature reports of 152° at Antalya, Turkey, on July 10, 1977; a temperature rise from 100° to 158° in two minutes near Lisbon, Portugal, on July 6, 1949; and an unbelievable 188° shade temperature apparently recorded in June 1967, at Abadan, Iran, where press reports said dozens of people died and asphalt streets liquefied. These reports from Portugal, Turkey, and Iran are almost certainly apocryphal; there appears to be no corroborating information aside from the original press reports themselves.

A verified heat burst at Kimberley, South Africa, raised the temperature from 67° to 110° in five minutes between 9:00 p.m. and 9:05 p.m. during a thunderstorm squall. The local weather observer stated that he believed the temperature actually rose higher than 110°, but his thermometer was not quick enough to register the highest point. The temperature had fallen back down to 67° by 9:45 p.m.

HEAT BURSTS IN NORTH AMERICA

Georgia

In 1860, *Scientific American* (vol.3:106) gave the following report of two heat flashes (no dates of occurrence were given):

> A hot wind extending about 100 yards in width, lately passed through middle Georgia, and scorched up the cotton crops on a number of plantations. A hot wind also passed through a section of Kansas; it burned up the vegetation in its track and several persons fell victim to its poisonous blast. It lasted for a very short period, during which the thermometer stood at 120° F.

Minnesota

The *Minneapolis Tribune* published the following story on July 10, 1879. "A blast of hot air passed from south to north through portions of New Ulm and Renville County last Sunday evening. It lasted only a minute or two, but so intense was the heat that people rushed out of their houses believing them to be on fire."

Oklahoma

A heat flash raised the temperature to 136° south of Cherokee, Oklahoma, at 3 a.m. on July 11, 1909. Crops in a small area at the center of the heat burst are said to have been instantly desiccated.

A heat burst at Gage on July 7, 1993, raised the temperature from 85° at 10:54 p.m. to 102° at midnight.

On the night of May 22, 1996, a heat burst emanating from collapsing thunderstorms near Ninnekah raised the temperature from the low 80°s to 105° between 11 p.m. and 3 a.m.

Texas

Just after midnight on the morning of June 15, 1960, a blast of hot wind estimated at 80–100 mph drove the temperature from 70° up to 140° on the northwest side of Lake Whitney, northwest of Waco. Cotton fields were reported to have been carbonized, leaving only burnt stalks standing.

Manitoba, Canada

Gretna, Manitoba experienced two heat bursts during the night of July 20–21, 1960, when the temperature rose from 80° to 96° for 15 minutes between 12:25 and 12:40 p.m. It fell back to a much lower temperature and then rose again to 97° for 15 minutes between 2:00 a.m. and 2:15 a.m.

South Dakota

Temperature rose from 85° at 1 a.m. to 104° at 3 a.m. in Pierre, South Dakota, the morning of June 20, 1989.

Montana

Great Falls reported a temperature rise from 67° to 93° in 15 minutes between 5:02 and 5:17 a.m. on September 9, 1994, tying the record high temperature for the date. The temperature then fell back to 68° by 5:40 a.m.

DROUGHTS AND DUST STORMS

Palmer Drought Severity Index

Palmer Index	Soil Moisture
Above +4	Extremely Moist
+3 to +4	Very Moist
+2 to +3	Moist
-2 to +2	Average, Normal
-2 to -3	Dry
-3 to -4	Very Dry
Below -4	Extremely Dry

A drought occurs when a month or more passes with less than 30% of normal precipitation. Severe droughts can last many months and even years. The method of measurement for drought is a scale called the Palmer Index where

-1.9 to +1.9 = normal or near normal soil moisture

-2.0 to -2.9 = moderate drought conditions

-3.0 to -3.9 = severe drought

-4.0 or less = extreme drought

A massive dust storm sweeps across Paradise Valley, Arizona, in August 1997. Wind-blown silts and clays are transported great distances from their origins when dry storm fronts lift them into the atmosphere and carry them across the earth's surface. (J. Trujillo/Weatherstock, Inc.)

Conversely, positive numbers indicate above-normal soil moisture with +4.0 or higher indicating extremely moist conditions.

Maps are issued weekly by the Climate Prediction Center of the National Oceanographic and Atmospheric Administration (NOAA) showing which areas of the United States are being affected by drought conditions. Droughts occur in all regions at one time or another and are not any more frequent in one place than another. They tend to be more devastating in the Midwest because of that region's dependence on non-irrigated agriculture (dry farming).

Running for cover: This classic photograph, taken in 1936, is a poignant illustration of the suffering endured by farmers in the Midwest during the Dust Bowl years of 1933–38. No other natural disaster has been more devastating to the country in terms of cost and lost livelihoods . (Library of Congress)

The worst drought in United States history accompanied the famous heat waves of the 1930s. Perhaps no other natural calamity in the history of the country was as devastating as this period of drought, coinciding as it did with the Great Depression. The worst periods of the drought were during the summers of 1934 and 1936, and during the fall of 1939. In each instance, the Palmer Index fell well below –5.0 in the Great Plains, even approaching –9.0 in 1934.

At its maximum extent, the drought area, called the Dust Bowl, encompassed over 50 million acres of the Plains and Midwest from Texas to Canada and from Colorado to Illinois. Windstorms blew the topsoil into roiling clouds of dust which often turned day into night.

In the Texas Panhandle there were 908 hours (the equivalent of almost 38 days) of dust storm in 1935 alone. Some individual storms lasted for three consecutive days. Dust from the Great Plains spread to and beyond the East Coast with deposits being reported on ships 300 miles out in the Atlantic.

Other periods of drought included 1893 to 1895, when the Palmer Index reached a low of –4.97 in Kansas. A severe drought once again affected the Midwest in 1953 and 1954. The Northeast had its worst drought from 1963 to 1965. Californians will remember the years from 1975 to 1977, when water rationing was instigated statewide. Sacramento received a paltry 6.63" of rain during the 1976–77 season and San Francisco 6.96", less than one-third of normal and the driest rainy season since records began in 1849.

More recently, one of the most extensive droughts in U.S. history occurred during the spring and summer of 2002. It was most severe in the West and Southwest where enormous wildfires burned over a million acres in Arizona and Colorado. San Diego and Los Angeles suffered through their driest rainy seasons in history with only 3.02" and 4.46" respectively, compared to normals of 10–15″.

REMARKABLE HIGH PLAINS DUST STORMS

At Tulia, Texas

On March 18, 1914, a strong wind, which raised much dust and sand, was blowing from the west until 4:30 p.m., when it became calm and sultry, with a temperature of 83°. The sky at the time was clear, except that a grayish bank of clouds or dust could be seen rising rapidly in the north. As it came nearer it was found to be composed of several spiral columns, apparently five feet in diameter and towering vertically upward to a height of three-quarters to one mile from the ground to the apex of the cloud. The phenomena [sic] was beautiful and awe-inspiring beyond description, and the effect was heightened by the rays of the sun reflecting from the storm cloud. It was traveling at the rate of probably 40 miles per hour, and about the time it crossed the Paledora Canyon there were two distinct flashes of lightning and two peals of thunder. There was no rain attending the phenomena. The storm struck here at 6:42 p.m., suddenly enshrouding the town in darkness and covering everything with dust and dirt. Wheat and oats were badly blown away and several windmills blown down, while at Happy, about 14 miles south, the Catholic Church and several small houses and windmills were wrecked. The oldest inhabitants consider it the worst storm in the last 23 years.

—*Climatological Data*, Texas Section, March 1914

In Oklahoma

During the morning of January 20, 1895, the sky was filled with cirrus clouds, very feathery and white. In the afternoon it became hazy, then dark, and looked like rain. Wind in puffs from the southwest. At nightfall the sky cleared, but somewhat hazy. At 8 p.m. the wind changed to the west, and a gale began. By 9 p.m. it was frightful. The dust passed along in columns fully 1,000 feet high and the wind rose to a speed of 35, then 45 miles per hour, with gusts reaching 55 miles, the temperature fell rapidly, and we saw for the first time (about 9 p.m.) flashes of light that apparently started from no particular place, but prevaded [sic] the dust everywhere. As long as the wind blew, till about 2 a.m., January 21, this free lightning was everywhere but there was no noise whatever. It was a silent electrical storm. This morning the sky is clear and except that dirt is piled up over books, windows, and all in the house, no one would know what a fierce raging of wind and sky we had.

—J. C. Neal of Oklahoma A&M College, in *Monthly Weather Review*, January 1895

In Kansas

The darkness was dust. The windows turned solid pitch; even flower boxes six inches beyond the pane were shut from view....Dust sifted into houses, through cracks around the doors and windows—so thick even in well-built homes a man in a chair across the room became a blurred outline. Sparks flew between pieces of metal, and men got a shock when they touched the plumbing. It hurt to breath[e], but a damp cloth held over mouth and nose helped for a while. Food on tables freshly set for dinner ruined. Milk turned black. Bed, rugs, furniture, clothes in the closets, and food in the refrigerator were covered with a film of dust. Its acrid odor came out of pillows for days afterward.

—Vance Johnson, a resident of western Kansas in the 1930s

GLOBAL WARMING

In a chapter devoted to extreme heat, it would be remiss not to touch upon the controversy surrounding global warming and its causes. There is little debate over the fact that over the past 25 years or so global temperatures have risen significantly, and the trend has escalated since 1990. Many factors might explain the rise of temperature averages around the world. For one thing, weather stations have increasingly been moved from sites within cities to their nearby airports, and the acres of airport asphalt absorb and then radiate more heat than do city rooftops or parks. This relocation would especially affect nighttime temperatures, an upward trend that is especially conspicuous. Urban areas, where most weather stations are located, have been growing steadily larger and therefore steadily warmer.

These are obviously not the only factors contributing to increases in ambient air temperature the world over. Other factors must be causing glaciers to melt and deserts to expand. Is this change the result of human actions (the burning of fossil fuels, destruction of rain forests, urbanization, etc.) or has it been caused by natural, cyclical, climate changes? Recent studies indicate that the increase in carbon dioxide in the atmosphere is inextricably associated with global temperature change. Thus it would appear that human activity is causing the planet to heat up.

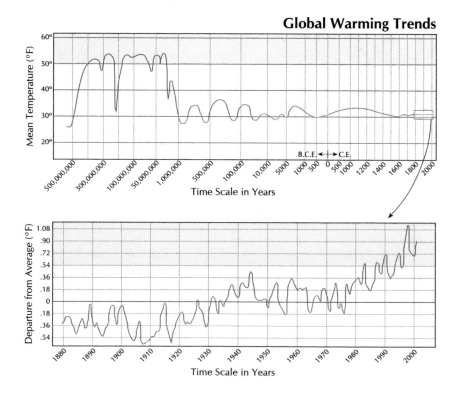

Ultimately, however, the sun is the planet's most important climate control system and even the smallest of perturbations of this vast heat engine would have catastrophic effects on our climate. Rapid changes in global climate have occurred many times over the course of climatic history and certainly will happen again regardless of human activity.

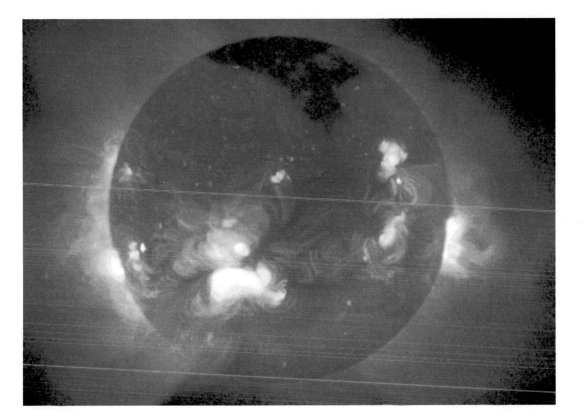

NATIONAL WEATHER SERVICE FORECASTS END OF THE WORLD

According to an urgent advisory issued Wednesday morning [December 17, 2003] by the National Weather Service, the Earth has broken out of its orbit and is careening toward a fiery end.

"Unusually hot weather has entered the region for December as the Earth has left its orbit and is hurtling towards the sun," read the message, posted on the National Oceanic & Atmospheric Administration's web site. "Unusually hot weather will occur for at least the next several days as the Earth draws ever nearer to the sun. Therefore, an excessive heat watch has been posted."

But don't assume this news is going to get you out of finishing your Christmas shopping. The release, signed "Heinlein," was only a test message, erroneously posted during a training session, according to a Weather Service spokesman. By midafternoon, the statement had been removed and a correction issued.

—Posted by David M. Ewalt on the *Weather Matrix* web site, December 19, 2003

chapter 2

COLD

Unlike the hottest regions of the United States, the coldest places in the country are sparsely inhabited. These areas include much of Alaska, northern Minnesota, North Dakota, northern Maine, and the higher elevations and valleys of the Rocky Mountains.

COLDEST PLACES IN THE UNITED STATES

The coldest major city in the lower 48 is, not surprisingly, Minneapolis–St. Paul, Minnesota. Weather records have been kept there since 1820 (beginning at Fort Snelling, prior to the establishment of the city itself), and an extreme low of -41° was reached in January of 1889 at St. Paul. The average temperature for the three months of winter is a bone-chilling 17.3°. One year, during 186 consecutive hours between New Year's Eve of 1911 and January 8, 1912, the thermometer did not once rise above zero. Then it rose above zero for four hours before falling under zero again for another 113 hours. Nowadays, residents can avoid the frigid temperatures in the downtown area by taking advantage of covered overpasses and underpasses that connect most of the city center.

15 Coldest Cities in the U.S.
Average Annual Temperatures
(period of record 1970–2000)

Location	Temperature
Fairbanks, AK	26.7°
Anchorage, AK	36.2°
International Falls, MN	37.4°
Duluth, MN	39.1°
Caribou, ME	39.2°
Butte, MT	39.5°
Sault. Ste. Marie, MI	40.1°
Grand Forks, ND	40.3°
Alamosa, CO	40.8°
Williston, ND	40.9°
Juneau, AK	41.5°
Fargo, ND	41.5°
St. Cloud, MN	41.8°
Bismarck, ND	42.3°
Kalispell, MT	42.6°

source: NCDC

15 Coldest Cities in the U.S.
Average Jan Minimum Temperatures
(period of record 1970–2000)

Location	Temperature
Fairbanks, AK	-19.0°
International Falls, MN	-8.4°
Grand Forks, ND	-4.3°
Alamosa, CO	-3.7°
Williston, ND	-3.3°
Fargo, ND	-2.3°
St. Cloud, MN	-1.2°
Duluth, MN	-1.2°
Bismarck, ND	-0.6°
Caribou, ME	-0.3°
Aberdeen SD	0.6°
Eau Claire, WI	2.5°
Huron, SD	3.5°
Wausau, WI	3.6°
Rochester, MN	3.7°

source: NCDC

Coldest Wintertime Regions in the United States

1	Yukon and Tanana River Valleys of Alaska (and anywhere inland from the Gulf of Alaska or Bering Sea)
2	Red River Valley of eastern North Dakota and Western Minnesota
3	Northern Maine and Northern New Hampshire
4	Yellowstone Plateau of Wyoming and Montana
5	High mountain Valleys of Colorado
6	West-central Wyoming from the Big Horns to the Salt River Range near Idaho border
7	Lake Superior region of NE Minnesota, North Wisconsin and Upper Peninsula of Michigan
8	Adirondack Mountains of New York
9	Sawtooth Mountains of Idaho
10	High plateau of Mono County, California near Nevada border

(opposite) Fairbanks, Alaska, is the coldest large city in the United States. Temperatures can be expected to fall below -40° for at least a few days every winter. (Patrick Endres, AlaskaStock.com)

Towns with Same Day Record Highs and Lows

Location	High	Low	Date
Alamosa, CO	91°	35°	7/2/1989
"	87°	30°	8/25/2002
"	88°	31°	8/26/2002
Astoria, OR	61°	28°	3/15/1988
Bakersfield, CA	75°	23°	1/3/1930
Coalville, UT	100°	37°	7/10/2003
Delta, UT	107°	42°	7/10/2003
Deeth, NV	87°	12°	9/21/1954
Eureka, CA	61°	28°	3/15/1988
Hilo, HI	91°	60°	5/25/2003
"	88°	60°	5/26/2003
Juneau, AK	56°	25°	4/16/2001
Kansas City, MO	76°	11°	11/11/1911
Las Vegas, NV	119°	48°	7/13/1972
(Sunrise Manor, Las Vegas, NV)			
Miami, FL	96°	70°	8/11/1984
Melbourne, FL	97°	68°	5/22/1998
Oklahoma City, OK	83°	17°	11/11/1911
Park City, UT	87°	36°	8/11/2002
"	89°	39°	8/15/2002
Paso Robles, CA	74°	21°	1/9/1999
Pine Valley, NV	97°	30°	7/30/1989
Pueblo, CO	101°	52°	7/15/1993
Rapid City, SD*	101°	39°	8/17/2002
*39° is only 1 degree from the monthly record low and 101° was only 5 degrees from the monthly record high			
Safford, AZ	105°	52°	8/25/2002
Sioux City, IA	91°	33°	5/16/1997
Springfield, MO	80°	13°	11/11/11

source: selected from various reports

The cities of North Dakota are even colder, with Bismarck, Fargo, and Grand Forks averaging 14.5°, 11.1°, and 9.7°, respectively, for the December-through-February period. The coldest town of all in North Dakota is Hansboro, on the Canadian border, which averages only 5.4° over the entire winter. In January the average nighttime low here is a rather unpleasant -10.4°. However, for the coldest of the cold spots in the lower 48 we have Tower, Minnesota, which averages a daily low of -13.6° during the typical January, and Roseau, Minnesota, which averages 4.5° over the winter months of December through February.

For coldest nights, Taylor Lake, Colorado, is second only to Tower, falling to -12.2° on an average January night. Taylor Lake also held the record-low temperature in the state of Colorado at -60°, until the village of Maybell dropped to -61° in February of 1985.

It is places like Maybell in the high mountain valleys and plateaus of the Rocky Mountains that endure some of the coldest winter nights in the United States outside of Alaska. The Yellowstone plateau of Wyoming and Montana is exceptionally cold in the winter. A record reading of -66° was measured on February 9, 1933, at the Riverside Ranger Station, located in the Montana section of the National Park. (This has caused some confusion among record keepers since the ranger station was later moved across the border into Wyoming. Wyoming's coldest official low of -63° was recorded at Moran in Grand Teton National Park that same night). However, daytime temperatures in the winter months tend to be warmer in the Rockies than in the coldest areas of the upper Midwest or northern New England.

An example of just how extreme the range in diurnal temperature can be is seen in the case of Gavilan, New Mexico, where during the bitter cold wave of January 1949, the following high and low temperatures were reported:

JANUARY 1949, Gavilan, New Mexico	DAY	HIGH	LOW	TEMPERATURE SPREAD
	January 5	29°	-40°	69°
	January 6	43°	-31°	74°
	January 7	43°	-22°	65°

An even more extreme example of diurnal temperature fluctuation was reported at Juniper Lake in Oregon on May 2, 1968, when the temperature purportedly rose from a morning low of zero to an afternoon high of 81°. Sometimes record-high and record-low temperatures are attained on the same day (see table). The dryness of the Western air and abundant sunshine account for such extremes.

Northern New England also ranks at the top of coldest U.S. places with Allagash, Maine, taking top honors with its average minimum temperature of -9.8° during January.

Only a handful of weather stations have recorded temperatures lower than -60° in the lower 48 states. These readings were taken in the states of Minnesota, North Dakota, Montana, Wyoming, Colorado, and Utah. Unofficial readings of -60° were reported at Rochford, South Dakota, on December 22, 1989 and -64° at Embarrass, Minnesota, on February 2, 1996.

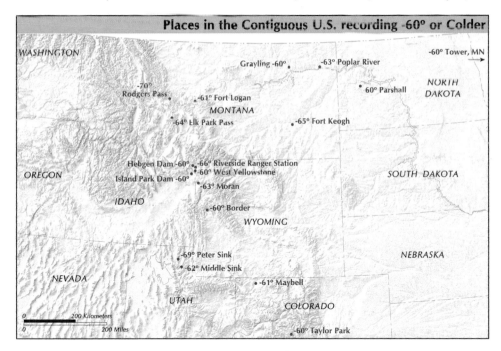

Places in the Contiguous U.S. recording -60° or Colder

Alaska

Of course, no location in the lower 48 can boast of frigid temperatures comparable to those existing over the Yukon and Tanana River Valleys of Alaska during the winter months. Fairbanks averages -9.7° during January, with a normal minimum of -19.0° that month. Intense cold waves send the mercury below -40° for days at a time. A cold wave in December of 1961 brought temperatures as shown in the table on p. 50.

(following pages) It is never too cold to snow.
This is proven every winter day somewhere in Alaska. (David Job)

Fairbanks, Alaska, Deep Freeze December 1961		
Date	High	Low
12/16	-15°	-42°
12/17	-38°	-48°
12/18	-14°	-46°
12/19	-14°	-40°
12/20	-36°	-45°
12/21	-44°	-49°
12/22	-43°	-51°
12/23	-48°	-53°
12/24	-51°	-54°
12/25	-50°	-56°
12/26	-53°	-56°
12/27	-48°	-58°
12/28	-55°	-62°
12/29	-55°	-62°
12/30	-15°	-57°
12/31	+9°	-24°

Note the *high* temperatures of -55° on December 28 and 29! Fairbanks's coldest reading ever was -66° on January 14, 1934.

Author John Murray recalls a mistake he made one evening after returning home from lecturing at the University of Alaska in Fairbanks some years ago. A cold snap had sent temperatures to -40° and because he forgot to put his gloves on while opening the front door of his house, his hand froze to the doorknob. Fortunately, his wife was home and was able to free him with a bit of warm water.

The town of Ambler, Alaska recorded a daily *maximum* temperature of -66° on January 27, 1989. The wind chill that day made it feel like -120°. During this cold wave the barometric pressure rose to 31.85″ at Northway. This was the highest air pressure ever measured in the U.S., and because most aircraft altimeters are unable to calibrate for such pressure, many flights had to be cancelled.

Prospect Creek in Alaska currently holds the all-time lowest temperature recorded in the U.S., with a reading of -79.8° on January 23, 1971. An unofficial reading of -82° was reported at the aptly named hamlet of Coldfoot. Fort Yukon, Alaska averaged -48.4° for the entire month of

---

MAKESHIFT ALASKAN THERMOMETERS

We had no thermometers in Circle City that would fit the case [reach low enough], until Jack McQueston invented one of his own. This consisted of a set of vials fitted into a rack, one containing quicksilver (mercury), one of the best whiskey in the country, one kerosene, and one Perry Davis's Pain-Killer. These congealed in the order mentioned, and a man starting on a journey started with a smile at frozen quicksilver, still went at whiskey, hesitated at the kerosene, and dived back in his cabin when the Pain-Killer lay down.

—Recorded by William Bronson, *The Last Grand Adventure*, describing Alaska in the Gold Rush era of the 1890s

COLDEST PLACE IN THE LOWER 48: EMBARRASS, MINNESOTA

Official National Weather Service stations are not always located in what may actually be the hottest, coldest, wettest, or snowiest locations in a state. And there isn't one in Embarrass, Minnesota, where on February 2, 1996, a weather record was broken.

The hamlet (pop. 691) is located in Minnesota's Arrowhead region about 10 miles southeast of Tower—the state's coldest "official" location. The diligent citizens of Embarrass have been keeping detailed weather records here since 1975, and their thermometers have consistently shown Embarrass to be even colder than Tower or any other location in the state.

When Tower dropped to -60° on February 2, 1996, Embarrass plunged to -64°. The thermometer registering this remarkable figure was checked for accuracy by Taylor Environmental Instruments and Watson Ridgeway Laboratory. The thermometer was deemed accurate. The village's average annual temperature of 34.4° is the coldest in the contiguous United States exclusive of high mountain locations.

December 1917, the coldest month observed in the United States.

In Alaska only the stations in the adjacent table have reported temperatures lower than -70°, although most locations in the interior of the state have recorded -60° or below at one time or the other. A thermometer left at an elevation of 15,000 feet on Mt. McKinley for 19 years recorded a low temperature of -100° at some time during its exposure, according to the U.S. Army Natick Laboratories in 1969.

The coldest location year-round in the United States (outside of Alaska where Barrow's 10.4° annual average takes the prize) is the summit of New Hampshire's Mt. Washington, with its annual average temperature of 27.2°. Of course, nobody actually lives year-round on the summit of Mt. Washington, and for that matter not many people live in the small towns of Hansboro (pop. 20) in North Dakota, or Tower (pop. 502) in Minnesota, either. The coldest large (relatively speaking) towns in the country are those listed on the tables on p. 45.

Places in Alaska Recording -70° or Colder

Temp	Location
-79.8°	Prospect Creek Camp
-79°	Circle
-78°	Fort Yukon
-77°	Manley Hot Springs
-76°	Tanana
-75°	Allakaket
-75°	Tanacross
-75°	McGrath
-74°	Ambler West
-74°	Chandalar Lake
-74°	Coldfoot Camp
-72°	Aniak
-72°	Chicken
-71°	Northway
-71°	Farewell Lake
-71°	Eagle
-71°	Tok
-70°	Galena
-70°	Bettles

COLDEST LOCATION BY STATE

State	Temp	Location
ALABAMA	25.5°	Hamilton
ALASKA	-33.1°	Arctic Village
ARIZONA	6.7°	Maverick
ARKANSAS	21.4°	Deer
CALIFORNIA	5.8°	Bodie
COLORADO	-11.2°	Taylor Park
CONNECTICUT	11.9°	Norfolk
DELAWARE	23.5°	Newark
FLORIDA	36.7°	De Funiak Springs
GEORGIA	24.7°	Blairsville
HAWAII	33.7°	Mauna Loa *(February)*
IDAHO	-1.8°	Grouse
ILLINOIS	7.5°	Mount Carroll
INDIANA	12.1°	Lowell
IOWA	2.0°	Sibley
KANSAS	11.4°	Atwood
KENTUCKY	19.0°	Ashland
LOUISIANA	31.4°	Plain Dealing
MAINE	-9.8°	Allagash
MARYLAND	16.5°	Frostburg
MASSACHUSETTS	6.9°	Chester
MICHIGAN	-3.7°	Strambaugh
MINNESOTA	-13.6°	Tower
MISSISSIPPI	26.4°	Iuka
MISSOURI	9.1°	Spickard 7 W
MONTANA	-4.9°	Westby
NEBRASKA	5.1°	Lynch
NEVADA	4.2°	Wild Horse Resevoir
NEW HAMPSHIRE	-2.5°	First Conn. Lake *(excl. Mt. Washnigton)*
NEW JERSEY	12.1°	High Point Park
NEW MEXICO	-1.1°	Gavilan
NEW YORK	1.6°	Wanakena Ranger School
NORTH CAROLINA	17.3°	Mount Mitchell
NORTH DAKOTA	-10.4°	Hansboro
OHIO	13.9°	Paulding
OKLAHOMA	16.7°	Turpin
OREGON	9.9°	Seneca
PENNSYLVANIA	9.9°	Pleasant Mount
RHODE ISLAND	16.5°	North Foster
SOUTH CAROLINA	27.6°	Caesars Head
SOUTH DAKOTA	-2.7°	Columbia 8 N
TENNESSEE	17.0°	Mt. Leconte
TEXAS	15.9°	Dalhart 6SW
UTAH	-1.0°	Scofield Dam
VERMONT	-1.3°	West Burke *(excl. Mt. Mansfield)*
VIRGINIA	16.8°	Big Meadows
WASHINGTON	5.8°	Stockdill Ranch
WEST VIRGINIA	13.0°	Marlinton
WISCONSIN	-6.0°	Big Falls Hydro
WYOMING	-10.1°	Darwin Ranch

ALASKA

source: NOAA

AVERAGE DAILY MINIMUM JANUARY TEMPERATURE

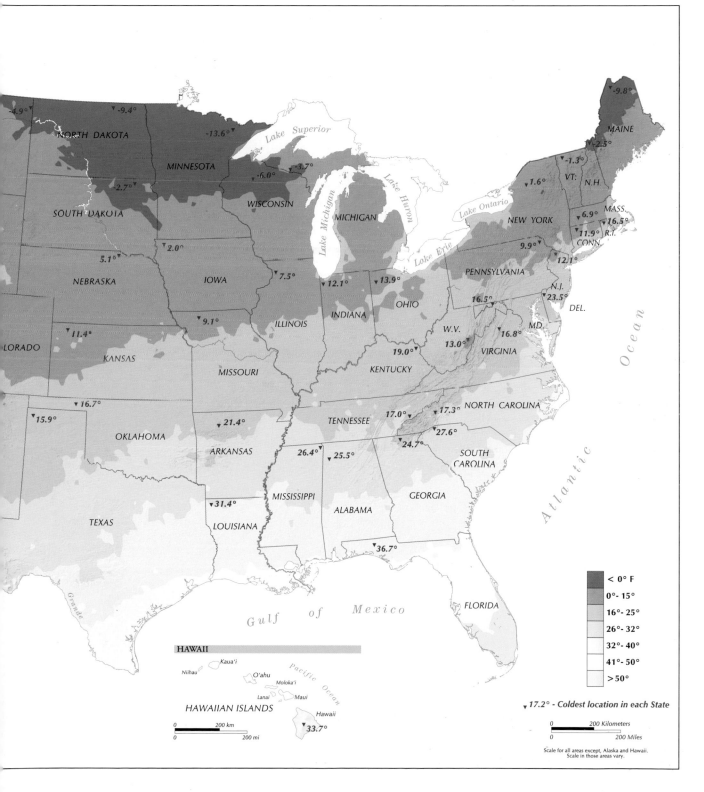

NORTH DAKOTA ▼-4.9° ▼-9.4°

▼-13.6° Lake Superior

MINNESOTA

▼-2.7° ▼-6.0° ▼-3.7°

▼-9.8° MAINE

SOUTH DAKOTA

WISCONSIN

MICHIGAN

Lake Michigan

Lake Huron

Lake Ontario

▼-2.5°

▼-1.3° VT: N.H.

▼1.6°

NEW YORK MASS.
▼6.9° ▼16.5°
▼11.9° R.I.
CONN.

▼5.1° NEBRASKA

▼2.0° IOWA

▼7.5° ILLINOIS

▼12.1° INDIANA

▼13.9° OHIO

PENNSYLVANIA
▼9.9°

▼12.1°

N.J.
▼23.5°
DEL.

▼11.4° COLORADO

KANSAS

▼9.1°

MISSOURI

KENTUCKY
▼19.0°

W.V. ▼16.8° MD.
▼13.0° VIRGINIA

Ocean

▼16.7°

▼15.9°

OKLAHOMA

▼21.4°

ARKANSAS

TENNESSEE ▼17.0°

▼17.3° NORTH CAROLINA

▼27.6°

▼26.4° ▼25.5° ▼24.7° SOUTH
CAROLINA

Atlantic

TEXAS

▼31.4°

LOUISIANA

MISSISSIPPI

ALABAMA

GEORGIA

▼36.7°

FLORIDA

Gulf of Mexico

Grande

HAWAII

Kaua'i

Niihau O'ahu

Moloka'i

Lanai Maui

Pacific Ocean

HAWAIIAN ISLANDS

Hawaii
▼33.7°

0   200 km
0   200 mi

	< 0° F
	0°- 15°
	16°- 25°
	26°- 32°
	32°- 40°
	41°- 50°
	>50°

▼17.2° - Coldest location in each State

0   200 Kilometers
0   200 Miles

Scale for all areas except, Alaska and Hawaii.
Scale in those areas vary.

LOWEST RECORDED TEMPERATURES BY STATE

State	Temp	Location	Date
ALABAMA	-27°	New Market	1/30/1966
ALASKA	-80°	Prospect Creek Camp	1/23/1971
ARIZONA	-41°	Hawley Lake	1/7/1971
ARKANSAS	-29°	Pond and Gavette	2/13/1905
CALIFORNIA	-45°	Boca	1/20/1937
COLORADO	-61°	Maybell	2/1/1985
CONNECTICUT	-37°	Norfolk (Valley Station)	2/16/1943
DELAWARE	-17°	Millsboro	1/17/1893
FLORIDA	-2°	Tallahassee	2/13/1899
GEORGIA	-17°	CCC camp F16 (Lafeyette)	1/27/1940
HAWAII	12°	Mauna Kea Obs. Sta.	5/17/1979
IDAHO	-60°	Island Park Dam	1/18/1943
ILLINOIS	-36°	Congerville	1/5/1999
INDIANA	-36°	New Whitehead	1/19/1994
IOWA	-47°	Elkader	2/3/1996
KANSAS	-40°	Lebanon	2/13/1905
KENTUCKY	-37°	Shelbyville	1/19/1994
LOUISIANA	-16°	Minden	2/13/1899
MAINE	-48°	Van Buren	1/19/1925
MARYLAND	-40°	Oakland	1/13/1912
MASSACHUSETTS	-40°	Chester	1/22/1984
MICHIGAN	-51°	Vanderbilt	2/9/1934
MINNESOTA	-60°	Tower	2/2/1996
MISSISSIPPI	-19°	Corinth	1/30/1966
MISSOURI	-40°	Warsaw	1/13/1905
MONTANA	-70°	Rogers Pass	1/20/1954
NEBRASKA	-47°	Camp Clarke	2/12/1899
NEVADA	-50°	San Jacinto	1/8/1937
NEW HAMPSHIRE	-47°	Mt. Washington	1/31/1934
NEW JERSEY	-34°	River Vale	1/5/1904
NEW MEXICO	-50°	Gavilan	2/1/1951
NEW YORK	-52°	Old Forge	2/18/1979
NORTH CAROLINA	-34°	Mt. Mitchell	1/21/1985
NORTH DAKOTA	-60°	Parshall	2/15/1936
OHIO	-39°	Milligan	2/10/1899
OKLAHOMA	-27°	Watts	1/18/1930
OREGON	-54°	Seneca	2/10/1933
PENNSYLVANIA	-42°	Smethport	1/5/1904
RHODE ISLAND	-25°	Greene	2/5/1996
SOUTH CAROLINA	-22°	Hogshead Mountain	1/21/1985
SOUTH DAKOTA	-58°	McIntosh	2/17/1936
TENNESSEE	-32°	Mountain City	12/30/1917
TEXAS	-23°	Seminole	2/8/1933
UTAH	-69°	Peter's Sink	2/1/1985
VERMONT	-50°	Bloomfield	12/30/1933
VIRGINIA	-30°	Mt. Lake Biology Station	1/22/1985
WASHINGTON	-48°	Mazama	12/30/1968
WEST VIRGINIA	-37°	Lewisburg	12/30/1917
WISCONSIN	-55°	Couderay	2/1/1996
WYOMING	-63°	Moran	2/9/1933

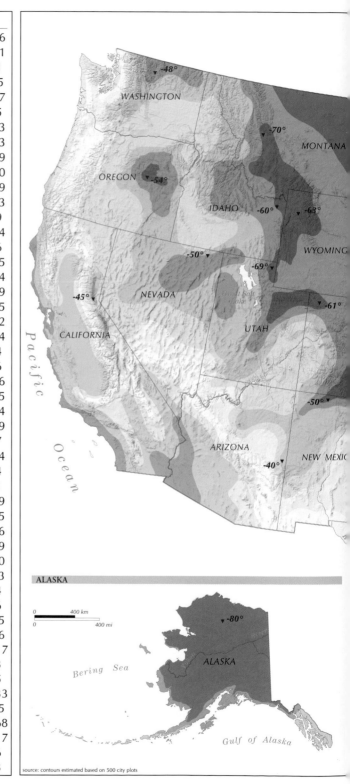

ALASKA

source: contours estimated based on 500 city plots

ABSOLUTE MINIMUM TEMPERATURE ON RECORD

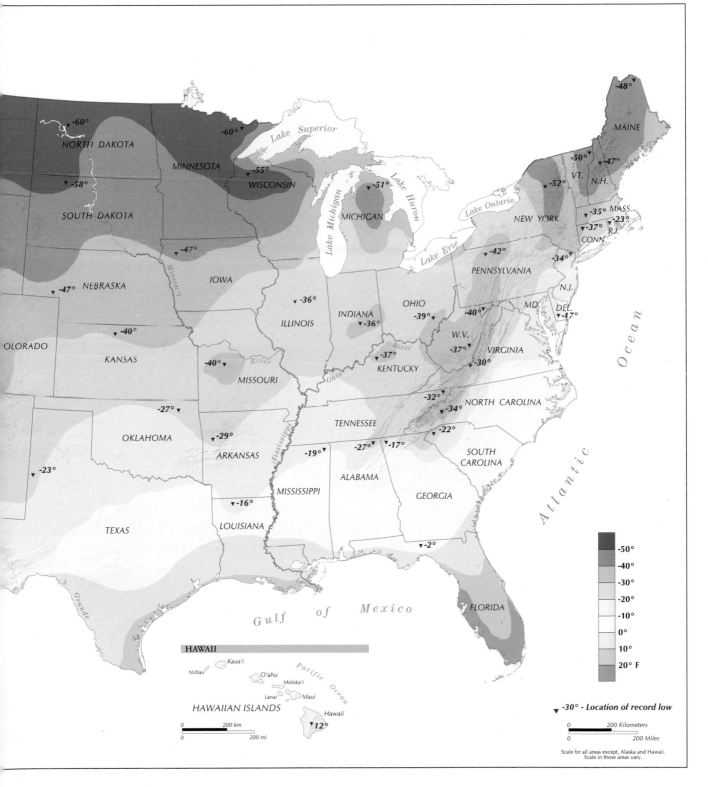

MAINE ▼ -48°

-60° ▼ NORTH DAKOTA

-60° ▼ MINNESOTA

-55° ▼ WISCONSIN

-58° ▼ SOUTH DAKOTA

-51° ▼ MICHIGAN

Lake Superior

Lake Huron

Lake Michigan

Lake Erie

Lake Ontario

-50° ▼
VT. -47° ▼
N.H.
-52° ▼
-35° ▼ MASS.
NEW YORK
-23° ▼
-37° ▼ R.I.
CONN.

-47° ▼ IOWA

-47° ▼ NEBRASKA

-36° ▼ ILLINOIS

-36° ▼ INDIANA

-39° ▼ OHIO

-42° ▼ PENNSYLVANIA

-34° ▼

N.J.

MD.

-40° ▼ COLORADO

-40° ▼ KANSAS

-40° ▼ MISSOURI

-37° ▼ KENTUCKY

-40° ▼ W.V.

-37° ▼ VIRGINIA

-30° ▼

DEL.

-17° ▼

Missouri River

Ohio River

-27° ▼ OKLAHOMA

-29° ▼ ARKANSAS

-19° ▼

-27° ▼ TENNESSEE

-17° ▼

-32° ▼ NORTH CAROLINA

-34° ▼

-22° ▼

-23° ▼

-16° ▼ LOUISIANA

MISSISSIPPI

ALABAMA

SOUTH CAROLINA

GEORGIA

TEXAS

Mississippi River

Rio Grande

-2° ▼

Gulf   of   Mexico

FLORIDA

Atlantic   Ocean

HAWAII

Kaua'i

Niihau

O'ahu

Moloka'i

Lanai   Maui

Pacific Ocean

HAWAIIAN ISLANDS

Hawaii

-12° ▼

0   200 km

0   200 mi

	-50°
	-40°
	-30°
	-20°
	-10°
	0°
	10°
	20° F

▼ -30° - Location of record low

0   200 Kilometers

0   200 Miles

Scale for all areas except, Alaska and Hawaii.
Scale in those areas vary.

World's Coldest Places
Lowest Temperatures Recorded

Country	Temp	Location
ANTARCTICA		
	-128.6°	Vostok*
coldest official temperature recorded in the world, Jul 21, 1983		
THE ARCTIC		
North Pole	-59°	North Pole
Greenland	-86.8°	Northice
ASIA		
Russia	-90°	Verkhoyansk
China	-62°	Mohe
Mongolia	-56°	Ulan Bator
Korea, North	-46°	Chunggangjin
Japan	-42°	Asahigawa, Hokkaido
Turkey	-40°	Erzurum
Afghanistan	-24°	Ghazni
India	-23°	Leh
Iran	-18°	Mashed
Iraq	-4°	Kurdistan
Pakistan	-3°	Quetta
Jordan	7°	Shoubak
Syria	9°	Aleppo
Lebanon	12°	Al Labwah
Saudi Arabia	12°	Turuyfa
Israel	16°	Har Meron
Taiwan	19°	Alishan
Sri Lanka	25°	Nuwara Eliya
Vietnam	28°	Muong Bu
Thailand	28°	Doi Inthanon
Kuwait	28°	Kuwait City
Laos	28°	Xiang Khoang
Burma	31°	Lashio
Hong Kong	32°	Victoria Peak
Oman	32°	Hibra
Yemen	32°	Sanaa
Bangladesh	33°	Jalpaiguri
Philippines	34°	Pangrango
Qatar	34°	Doha
Malaysia	36°	Cameron Highlands
U.A.E.	37°	Sharjah
Bahrain	37°	Manama
Indonesia	48°	Sabang
Cambodia	49°	Siem Reap
Singapore	66°	Singapore
NORTH AMERICA		
Canada	-81.4°	Snag, Yukon
U.S.	-79.8°	Prospect Creek, Alaska
Mexico	5°	Chihuahua
Cuba	33°	Bainoa

source: various and weatherbase.com

At an elevation of 9,300 feet, Antarctica's Amundsen-Scott South Pole Station is not the coldest spot on the continent but it's close. The Russian station of Vostok, 600 miles from the Pole itself but 2,000 feet higher than the Amundsen-Scott Station, takes that honor. (Galen Rowell)

WORLD'S COLDEST PLACES

The continent of Antarctica is the coldest region on the planet. The lowest temperature ever recorded, -128.6°, was measured at the Russian research station of Vostok near the South Pole at an elevation of 11,220 feet. An even lower reading of -132° was reported from Vostok during the winter of 1997, but has never been confirmed. Only one reading above zero has ever been recorded here, when 4° was attained on a summer afternoon. The South Pole itself has warmed up to a balmy 7.5°.

The average winter (June–August) temperature at the South Pole is about -72°. The lowest average temperature for any single month was recorded at the Plateau Station (elev. 11,890 feet) during July 1968, when the month averaged -99.8° degrees. On July 4, 2003, a remote-recording weather station located on Queen Maud Land in east Antarctica reported an air temperature of -94° and a wind speed of 75 mph. The resulting wind chill of -150° is perhaps the coldest ever on earth. And a good thing it was that the recording was operated by remote control since it is difficult to imagine how a human could endure such conditions.

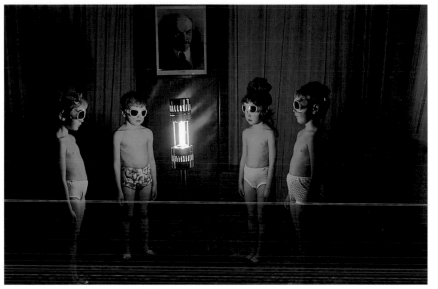

These children are undergoing ultraviolet treatment to make up for the lack of sunshine during the long, dark, and cold Siberian winters. Siberia is the coldest permanently inhabited region on earth.
(Mark S. Wexler)

The icecap of Greenland is another center of cold, where temperatures as low as -87° have been registered by scientific expeditions.

The coldest permanently inhabited region in the world is in Russia's far east. Specifically, the so-called "Pole of Cold" region in Siberia's northeastern interior surrounding the city of Verkhoyansk. During January, Verkhoyansk averages -58°F with a normal daily range of -54° to -63°. Its coldest reading ever was an almost unbelievable -90° on the nights of February 5 and 7, 1892.

Unofficial temperatures as low as -108° have been reported from Oymyakon, situated in a river valley some 300 miles southeast of Verkhoyansk and considered to be the very coldest inhabited town on earth. Oymyakon has never seen the temperature rise above zero between December 1 and March 1, its warmest winter reading being -4° in February. Supply trucks servicing this area must never turn off their engines during the core of the winter season. If the engine block freezes, truckers light a fire underneath to warm it up. When the temperature falls below -58°F (-50°C) a phenomenon known as the "whispering of the stars" is said to occur, when ice crystals in the atmosphere make a light swishing sound.

World's Coldest Places
Lowest Temperatures Recorded

Country	Temp	Location
EUROPE		
Russia	-67°	Ust'Shchugor
Sweden	-66°	Ultima
Norway	-62°	Karosjok
Finland	-61°	Kittila and Pokka
Germany	-51°	Funtensee
Estonia	-46°	Jogeva
Czech Republic	-44°	Litvinovice
Switzerland	-43°	La Brevine
Poland	-41°	Zywiec
Bulgaria	-37°	Veliko Turnova
Iceland	-36°	Reykjahllo
Latvia	-36°	Daugavpits
Yugoslavia (former)	-36°	Mt. Onul and Sjenica
Austria	-35°	Saalfelden
Romania	-31°	Virfu-Omul
Hungry	-29°	Debreen
France	-27°	Langres
Denmark	-24°	Silkeborg
Slovakia	-22°	Kesmarok
Spain	-22°	Calamocha (Teruel)
Great Britain	-17°	Braemer, Scotland
Italy	-13°	Tarvisio
Greece	-13°	Kavalla
Netherlands	-13°	De Bilt & The Hague
Belgium	-11°	Valender
Luxembourg	-10°	Larochette
Ireland	-3°	Omagh, Tyrone
Albania	1°	Korce
Portugal	3°	Penhas da Sude
Holy See	19°	Vatican City
Cyprus	21°	Nicosia
SOUTH AND CENTRAL AMERICA		
Argentina	-38°	Pato Superior Valley
Chile	-23°	Cristo Redentor
Peru	-13°	Imata
Bolivia	-10°	Charana
Brazil	1°	Sao Jose dos Ausentes
Costa Rica	16°	Mt. Chirripo
Uruguay	20°	Melo
Paraguay	23°	Mariscal Estigarribia
Guatemala	37°	Coban
Honduras	41°	Isla del Cisne
El Salvador	45°	San Salvadore
Belize	52°	Belize City
French Guiana	54°	Saul
Guyana	60°	Mazaruni
Suriname	60°	Paramaribo
Panama	63°	Balboa Heights

World's Coldest Places
Lowest Temperatures Recorded

Country	Temp	Location
AFRICA		
Morocco	-11°	Ifrane
South Africa	3°	Bethlehem
Algeria	12°	Batna
Namibia	15°	Gobabis
Zimbabwe	16°	Matopos
Canary Islands	16°	Izana
Madagascar	17°	Nanokely
Angola	18°	Chitembo
Egypt	18°	El Nakhl
Libya	19°	Hun
Tunisia	21°	Gafsa
Botswana	23°	Francistown
Malawi	26°	Lilongwe
Somalia	27°	Erigavo
Zambia	27°	Livingston
Ethiopia	28°	Addis Ababa
Mozambique	28°	Maputo
Sudan	28°	Wadi Halfa
Zimbabwe	28°	Bulawayo
Niger	29°	Magaria
Tanzania	30°	Mbfya
Niger	30°	Agadez
Nigeria	31°	Hadeijal
Kenya	32°	Eldoret
Eritrea	32°	Asmara
Mali	34°	Timbuctu
Congo	33°	Lubumbashi
Chad	36°	N'Djamena
Mauritania	37°	Kifa
Somalia	37°	Hargeysa
West Sahara	37°	El Aaiun
Senegal	41°	Kolda
Benin	43°	Kandi
Burkina Faso	44°	Dori
Gambia	45°	Banjul
Sierra Leone	45°	Daru
Ivory Coast	46°	Dimbokro
Cameroon	46°	Ngaoundere
Central African Rep	46°	Bangui
Togo	48°	Lome
Cape Verde	50°	Mindelo
Ghana	52°	Bole
Dem Rep of Congo	52°	Brazzaville
Liberia	59°	Monrovia
Djibouti	61°	Djibouti
OCEANIA		
Australia	-9.4°	Charlotte Pass
New Zealand	-6.9°	Ophir

After Siberia, the Yukon and Northwest Territories of Canada comprise the world's coldest inhabited region. The airport at Snag in the Yukon holds the record for the coldest temperature ever recorded in North America, with a reading of -81.4°F (-63°C) reported on February 3, 1947. On January 11, 1911, the temperature fell to -78°F at Fort Vermilion in northern Alberta. More recently, temperatures of -96° and -92° have been recorded—these on January 7, 1982—by two weather stations near Fort Nelson, British Columbia, that had been established as part of a permafrost study. These readings have not been accepted as official by Canada's weather bureau, the Atmospheric Environment Service.

Readings as low as -73°F were recorded not far from the U.S. border in the town of Iroquois Falls, Ontario, in January of 1935. Likewise, a temperature of -53°F was reported at Sisson Dam, New Brunswick, just 45 miles northeast of Van Buren, Maine. (No temperature lower than -48°F has ever been reported from Maine.) For enduring cold, nowhere in North America can beat Eureka in the Northwest Territories, where an average of -54°F in February of 1979 was recorded, the coldest month ever measured on the continent.

One would have to be resolute to endure the winter of 1948–1949 in Resolute, also in the Northwest Territories, where the temperature averaged -31.4°F for the entire winter, a continental record for the coldest winter ever (winter being considered the three months of December, January, and February).

The coldest wind-chill temperature ever recorded in North America occurred at Pelly Bay, Northwest Territories, where on January 13, 1975, a -60°F air temperature was accompanied by a sustained 35-mph wind, resulting in a wind-chill effect of -135°. (This is based on an earlier method of calculating the wind-chill factor and would be the equivalent of about -110° using today's standard.)

Europe's coldest area outside of Russia is the island of Spitsbergen, Norway, at a latitude of some 78°N. The lonely outpost of Gronfjorden has reported temperatures as low as -57°F and averages -4°F during March, its coldest month. On the mainland, the northern Norwegian town of Karasjok is Europe's coldest town, averaging about 2°F during February. Its absolute minimum is -37°. The northern part of Finland is subjected to similar temperatures.

WHAT'S IT LIKE AT -80°?

The coldest temperature ever officially recorded in the United States was at a camp along the Alaska pipeline. The camp was at Prospect Creek, just 20 miles north of the arctic circle along the James Dalton Highway. It was here that a reading of -79.8° was registered on January 23, 1971.

The hamlet of Coldfoot, another 40 miles north of Prospect Creek reported a reading of -82°, which is probably credible but not official for the various reasons temperatures are not considered "official" (exposure of thermometer, calibration, etc.).

Little information about the reading at Prospect Creek seems to be available beyond the measurement itself. However, after the temperature at Snag airport in the Yukon of Canada dropped to an "official" -81.4° on February 3, 1947, the 16 airport employees provided accounts of the experience.

Two of the weather observers, Wilf Blezard and Gordon Toole, made the following observation:

> We threw a dish of water high into the air, just to see what would happen. Before it hit the ground, it made a hissing noise, froze, and fell as tiny round pellets of ice the size of wheat kernels.
>
> Spit also froze before hitting the ground. Ice became so hard the ax rebounded from it. At such temperatures, metal snapped like ice; wood became petrified; and rubber was just like cement. The dogs' leather harness couldn't bend or it would break...It was unique to see a vapor trail several yards long pursuing one as he moved about outside. Becoming lost was of no concern. As an observer walked along the runway each breath remained as a tiny motionless mist behind him at head level. These patches of human breath fog remained in the still air for three or four minutes before fading away. One observer even found such a trail still marking his path when he returned along the same path 15 minutes later.

When breath freezes at this temperature it makes a hissing or swishing sound. When ice breaks it sounds like breaking glass. Extremely cold air makes radio static in the same way lightning does. Observer Wilf Blezard added that at -70° or lower, "It was easy to freeze your nose without even knowing it. At -30° you feel it coming."

David Phillips writes in his book *Blame it on the Weather*, an account of the event:

> There were other cold-weather experiences mentioned by the observers at Snag. For days, a small fog or steam patch would appear over the sled dogs at a height of about 20 feet. It would disappear only in the warm part of the day when the temperature warmed up to -60°. A chunk of ice was so cold that when brought into a warm room, it took five full minutes before there was a trace of moisture, even when held in the hand.

HISTORIC U.S. COLD WAVES OF THE 19TH CENTURY

Eastern Cold Waves of January and February 1835

No cold wave has rivaled the arctic outbreak of 1899 (see following), although earlier in the 19th century there were two freezes that may have been just as, or more, severe. The first of these was in January 1835. Mercury thermometers froze throughout the Northeast during this cold snap, so the absolute minimum temperatures reached will never be known. Mercury froze at -40° in Bangor and Bath, Maine, as well as Montpelier and White River in Vermont. Hartford's -27° and New Haven, Connecticut's, -23° have never been equaled since. The same was true in Massachusetts, where Williamstown hit -30° and Pittsfield -32°. In the Southeast, another cold wave one month later dropped the temperature to zero in Savannah, Georgia, a full eight degrees colder than the figure reached in 1899. Charleston, South Carolina, dropped to 2° and at Fort Marion in St. Augustine, Florida, a low of 7° and high of just 21° were measured on February 8.

With its depictions of horse-drawn sleighs, snow drifts, and ice skaters, this Currier & Ives print of an old-fashioned winter's day in New York City's Central Park (circa 1860) may have been inspired by the cold waves of 1857 and 1859.

NEW YORK CITY'S COLDEST DAYLIGHT EVER:
JANUARY 10, 1859:

The coldest daytime ever experienced in New York City (and throughout New England for that matter) occurred on January 10, 1859. Accurate thermometers were commonplace and well distributed by this time although most of them were not self-registering, meaning that observations had to be made visually, usually three times daily at 7 a.m., 2 p.m., and 9 p.m. Consequently, the daily absolute minimum or maximum temperatures are not precisely known. In any case, the day of January 10 was most likely the coldest ever, based on observations from Montreal, Canada, to New York City. Montreal reported a 7 a.m. reading of -43.6°F on this day, a reading some 15 degrees lower than its modern record minimum of -29°F recorded in 1933. Professor Petty of the University of Vermont in Burlington recorded -31.5° at 7 a.m. and a bone-chilling 2 p.m. temperature of only -26°. A spirit thermometer in Woodstock, Vermont, measured -45°. The campus thermometer at Harvard in Cambridge, Massachusetts, registered only -4.5° at 2 p.m. and dropped to -18° by the morning of the 11th, the lowest reading and coldest day ever in the Boston area. Even Nantucket Island, 20 miles off the coast of Cape Cod, measured -12° according to Smithsonian thermometers, some six degrees lower than the coldest modern official reading. In New York the official thermometer located at Erasmus Hall in Brooklyn registered the following temperature range that day:

7 a.m.	-3.7°
11 a.m	-7.5°
2 p.m.	-3.8°
9 p.m.	-8.0°

A Brooklyn Heights thermometer reported a reading of -9° at noon on January 10. Union Hall in the Jamaica section of Queens reported -12° at midnight on the 10–11th. White Plains, just north of the city, reported readings of -13° at 7 a.m., -10° at 2 p.m., and -15° at 9 p.m. This was probably the only day in New York City history when the temperature failed to rise above zero degrees. Lower nighttime readings have been observed since; -15° on the morning of February 9, 1934, is the city's lowest official reading as measured in the Central Park weather office.

New England Cold Wave of January 1857

January 1857 was the coldest month ever recorded in New England. The average temperatures of 16.7° at New Haven, 16.8° at Boston, and 19.6° in New York City remain the coldest months on record. The worst of the month's cold waves descended on New England on January 22, with the 23rd probably being the coldest day ever throughout the region.

Readings of -52° at Bath on the Maine coast; -51° in Franconia, New Hampshire; and -50° at both Montpelier and St. Johnsbury, Vermont, were reported by the press, ostensibly on spirit thermometers, since mercury freezes at -40°. The lowest official reading on a Smithsonian thermometer was -44° at Norwich, Vermont, lending credibility to the other readings in normally colder locations.

The temperature never rose above zero in Boston on this day, with Cambridge reporting a reading of -8° at 2 p.m. The Boston suburbs of Malden and West Newton both recorded -30° overnight. Incredibly, Nantucket Island was cut off from the mainland by ice, and the temperature at this usually moderate location dropped to -6.5°, with unofficial reports of -11° being noted. Even on Cape Cod temperatures as low as -24° occurred in Wareham and -19° at Yarmouth. In New York City, the Smithsonian thermometer at Erasmus Hall in Brooklyn reached a high temperature of zero, and the Hudson River froze over solidly enough for people to walk to Hoboken.

Cold Wave of February 1899

-32° - Coldest recorded low in state during February 1899

-32°* - State record low still standing

• -49° Billings - City record low still standing

Lowest observed temperature during cold wave

source: Monthly Weather Review Vol. 27 no. 2

A rare photograph of the three to four inches of snowfall that came down in New Orleans on February 14, 1899. The temperature dropped to 6.8° just prior to the snowfall on February 13, the coldest day ever reported in the city. (Historic New Orleans Collection)

Great Nationwide Cold Wave of February 1899

The greatest cold wave in modern U.S. history (modern meaning during the period of U.S. Weather Bureau record keeping) occurred during the first two weeks of February in 1899. For the first and only time, temperatures fell below zero in every state of the Union, including Florida, where Tallahassee dropped to an unbelievable -2°. The Weather Bureau reported unofficial readings as low as -4° at other western Florida locations. In Mobile, Alabama, the reading of -1° was a full 13 degrees colder than the previous record low of 12°. Even tropical Havana, Cuba, was affected on February 14, when the temperature did not rise above 54°.

In New Orleans the lowest reading was 6.8°, and ice floes choked the Mississippi River and even drifted into the Gulf of Mexico. Three inches of snow fell in New Orleans and ice two inches thick formed on standing freshwater. Snow showers were reported by ships in the Gulf of Mexico.

The arctic outbreak spread south out of Canada on February 7, dropping temperatures to -61° in Montana, -59° in Minnesota, (a state record that stood for almost a century, until the village of Tower recorded -60° on February 1, 1996), -51° in Wyoming, and -50° in Wisconsin.

On February 8, 1899, the maximum temperature climbed to only -39° in Roseau. Minnesota, the second-coldest maximum temperature ever recorded in the lower 48 states. (The coldest maximum temperature ever recorded in the lower 48 was -44° in Glasgow, Montana, on January 12, 1916). By mid-February 1899, the Mississippi River had frozen 16 inches thick at St. Louis. Washington, D.C., reported -15° on February 11, the coldest modern temperature ever recorded in the nation's capital. Record coldest temperatures for over 40 major cities and four states still stand today.

HISTORIC U.S. COLD WAVES OF THE 20TH CENTURY

Great Plains Winter of 1935–36

The Great Plains experienced its coldest winter in 1935–1936. Most of Montana averaged 20 degrees below normal for the entire month of February 1936. The town of Malta averaged an incredible 31.4 degrees below normal and had four days when the temperature fell below -50°. In North Dakota, the town of Langdon remained below zero for 41 consecutive days from January 11 to February 20, the

Langdon, N.D. Deep Freeze of Winter 1935-1936

Minimum below 32°	176 days	Oct 17, 1935 - Apr 10, 1936
Maximum below 32°	92 days	Nov 30, 1935 - Feb 29, 1936
Minimum below 0°	67 days	Dec 31, 1935 - Mar 6, 1936
Maximum below 0°	41 days	Jan 11, 1936 - Feb 20, 1936

longest such stretch in the history of the U.S. outside of Alaska. The temperature never rose above 15° during the entire month of February.

Fort Berthold Agency, North Dakota, went for seven consecutive days with a maximum temperature of -20° or lower, and for two full weeks remained below -10°. Parshall, North Dakota, dropped to -60° on February 15 for the state's coldest reading ever. Turtle Lake, North Dakota, had a mean temperature of -19.4° in February, which constitutes the coldest monthly average temperature ever recorded in the lower 48 states. The village of Amenia, also in North Dakota, registered a maximum temperature of only 10° for the entire month of January, perhaps the lowest monthly maximum temperature ever measured in the lower 48 states.

42 BELOW FEB, 16. 36

DOWN & DOWN SHE GOES.
AND WHERE SHE STOPS.
NO BODY KNOWS.

February 1936, was the coldest month ever, from Montana to Minnesota. For this gentleman in Otter Tail County, Minnesota, however, -42° wasn't all that chilly. He's taken his hat off and appears ready to fan himself with it. (Otter Tail County Historical Society)

In January 1949, blizzards isolated rural homesteads from Nevada to Nebraska. In many cases supplies had to be airlifted to stranded residents. (Nebraska State Historical Society)

Western U.S., January 1949

The West, from California to Colorado and north to Idaho, experienced its worst cold wave during January 1949. Spokane, Washington, and Boise, Idaho, both had their coldest months ever with averages of 8.5° and 10.3° respectively. This was the coldest winter at almost every weather station in California, Nevada, Idaho, and Oregon, with January being the coldest single month. The cold was accompanied by severe blizzards, which paralyzed the Great Basin region and isolated ranches in Wyoming for weeks at a time. In response, the U.S. government initiated "Operation Hay Lift." It was run by the army, which distributed hay and food to isolated ranches throughout Nevada, Wyoming, Colorado, and South Dakota.

Las Vegas, Nevada received an unprecedented and unequaled 16.7 inches of snowfall during the month of January.

Wind Chill Chart

Wind Speed (MPH)	40°	35°	30°	25°	20°	15°	10°	5°	0°	-5°	-10°	-15°	-20°	-25°	-30°	-35°	-40°	-45°
5	36°	31°	25°	19°	13°	7°	1°	-5°	-11°	-16°	-22°	-28°	-34°	-40°	-46°	-52°	-57°	-63°
10	34°	27°	21°	15°	9°	3°	-4°	-10°	-16°	-22°	-28°	-35°	-41°	-47°	-53°	-59°	-66°	-72°
15	32°	25°	19°	13°	6°	0°	-7°	-13°	-19°	-26°	-32°	-39°	-45°	-51°	-58°	-64°	-71°	-77°
20	30°	24°	17°	11°	4°	-2°	-9°	-15°	-22°	-29°	-35°	-42°	-48°	-55°	-61°	-68°	-74°	-81°
25	29°	23°	16°	9°	3°	-4°	-11°	-17°	-24°	-31°	-37°	-44°	-55°	-58°	-64°	-71°	-78°	-84°
30	28°	22°	15°	8°	1°	-5°	-12°	-19°	-26°	-33°	-39°	-46°	-53°	-60°	-67°	-73°	-80°	-87°
35	28°	21°	14°	7°	0°	-7°	-14°	-21°	-27°	-34°	-41°	-48°	-55°	-62°	-69°	-76°	-82°	-89°
40	27°	20°	13°	6°	-1°	-8°	-15°	-22°	-29°	-36°	-43°	-50°	-57°	-64°	-71°	-78°	-84°	-91°
45	26°	19°	12°	5°	-2°	-9°	-16°	-23°	-30°	-37°	-44°	-51°	-58°	-65°	-72°	-79°	-86°	-93°
50	26°	19°	12°	4°	-3°	-10°	-17°	-24°	-31°	-38°	-45°	-52°	-60°	-67°	-74°	-81°	-88°	-95°
55	25°	18°	11°	4°	-3°	-11°	-18°	-25°	-32°	-39°	-46°	-54°	-61°	-68°	-75°	-82°	-89°	-97°
60	25°	17°	10°	4°	-4°	-11°	-19°	-26°	-33°	-40°	-48°	-55°	-62°	-69°	-76°	-84°	-91°	-98°

Frostbite Times ☐ 30 Min ☐ 10 Min ☐ 5 Min   Temperature °F

Ohio Valley and Eastern U.S., January 1977

The greatest nationwide cold wave of the 20th century occurred in January of 1977, when 69 of the 302 stations listed in the weather tables of this book recorded their coldest month on record. The core of the cold air extended from Florida to New Hampshire and west to Iowa and Missouri. At the very heart of the cold air mass was Ohio, where every station measured its coldest month on record, and Cincinnati dropped to -25°, its coldest temperature ever (especially remarkable because the length of its record stretches back to 1820). For the first time, snow was reported as far south as Miami. During the entire month of January, from eastern Iowa to western Pennsylvania and points north, temperatures failed to rise above freezing, even for a day—an unprecedented event. At Minneapolis the wind-chill temperature dropped to -78° on the morning of January 28, perhaps the lowest on record in that city.

Buffalo, New York, was slammed by its worst blizzard ever during the last week of January, when near hurricane-force winds created white-out conditions for almost three days. Along with the wind and snow, temperatures hovered near zero with wind-chills of -60°. Snowdrifts up to 30 feet deep literally buried entire homes with their occupants.

The blizzard of January 1977, in and around Buffalo, New York, was so severe that winds up to 73 mph broke windows in homes, which were then quickly filled with blowing and drifting snow. (Dino Innandrea, whitedeath.com)

TEMPERATURE ANTICS IN THE BLACK HILLS & GREAT PLAINS

The Indians called chinooks "snow eaters." These warm and wild winds arrive in the frigid Great Plains from the southwest in the middle of winter. Today, a chinook is still a welcome winter visitor, melting snow so rapidly that a deep cover could disappear in a single day. Chinook winds occur when a stationary front retreats northward raising temperatures dramatically along and to the east of the Rocky Mountain Front Range.

SO COLD; THEN SO WARM!

Kipp, Montana, on December 1, 1896, saw the temperature rise 80 degrees in about 15 hours, entirely melting 30" of snow. In Granville, North Dakota, the temperature rose from a morning low of -33° to an afternoon high of 50° on February 21, 1918, an astonishing 83-degree rise, as a result of a chinook wind. Even more dramatic was the 42-degree increase in temperature in just 15 minutes that occurred at Fort Assiniboine, Montana, on January 19, 1893. More recently, Helena, Montana experienced a chinook that raised the temperature from -36° to 54° in 50 hours during January of 1954. And topping all the above records, in Loma, Montana, the temperature rose from -54° to 49° on January 14–15, 1972, the 103° change being the greatest ever recorded in the world for a 24-hour period.

SO WARM; THEN SO COLD!

Conversely, a stationary front can retreat southward as a cold front, and cause a sudden drop in temperature. Browning, Montana, saw the thermometer plummet 100 degrees in 24 hours on January 23–24, 1916, from a relatively mild 44° to a bone-chilling -56°. This is generally credited as being the greatest temperature drop ever reported in the world for a 24-hour period.

In Fairfield, Montana, the temperature once dropped 84 degrees in just 12 hours, from 63° at noon on December 24, 1924, to -21° by midnight that same day, the record for a 12-hour temperature change. Huron, South Dakota, once suffered through a precipitous drop of temperature from a pleasant 71° to –8° on March 17–18, 1883.

The famous cold front passage of January 10–11, 1911, was responsible for the sharpest drop in temperature ever across a wide area of the Midwest. Kansas City saw the thermometer fall from a balmy 76° around noon on January 11, to 11° by midnight. Oklahoma City's temperature dropped from 83° to 17°. Most dramatic of all was the 75-degree drop in temperature from 62° to -13°, at Rapid City, South Dakota, in just two hours on January 10 between 6:00 and 8:00 a.m.

SLOSHING TEMPERATURES

The Black Hills of South Dakota occasionally experience a unique phenomenon when a stationary front separating very cold air from mild air stalls over its mountains and valleys. When this happens, cold or warm pockets of air become trapped in the valleys and slosh back and forth like water in a shallow bowl. This can result in dramatic changes of temperature measured in minutes rather than hours. Spearfish saw its temperature warm from -4° at 7:30 a.m. to 45° at 7:32 a.m., a 45-degree rise in just two minutes during the morning of January 22, 1943. By 9:00 a.m. the temperature had risen gradually to 54° when it suddenly dropped again to -4° over the next 27 minutes. Dressing for the day must have been problematic for Spearfish's residents. Rapid City suffered the same effect a few hours later as evidenced by the thermograph trace below.

The town of Lead, up in the hills, experienced such a shocking change in temperature that plateglass windows cracked. At one point, the town of Deadwood, in a canyon 600 feet lower than Lead but only one and a half miles away, had a temperature of -16° at the same time that it was 52° in Lead. Wind gusts of 40–50 mph were whipping through the region. Motorists found it difficult to drive as their windshields would instantly frost over as they drove from a warm pocket to a cold one.

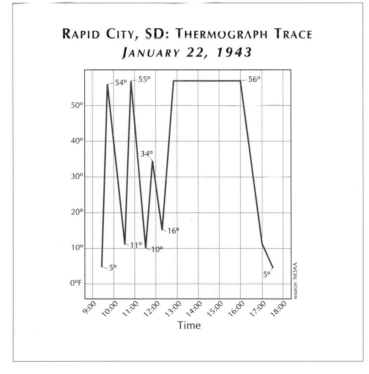

RAPID CITY, SD: THERMOGRAPH TRACE
JANUARY 22, 1943

source: NOAA

chapter 3
SNOW & ICE

It's hard to believe that cities argue over which may call itself the "snowiest in America," but such wrangles do occur. It's not always clear which city can claim to be snowiest because there are different ways to define a city and different places within or just outside a city to measure snowfall. Some towns too small to truthfully call themselves cities have claimed to be the "snowiest cities" in America, much to the dismay of their more populous but less snowy big sisters. So, for the sake of this book, the word "city" will refer only to an urban area of 10,000 citizens or more.

SNOWIEST CITIES IN THE UNITED STATES

Whether a city can claim it's the snowiest may well depend on the location of its weather station. Take the situation of Marquette, Michigan. After its official weather station was moved some years ago from downtown to the airport, the city's annual average snowfall increased dramatically. This is because the airport station is located about 500 feet higher than, and eight miles west of, the town and on terrain more exposed to the effects of lake-effect snowfalls. When the snow squalls blowing off Lake Superior hit the hills a few miles inland from the lakeshore, they drop copious amounts of snow. The difference in seasonal snowfall totals between the city location and the airport location is more than 60 inches in a typical winter. This now makes Marquette the second, rather than the sixth, snowiest city in the United States. All of the "major" U.S. cities (100,000 people or more) on the list are in New York State and record consistently high snowfalls because of their proximity to Lake Erie or Lake Ontario.

Among the snowiest places in the U.S., and for that matter, the world, are the higher elevations of Alaska's coastal mountain ranges, where more than 600 inches of the white stuff accumulate each winter. (Calvin W. Hall, AlaskaStock.com)

Snowiest U.S. Cities

Snowfall	Location	Population
203.4"	Truckee, CA	13,800
179.8"	Marquette, MI (airport)	
118.2"	Marquette, MI (city)	19,600
173.3"	Steamboat Springs, CO	9,000
153.3"	Oswego, NY	18,000
131.2"	Sault Ste. Marie, MI	16,600
120.2"	Syracuse, NY	147,300
111.2"	Meadville, PA	13,700
111.1"	Flagstaff, AZ	53,000
110.8"	Watertown, NY	26,700
105.9"	Muskegon, MI	40,100
99.5"	Rochester, NY	219,800
98.5"	Utica, NY	60,600
97.9"	Montpelier, VT	8,100
96.8"	Traverse City, MI	14,500
95.7"	Buffalo, NY	292,700
95.3"	Juneau, AL	30,700
93.2"	Presque Isle, ME	9,600
91.3"	Cortland, NY	18,700
85.6"	Casper, WY	49,700
83.3"	Duluth, MN	87,000
82.6"	Berlin, NH	10,300
81.6"	Burlington, VT	38,900

Population figures based on 2000 census
Source: NCDC and Regional Climate Centers

A few cities with populations under 10,000 make the list above. Steamboat Springs, Colorado, is a rapidly growing community and probably by now has more than 10,000 residents. Montpelier is included, despite its small size, as it is the capital of Vermont.

Flagstaff, Arizona, located near some of the country's hottest cities, is America's eighth snowiest city. This is a result of its location, of course—at an altitude of 7,000 feet. Seasonal snowfall in Flagstaff, however, may vary widely from year to year depending on such factors as the occurrence and strength of an El Niño and the alignment of the jet stream during winter months. In Flagstaff as little as 32″ of snow (in 1983–1984) and as much as 210″ of snow (in 1972–1973) have fallen during the winter season. The extreme range for Juneau, Alaska is even more impressive, from a low of 24″ (1987–1988) to a high of 246″ during the winter of 1917–1918.

Snowfall averages not only vary widely from one season to the next but also may be changing over time. Data seems to suggest that the cities around the Great Lakes are becoming snowier, while some places, like Juneau, are becoming less so. Juneau's seasonal snowfall for the period of 1943–1974 averaged 109.1″. Since then (for the period 1970–2000) it has decreased to only 95.3″. However, other factors than climate change could be at work. Instruments may have been moved from one location to another, and Juneau is subject to a staggering diversity of microclimates. A difference in elevation of only 100 to 200 feet, say, from the harbor to the top of 6th street (a distance of only about 1,000 yards), can be the difference between no snowfall and a 10″ accumulation. Likewise, the annual total precipitation varies from just 53″ in the town center to 93″ at the airport eight miles away.

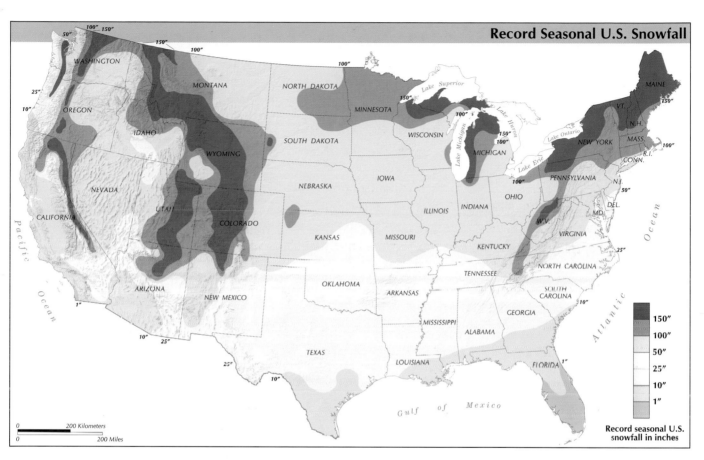

Record Seasonal U.S. Snowfall

Record seasonal U.S.
snowfall in inches

150"
100"
50"
25"
10"
1"

SNOWIEST PLACES IN THE UNITED STATES

Western, Coastal Mountain Ranges

The high mountain ranges of the West Coast of America are the snowiest places in America, and possibly, in the world. Moisture flowing off the Pacific Ocean piles up against the slopes of the Chugach Mountains of Alaska, the Coast Mountains of British Columbia, the Cascades of Washington and Oregon, and the Sierra Nevada of California. Moisture-laden clouds are lifted and condensed, squeezing out precipitation that during the winter months falls almost exclusively in the form of snow. The elevations receiving the greatest accumulations range between 4,000 and 9,000 feet in Washington; 5,000 and 9,000 feet in Oregon; and between 7,000 and 9,000 feet in California. (Above 9,000 feet, precipitation tends to decrease as available moisture decreases with altitude.)

(opposite) The Grand Canyon dramatically demonstrates the effect elevation has on snowfall. At Phantom Ranch, at the bottom of the canyon (elev. 2,570 feet), almost no snow falls. On the North Rim, about 6,000 feet higher but only a few miles away, 130 inches normally accumulates each winter. (Kerrick James)

SNOWIEST LOCATIONS BY STATE

State	Snowfall	Location
ALABAMA	5.3"	Valley Head
ALASKA	551.5"	Thompson Pass
ARIZONA	243.0"	Sunrise Mountain
ARKANSAS	16.2"	Gravette
CALIFORNIA	470.7"	Soda Springs
COLORADO	435.6"	Wolf Creek Pass
CONNECTICUT	106.8"	Norfolk
DELAWARE	18.7"	Wilmington
FLORIDA	0.2"	Milton Exp. Station
GEORGIA	6.1"	Clayton
HAWAII	0"	
IDAHO	283.5"	Mullan Pass
ILLINOIS	42.8"	Chicago (Midway)
INDIANA	76.6"	South Bend
IOWA	43.8"	Dubuque
KANSAS	46.0"	McDonald
KENTUCKY	23.6"	Covington
LOUISIANA	1.9"	Plain Dealing
MAINE	118.0"	Rangeley
MARYLAND	95.9"	Oakland
MASSACHUSETTS	85.0"	West Cummington
MICHIGAN	235.8"	Herman
MINNESOTA	83.3"	Duluth
MISSISSIPPI	4.9"	Cleveland
MISSOURI	25.5"	Bethany
MONTANA	305.5"	Kings Hill
NEBRASKA	59.3"	Mullen
NEVADA	241.0"	Marlette Lake
NEW HAMPSHIRE	315.4"	Mt. Washington
NEW JERSEY	39.8"	Sussex
NEW MEXICO	164.9"	Red River
NEW YORK	226.7"	Old Forge
NORTH CAROLINA	57.8"	Grandfather Mountain
NORTH DAKOTA	56.5"	Bowman
OHIO	97.0"	Chardon
OKLAHOMA	31.6"	Boise City
OREGON	529.9"	Crater Lake
PENNSYLVANIA	121.6"	Corry
RHODE ISLAND	55.5"	North Foster
SOUTH CAROLINA	8.7"	Caesars Head
SOUTH DAKOTA	193.0"	Lead
TENNESSEE	16.7"	Tazewell
TEXAS	23.9"	Borger
UTAH	516.3"	Alta
VERMONT	222.0"	Mt. Mansfield
VIRGINIA	51.8"	Burkes Garden
WASHINGTON	680.0"	Rainier Paradise R.S.
WEST VIRGINIA	157.8"	Terra Alta
WISCONSIN	138.7"	Gurney
WYOMING	285.0"	Bechler Ranger Station

source: NCDC

AVERAGE ANNUAL SNOWFALL

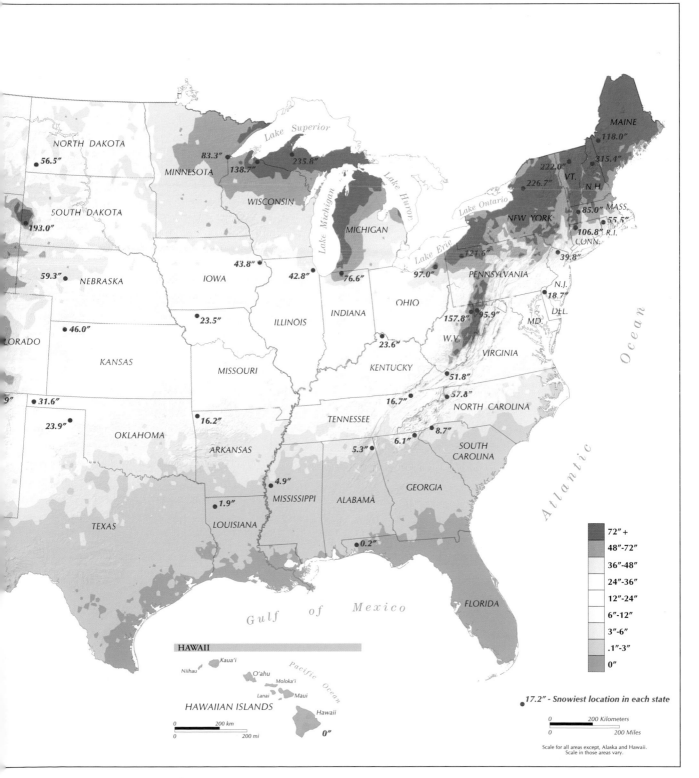

Legend:

	72" +
	48"–72"
	36"–48"
	24"–36"
	12"–24"
	6"–12"
	3"–6"
	.1"–3"
	0"

17.2" - Snowiest location in each state

0   200 Kilometers
0   200 Miles

Scale for all areas except, Alaska and Hawaii.
Scale in those areas vary.

NORTH DAKOTA ● 56.5"
SOUTH DAKOTA ● 193.0"
MINNESOTA 83.3" 138.7"
WISCONSIN 235.8"
MICHIGAN 76.6"
NEBRASKA ● 59.3"
IOWA ● 43.8"
ILLINOIS ● 23.5"
INDIANA 42.8"
OHIO 97.0"
● 46.0"
KANSAS
MISSOURI
KENTUCKY 23.6"
● 51.8"
● 31.6"
23.9"
OKLAHOMA
ARKANSAS ● 16.2"
TENNESSEE 16.7"
NORTH CAROLINA 57.8"
8.7"
6.1"
5.3"
SOUTH CAROLINA
MISSISSIPPI ● 4.9"
ALABAMA
GEORGIA
LOUISIANA ● 1.9"
0.2"
TEXAS
FLORIDA
MAINE ● 118.0"
315.4"
222.0"
226.7" VT.
N.H.
NEW YORK
85.0" MASS.
55.5"
106.8" R.I.
CONN.
121.6"
39.8"
PENNSYLVANIA
N.J. ● 18.7"
157.8" 95.9"
DEL.
W.V.
MD.
VIRGINIA

Lake Superior
Lake Michigan
Lake Huron
Lake Ontario
Lake Erie
Ocean
Atlantic
Gulf of Mexico

HAWAII
Kaua'i
Niihau
O'ahu
Moloka'i
Lanai   Maui
Hawaii
HAWAIIAN ISLANDS   0"
Pacific Ocean
0   200 km
0   200 mi

The snowiest places of all are probably located somewhere in the coastal mountains of Alaska, but no weather stations exist there to record what may amount to more than 1,000″ of annual snowfall. Thompson Pass, located at 2,600 feet and about 20 miles inland from the port of Valdez, receives an average of 600″ of snowfall per season. During the winter of 1952–1953 a staggering 975″ fell, with 346″ once being recorded in a single month. During a five-day period in December of 1955, a single snowstorm deposited 175.4″.

Valdez (pop. 4,068), at sea level, is Alaska's snowiest town with 328″ falling during an average season. It can, therefore, be safely assumed that somewhere in the mountains of this region far greater amounts fall. This is also true for British Columbia's Coast Mountains where 100–150″ of rain fall along the coastline every year, with most of this precipitation falling as snow in the mountains above 2,000 feet. The isolated settlement of Kemano, 40 miles inland from Princess Royal Island and located at an elevation of 4,500 feet, recorded 888″ of snow during the 1956–1957 season, a Canadian record until Mt. Copeland, near Revelstoke, also in British Columbia, reported 963″ the winter of 1971–1972.

Valdez, Alaska is the snowiest sea-level location in the world, averaging more than 300 inches of snowfall a year. (Don Pitcher)

The snowiest location in North America is the Paradise Lodge and Ranger Station located at 5,430 feet on the slopes of Mt. Rainier in Washington. An average of 680 inches falls each winter and this has accumulated to a depth of as much as 30 feet on occasion. This photograph, taken in 1927, shows the lodge buried under a 27-foot accumulation. (NOAA)

U.S. Pacific Northwest

In the United State's lower 48, ranger stations and ski resorts maintain snow-depth stations high on the slopes of the snowiest mountains, and thus a more complete record is available than in Alaska. The Mt. Baker Lodge at about 5,000 feet in Washington recorded 1,140" (95 feet!) of snow during the winter of 1998–1999. This is now considered the heaviest seasonal snowfall recorded on earth. The previous record holder was the Paradise Ranger Station at 5,500 feet on the slopes of nearby Mt. Rainier, where 1,122" was recorded during the season of 1971–1972. Another 1,000"-plus season occurred in 1955–1956 when a record snow depth of 367", over 30 feet, was on the ground by March. The last of this massive accumulation did not melt until July 29, 1956.

Further south, Crater Lake, Oregon's snowiest location, had 903" of snowfall in 1949–1950. But the record for the most snow in a single month, and the deepest snow depth ever recorded in North America, belongs to Tamarack, California. Here, near Yosemite's Tuolumne Meadows, 390" fell in January 1911. This led to a level snow depth of 451" (37.5 feet) by March of that

20 Snowiest Locations in the Western U.S.

Snowfall	Location
680″	Paradise R.S., Mt. Rainier, WA
552″	Thompson Pass, AK
530″	Mt. Baker Lodge, WA
530″	Crater Lake, OR
516″	Alta, UT
471″	Soda Springs, CA
445″	Tamarack, CA
442″	Stampede Pass, WA
436″	Wolf Creek Pass, CO
429″	Silver Lake Brighton, UT
395″	Twin Lakes, CA
328″	Valdez, AK
305″	Kings Hill, MT
285″	Bechler River Ranger Station, WY
283″	Mullan Pass, ID
276″	Snake River, WY
271″	Climax, CO
265″	Silver Lake, CO
257″	Government Camp, OR
254″	Holden Village, WA

year. Tamarack also holds California's greatest seasonal catch of 884″ in the notoriously wet winter of 1906–1907. Tamarack normally catches 445″ of snow per season, one of the highest such figures in California.

Rockies and Black Hills

Outside of the Pacific states, the snowiest places are in the Wasatch Mountains of Utah and the San Juan Mountains in southwestern Colorado. Higher elevations here normally receive over 400″ of snowfall a season. Wyoming's Pitchstone Plateau in Yellowstone and the spine of the Bitterroot Mountains between Idaho and Montana are the West's next snowiest places, averaging 250–300″ of snowfall. A surprisingly snowy location is the Black Hills of South Dakota. The town of Lead (pop. 3,062) receives 193″ of snow on average each winter and is often blasted by storms that produce phenomenal accumulations. One such storm buried the town under almost 115″ of snow in just seven days in 1998. Easterly

SNOW IN THE SIERRA NEVADA

Some of the greatest accumulations of snow on earth occur in California's Sierra Nevada above the 5,000-foot level almost every winter. In the 1860s, the construction crews on the Central Pacific Railroad, found the deep snows a major obstacle. The line's first winter season, 1866–1867, was an exceptionally snowy one with a level accumulation reaching 15 feet by mid-April. Worse yet, the snow drifted into some cuts to the enormous depth of 60 feet! In these stretches dynamite was used to clear the packed snow away.

The worst season of all was that of 1879–1880. A single storm dropped 10 feet of snow in two days (January 9 and 10), blocking the rail line and stranding train passengers for days. They included Nellie Bly, a New York newspaper woman who was trying to best the record of the heroes in Jules Verne's Around the World in Eighty Days. In the end, she detoured to the south via the Southern Pacific route and made her challenge by arriving back in New York in under 73 days.

In April of that season, an even stronger storm dropped a world-record 194" (over 16 feet) in just four days on Norden, a depot near Donner Summit, the greatest amount ever measured from a single storm anywhere in the world.

The 1879–1880 season total amounted to 783" (over 65 feet) at the Summit station located at 7,017 feet.

Even greater accumulations have been recorded at Tamarack near Yosemite's Toulumne Meadows. A North American–record snow depth of 451" (over 37 feet) was attained here in March of 1911, and a state-record seasonal total of 884" accumulated here during the winter of 1906–1907. More recently, Donner Summit, along Interstate Highway 80, measured 815" during the winter of 1951–1952. The Southern Pacific Railroad's flagship, City of San Francisco, was trapped by an avalanche for three days during a blizzard near Yuba Pass beginning January 13. The train lost power, and food supplies for the 226 passengers almost ran out before a rescue party arrived.

The incident was reminiscent of the ordeal suffered by the Donner Party some hundred years earlier when they were trapped below the pass later named after them for almost the entire winter of 1846-1847. Of the 83 pioneers, only 45 survived.

(opposite) Although the mountains of the Pacific Northwest receive heavier annual snowfalls, the Sierra Nevada of California holds the record for the greatest accumulated depth. Keeping the rail line open over Donner Summit has always been a chore, as this photo of Blue Canyon in 1917 illustrates. (NOAA)

Selected Record Point Snowfalls

Inches	Time	Location	Date
2.4″	15 minutes	Oswego, NY	1/26/1972
4.8″	30 minutes	Oswego, NY	1/26/1972
9.1″	1 hour	Oswego, NY	1/26/1972
17.5″	2 hours	Oswego, NY	1/26/1972
27.0″	3.5 hours	Oswego, NY	1/26/1972
36.0″	9 hours	Adams, NY	1/18/1993
40.0″	12 hours	Montague Township, NY	1/11-12/1997
70.9″*	15 hours	Dartmoor, Great Britain	2/16/1929
51.0″	16 hours	Bennett Ridge, NY	1/17/1959
67.8″	19 hours	Bessans, France	4/5-6/1959
62.0″	22 hours	Freye's Ranch, CO	4/14-15/1927
84.0″*	24 hours	Crestview C.H. Dept, CA	1/14-15/1952
77.0″*	24 hours	Montague Township, NY	1/11-12/1997
75.8″	24 hours	Silver Lake, CO	4/14-15/1927
68.2″	*24 hours*	*Tsukayama, Japan*	*12/30-31/1960*
68.0″	*24 hours*	*Adams, NY*	*1/9/1976*
68.0″	*24 hours*	*Squaw Valley, CA*	*1/1/1997*
67.0″	*24 hours*	*Echo Summit, CA*	*1/4/1982*
65.0″	*24 hours*	*Crystal Mountain, WA*	*1/23-24/1994*
62.0″	*24 hours*	*Thompson Pass, AK*	*12/29/1955*
58.6″	*24 hours*	*Takada, Honshu, Japan*	*2/8/1927*
57.1″	*24 hours*	*Tahtsa Lake, BC, Canada*	*2/11/1999*
56.0″	*24 hours*	*Randolf, NH*	*11/23-24/1943*
55.5″	*24 hours*	*Alta, UT*	*1/5-6/1994*
54.4″	*24 hours*	*Montevergine, Italy*	*2/22/1929*
87.0″	27.5 hours	Silver Lake, CO	4/14-15/1927
95.0″	32.5 hours	Silver Lake, CO	4/14-15/1927
99.7″	68 hours	Alta, UT	1/15-17/1995
129.0″	3 days	Laconia, WA	2/24-26/1910
194.0″*	4 days	Norden, CA	4/20-23/1880
189.0″	6 days	Mount Shasta, CA	2/13-19/1959
424.0″	30 days	Tamarack, CA	1-2/1911
390.0″	1 month	Tamarack, CA	1/1911
363.0″	*1 month*	*Paradise R.S., WA*	*1/1925*
362.2″	*1 month*	*Takada, Honshu, Japan*	*1/1945*
504.0″	2 months	Paradise R.S., WA	1-2/1925
503.0″	*2 months*	*Summit, CA*	*12/1894-1/1895*
710.5″	3 months	Mount Baker, WA	12/1998-2/1999
1107.0″	6 months	Mount Baker, WA	10/1998-4/1999
1140.0″	12 months	Mount Baker, WA	1998-1999
1122.0″	*12 months*	*Paradise R.S., WA*	*1971-1972*
974.4″	*12 months*	*Thompson Pass, AK*	*1952-1953*
963.2″	*12 months*	*Revelstoke Mt., BC Canada*	*1971-1972*

*unofficial measurements.
Italicized text indicates not record but other notable snow accumulations

winds accomplish this by carrying moisture all the way from the Gulf of Mexico, colliding with arctic air pouring down from Canada, and condensing the moisture into snowfall over the rugged terrain of the Black Hills.

Midwest and East

East of the Mississippi, the snowiest locations are Michigan's Upper Peninsula, the lake-effect snow belts of upstate New York, and the mountain tops of northern New England.

Portions of Michigan's Upper Peninsula normally receive the heaviest lowland snowfalls east of the Rockies. The heavy snow is a result of intense snow squalls blowing off Lake Superior. The Keweenaw Peninsula on the Upper Peninsula is the snowiest location of all, with a dot on the map called Herman receiving an average 236″ accumulation every winter (see table).

For a mountainous location, New Hampshire's Mt. Washington (elev. 6,288 feet) tops the list with over 300″ of snowfall each winter. The observatory on the mountain's summit recorded 566″ in the season of 1968–1969, the most ever recorded anywhere east of the Rocky Mountains. The snowiest low-level location in the Northeast is Old Forge, New York (pop. 1,061), where lake-effect snow squalls off Lake Ontario contribute to its 227″ annual average. Some of the most intense snowfalls recorded anywhere in the world occur in this region of the state. Montague Township reported 77″ of snowfall in 24 hours on January 11–12, 1997. The fall is not considered official, however, because measurements were made too frequently over the period of snowfall. The village of Adams holds New York state's official 24-hour snowfall record, when an equally impressive 68″ fell on January 9, 1976.

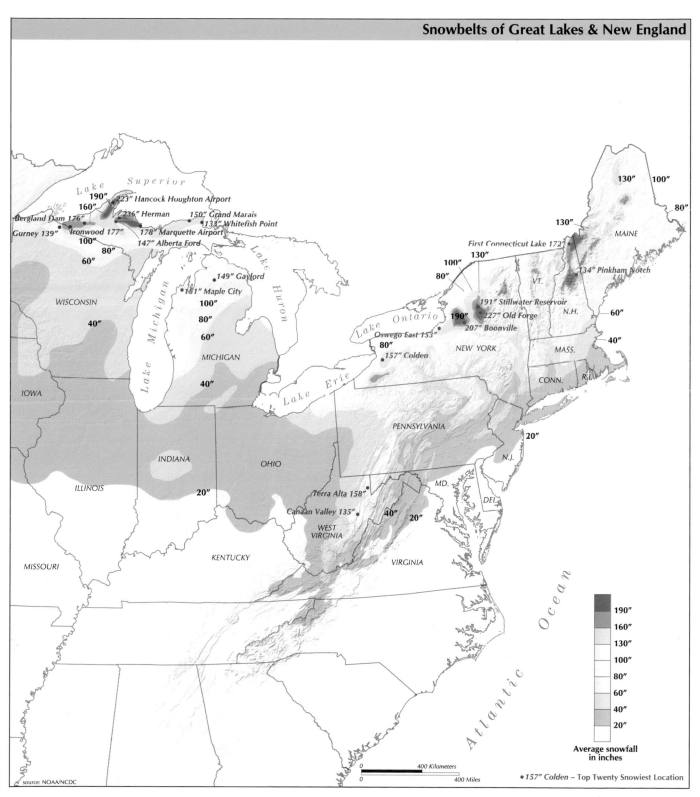

Snowbelts of Great Lakes & New England

Lake Superior

190" 223" Hancock Houghton Airport
160"
236" Herman   150" Grand Marais
Bergland Dam 176"   133" Whitefish Point
Gurney 139"   Ironwood 177"   178" Marquette Airport
100"   147" Alberta Ford
80"
60"

130" 100"
MAINE   80"

130"

First Connecticut Lake 172"

149" Gaylord
151" Maple City
100"
80"
60"

WISCONSIN

40"

Lake Michigan

Lake Huron

MICHIGAN

40"

IOWA

100"
80"

134" Pinkham Notch

VT.

N.H.   60"

191" Stillwater Reservoir
Lake Ontario   190"   227" Old Forge
Oswego East 153"   207" Boonville
80"
157" Colden   NEW YORK

MASS.   40"

CONN.   R.I.

Lake Erie

PENNSYLVANIA

20"

N.J.

MD.

DEL.

INDIANA   OHIO

ILLINOIS

20"

Terra Alta 158"
Canaan Valley 135"
40"   20"
WEST
VIRGINIA

MISSOURI   KENTUCKY

VIRGINIA

Atlantic Ocean

190"
160"
130"
100"
80"
60"
40"
20"

**Average snowfall
in inches**

● *157" Colden* – Top Twenty Snowiest Location

0   400 Kilometers
0   400 Miles

source: NOAA/NCDC

Snow in New England

Every form of severe weather occurs almost annually somewhere in New England—blizzards and cold waves in the winter and spring, severe thunderstorms, heat waves, and occasionally tornadoes in the summer, and droughts and floods every decade.

This extreme climate is the result of New England's being at the crossroads of diverse weather systems—cold, dry air flowing in from the Canadian arctic, moist maritime air pushing inland from the Atlantic Ocean, tropical heat and moisture flowing up from the Gulf of Mexico, and storms and air masses carried all the way across the country riding the jet stream.

Since weather generally moves from west to east, the moderating influences of the ocean are usually subsumed by the movement of air masses over the long and not-at-all moderating fetch of the continental mainland.

It is when two of these air masses collide along the East Coast—dry polar air from Canada hitting moist subtropical air from the Gulf of Mexico, along the route of a strong and bending jet stream—that New England's famous nor'easters occur.

In winter these storms produce New England's most intense blizzards. Often the storms' easterly winds draw warm air from the Atlantic into their circulation and turn the snow into rain along coastal areas. As a result, heavy snow is usually confined to areas at least 50 miles from the shoreline. Average annual snowfall in New England ranges from 30" along the immediate coastline of Connecticut, Rhode Island, and Massachusetts to over 100" in most of the interior regions.

Paul J. Kocin and Louis W. Uccellini published a study of New England snowstorms in their book *Snowstorms Along the Northeastern Coast of the United States: 1955–1985,* and calculated the total number of 5" or greater snowfalls for each of 11 different New England regions.

Statistically, the biggest snows occur in southern and central Vermont; inland in southern, and central New Hampshire; and along a broad strip through the interior of central Maine where 200–220 such storms occurred between 1955—1985, an average of seven to eight per year. Areas to the north of these regions are often too far from the coastal storm-track to collect the heaviest snowfalls, although more frequent, smaller snows contribute to an overall greater seasonal amount. Western Massachusetts and the northwest corner of Connecticut are the next most likely areas to receive heavy snowfalls (they average six to seven a year).

Coastal Connecticut, coastal Rhode Island, and Cape Cod and its islands are the least likely to have a big snowstorm, with an average of only two to three 5" snowfalls a season.

The Boston area is between these two extremes, averaging about four 5"-or-greater snowfalls each winter. February is the month most likely to have big snowstorms across all the regions.

If you are a snow lover and want to know which towns report the heaviest annual average snowfall in each state (excluding mountaintops like Mt. Washington in New Hampshire or Mt. Mansfield in Vermont), these would be your localities of choice:

STATE	TOWN	AVERAGE SEASONAL SNOWFALL
New Hampshire	First Connecticut Lake	172″
Maine	Rangely	118″
Vermont	Somerset	114″
Connecticut	Norfolk	107″
Massachusetts	West Cummington	85″
Rhode Island	North Foster	55″

Of course, if you prefer a rural setting, any location above 2,000 feet will most likely receive over 100″ of snow each winter. Elevation adds on average an additional one inch of snow per season for every 25 feet of increased elevation. Your author experienced a snowstorm in Bennington, Vermont, where the town received only 5″ of slush, while the home he was visiting, located on a ridge only five miles outside of town but 1,200 feet higher, received 20″ of wind-driven snow.

Alternatively, if you like your snow in the city, Burlington, Vermont, has more snow than any other large New England city, with a catch of 82″ on average per season, about two and a half feet more than Boston or Hartford, and one foot more than Portland.

If you hate snow you will be happiest living on Cape Cod, Nantucket, Martha's Vineyard, or Block Island, all of which average less than 30″ of snow per winter, about the same as New York City, and some winters virtually no snow falls. On rare occasions, however, a powerful northeaster will stay far enough offshore to spare the big cities heavy snow but close enough to put the islands and Cape Cod in the path of the heaviest snowfall. One of the worst such storms occurred February 27–29, 1952, when 23″ buried Nantucket and 20″ fell at Hyannis on Cape Cod. Winds of near hurricane force created drifts to 20 feet deep. Boston and Providence, however, received only 5″ of snow from this storm.

Bennett Ridge, 30 miles east of Oswego and considered one of the snowiest locations in America, is also famous for its "snowbursts." On January 17, 1959 an astounding 51" of snow fell in just 16 hours! Oswego itself received 27" of snow in only three and a half hours on the afternoon of January 26, 1972, and more recently, Adams accumulated 36" in just nine hours on January 18, 1993.

The heaviest lake-effect snows are normally very fluffy, with snow-to-water ratios as high as 40" of snow to one inch of melted precipitation. Snow that falls at a temperature of 32° normally has a ratio of 10" of snow to one inch of liquid precipitation. The snow belts around the Great Lakes that get the heaviest falls tend to be located not right on the lake shores, but rather places inland 20 or so miles, where hills provide the orographic lift necessary to squeeze moisture from the onshore winds. The heaviest squalls normally occur in early winter while the lakes are still relatively warm and the first cold winds of winter blow over them drawing moisture from the lake's surface. The longer the fetch over the water, the heavier the snowbursts are likely to be.

Another exceptionally snowy location in the East is along the high mountain ridges and plateaus of West Virginia. Above the 3,000-foot level, snowfall averages well over 100" per season in places such as Pickens and Terra Alta. Krumbrabow State Forest once recorded 301" during the season of 1959–1960. Of this, 100" fell in the month of March alone.

This photograph was taken during the remarkable snow burst at Montague Township, New York, in January 1997, when an unofficial world-record 77 inches of snow fell in just 24 hours. (courtesy Cheryl Boughton)

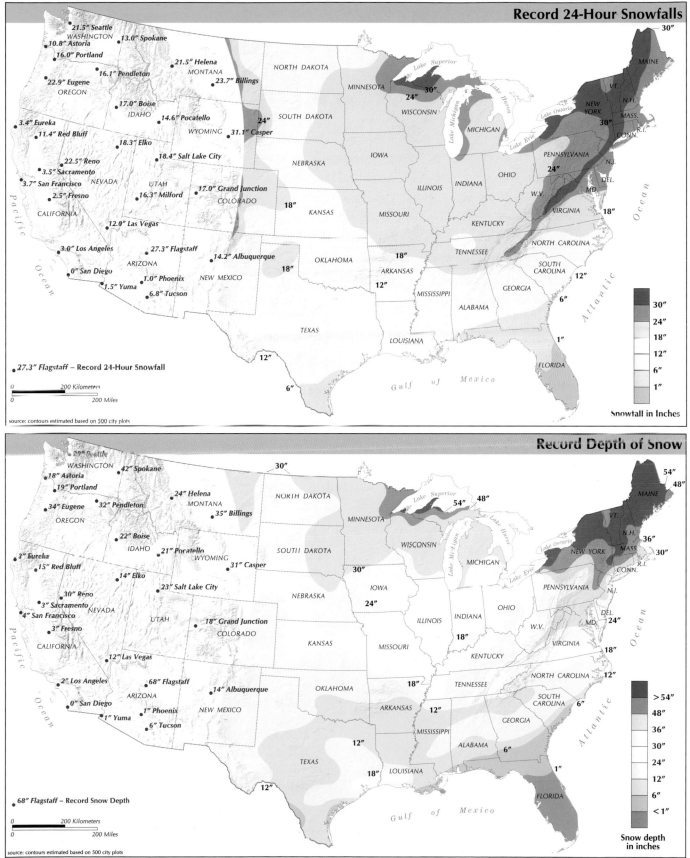

Record 24-Hour Snowfalls

21.5" Seattle
WASHINGTON
13.0" Spokane
10.8" Astoria
16.0" Portland
16.1" Pendleton
21.5" Helena
MONTANA
22.9" Eugene
OREGON
23.7" Billings
17.0" Boise
IDAHO
14.6" Pocatello
3.4" Eureka
11.4" Red Bluff
18.3" Elko
WYOMING
31.1" Casper
18.4" Salt Lake City
22.5" Reno
3.5" Sacramento
NEVADA
UTAH
17.0" Grand Junction
3.7" San Francisco
16.3" Milford
COLORADO
2.5" Fresno
CALIFORNIA
12.0" Las Vegas
3.0" Los Angeles
27.3" Flagstaff
14.2" Albuquerque
0" San Diego
ARIZONA
1.0" Phoenix
NEW MEXICO
1.5" Yuma
6.8" Tucson

NORTH DAKOTA
MINNESOTA
SOUTH DAKOTA
NEBRASKA
KANSAS
OKLAHOMA
TEXAS
WISCONSIN
IOWA
MISSOURI
ARKANSAS
LOUISIANA
MICHIGAN
ILLINOIS INDIANA OHIO
KENTUCKY
TENNESSEE
MISSISSIPPI ALABAMA GEORGIA
MAINE
VT. N.H.
NEW YORK MASS.
CONN. R.I.
PENNSYLVANIA N.J.
W.VA. MD. DEL.
VIRGINIA
NORTH CAROLINA
SOUTH CAROLINA
FLORIDA

Lake Superior
Lake Michigan
Lake Huron
Lake Ontario
Lake Erie

30"
24"
30"
24"
18"
18"
18"
12"
12"
6"
6"
1"
18"
12"

Pacific Ocean
Atlantic Ocean
Gulf of Mexico

27.3" Flagstaff – Record 24-Hour Snowfall

0   200 Kilometers
0   200 Miles

source: contours estimated based on 500 city plots

Snowfall in Inches

30"
24"
18"
12"
6"
1"

Record Depth of Snow

22" Seattle
WASHINGTON
42" Spokane
18" Astoria
19" Portland
32" Pendleton
24" Helena
MONTANA
34" Eugene
OREGON
35" Billings
22" Boise
IDAHO
21" Pocatello
2" Eureka
15" Red Bluff
14" Elko
WYOMING
31" Casper
23" Salt Lake City
30" Reno
3" Sacramento
NEVADA
UTAH
18" Grand Junction
4" San Francisco
COLORADO
3" Fresno
CALIFORNIA
12" Las Vegas
2" Los Angeles
68" Flagstaff
14" Albuquerque
0" San Diego
ARIZONA
1" Phoenix
NEW MEXICO
1" Yuma
6" Tucson

NORTH DAKOTA
MINNESOTA
SOUTH DAKOTA
NEBRASKA
KANSAS
OKLAHOMA
TEXAS
WISCONSIN
IOWA
MISSOURI
ARKANSAS
LOUISIANA
MICHIGAN
ILLINOIS INDIANA OHIO
KENTUCKY
TENNESSEE
MISSISSIPPI ALABAMA GEORGIA
MAINE
VT. N.H.
NEW YORK MASS.
CONN. R.I.
PENNSYLVANIA N.J.
W.VA. MD. DEL.
VIRGINIA
NORTH CAROLINA
SOUTH CAROLINA
FLORIDA

Lake Superior
Lake Michigan
Lake Huron
Lake Ontario
Lake Erie

30"
54" 48"
54" 48"
48"
36"
30"
30"
24"
24"
18"
18"
18"
12"
12"
12"
6"
6"
6"
1"
12"
18"
12"

Pacific Ocean
Atlantic Ocean
Gulf of Mexico

68" Flagstaff – Record Snow Depth

0   200 Kilometers
0   200 Miles

source: contours estimated based on 500 city plots

Snow depth in inches

>54"
48"
36"
30"
24"
12"
6"
<1"

SNOWY PLACES AROUND THE WORLD

Japan

Outside of North America the snowiest places in the world are in the Japanese Alps of Honshu Island. The greatest snow depth ever measured on earth was an incredible 466″ (38.8 feet) recorded in 1927 on Mt. Ibuki at the 5,000-foot level. Nearby, along the coast of the Sea of Japan, is Takada, the snowiest sea-level location in the world outside of Valdez, Alaska. The city of Takada averages over 262″ of snowfall a season, and once during a single month (January 1945), recorded 362″ (30.1 feet). The snow on the ground accumulated to a level depth of 148″ (12.3 feet) in February of that year. In just 24 hours on February 8, 1927, 58.6″ once fell.

The houses in this region have long eaves called "gangis" that overhang the sidewalks to keep them clear of the snow. The deep snow cover must be swept off the roofs and into the streets, and these roadways have been equipped with underground sprinklers that spray warm water to melt the snow.

Snowfall in the towns along the Sea of Japan in Niigata Prefecture are so deep that artificial snow-melting systems must be employed. These warm-water sprinklers, which run down the middle of a street in Imokawa, keep the community functioning. (M. Ishii)

Tokyo, by contrast, has never had more than 18″ of snow on the ground—even though it is at the same latitude as Takada and only 150 miles to the east. The spine of the Japanese Alps, running through the heart of Honshu Island, blocks the moisture blowing in off the Sea of Japan. It is these same mountains that are responsible for the enormous snowfalls blanketing their northern and western flanks. The deepest low-elevation snow depth in this area was measured at Makawa (Toyama Prefecture) when 295″ (24.6 feet) accumulated. A near world record, 68.2″ of snowfall in 24 hours, fell at Tsukayama, about 50 miles east of Takada, on December 30–31, 1960.

(opposite) The snowiest region of the world outside North America's western mountain ranges is the Japanese Alps of Honshu Island. The mountains of Hokkaido Island, where this photograph was taken, also receive phenomenal snowfalls each season. (Michael Yamashita)

Europe

Snowfall statistics for Europe are sketchy, because of the method of measurements employed. Snowfall usually is not measured directly, but rather its liquid content. In any case, the Alps of France and Switzerland are probably the snowiest inhabited region of the world after Japan and North America. Snowfalls on the southern and western slopes of the Alps above 6,000 feet (1,500 meters) can be as much as 200–600" each winter. Santis, at 8,200 feet in the Swiss Alps, receives an average of 570" each winter. The record 24-hour snowfall here can be estimated at around 50" based on liquid precipitation reports. Col du Tonale bordering Italy and France received 667" during the winter season of 1950–1951. Bessans, in the French Alp region of Savoie, also near the Italian border, once received 67.8" of snow in a 19-hour period, one of the most intense snowfalls on record anywhere in the world. The valley in which Bessans is located receives the full effect of a local wind phenomenon known as "la lombarde," a strong southeasterly wind that funnels moisture into the valley.

The mountains of southern Italy can receive huge snowfalls. Montevergine, in the southern Apennines once had 54.4" of snow in 24 hours, and the mountain village of Floresta in Sicily recorded 51.2" in 24 hours on February 22, 1929.

Great Britain's greatest snowstorm brought an astonishing, (and near world record) snowfall of 70.9" (180 cm) in 15 hours to Dartmoor on February 16, 1929. London's greatest snowstorm probably occurred in February of 1579 when 24" fell in a single storm. In Scandinavia the greatest snowfalls occur in Norway's Jostedalsbreen Mountains in the southern interior of the country. Fanaraken, at 6,600 feet, receives over 400" of snow annually, with 24-hour falls over 40" common. Another of Scandinavia's snowiest regions is Sweden's Kebnekaise Mountains where accumulations of 10 to 20 feet pile up during particularly snowy winters.

Russia

Russia, of course, is inevitably associated with snow, and indeed the entire country experiences some snow during a typical winter. The snowiest region of Russia is not Siberia or the Ural Mountains, as might be expected, but the Western Great Caucasus near Turkey and the Black Sea. Snow depths at Achishko (elevation 6,200 feet) reach 150–200" every winter and have been known to be as high as 315" during their snowiest winter. Most of Siberia is so dry during the winters that often no more than 20–30" fall in the interior and coldest region. Only the Siberian Pacific Coast, in such places as the Kamchatka Peninsula and Sakhalin Island, receive exceptionally heavy snowfalls.

Heavy snowstorms in the Austrian Alps during February 1999 resulted in numerous avalanches such as this one descending upon the town of Zurs. (Gamma Stock Agency)

Remote Snowy Places

Prodigious snowfalls occur in high mountain areas all over the world, but by and large these places are uninhabited. Particularly snowy mountains include the Alps of the South Island of New Zealand above 3,000 to 4,000 feet. The southern tip of the Andes near Tierra del Fuego and Patagonia, in Chile and Argentina, experiences tremendous snow accumulations above the 3,000- to 4,000-foot level as do the southern flanks of the high Himalayas east of the 80° longitude.

Surprisingly, the arctic and antarctic receive very small amounts of snowfall due to lack of atmospheric moisture. It is estimated, in fact, that the South Pole is one of the driest places on earth. It is impossible to actually measure precipitation here because of the high winds, but less than one-tenth of an inch of precipitation (just one or two inches of snow) probably falls on an annual basis.

HISTORIC U.S. SNOWSTORMS

New England's Greatest Snowstorm: Great Snow of 1717

Between February 27 and March 9 of 1717, four snowstorms pounded the Northeast leaving snow depths from the Philadelphia outskirts to New Hampshire of three to five level feet. As David Ludlum explained in his book *Early American Winters, 1680–1820,* "There was probably no event of a non-political nature in New England's history that acquired such a reverential status as the Great Snow of 1717." Snowdrifts up to 25 feet were reported in the Boston area. The famous Rev. Cotton Mather expounded in his diary on March 11, 1717, "As mighty a snow, as perhaps has been known in the memory of man, is at this time lying on the ground." Some

accounts estimated that 95% of the deer population perished following the storms. John Winthrop reported that a herd of sheep, buried under the snow for 28 days, was dug out alive. (How the sheep managed to survive remains unclear.)

A block print of the Great Snow of 1717 shows the snow almost burying fenceposts. An account of the Great Snow by Cotton Mather was one of the first publications of the Massachusetts Historical Society.

A LUMINOUS SNOWSTORM IN NEW ENGLAND

On the night of January 17, 1817, a severe snowstorm raged across New England. The storm was attended by lightning and thunder, and many residents in Massachusetts and Vermont reported the phenomenon of "St. Elmo's fire" lighting up objects as diverse as treetops, fence posts, house roofs, and even people's bodies. A professor from Harvard in Boston related some of these observations in a memoir.

I am informed also by another observing and intelligent gentleman [a professor Dewey of Williams College] who had collected a number of facts relating to this extraordinary phenomena [sic], from persons in whom he placed the highest confidence, that at Williamstown [Vermont], it was seen by a physician on the ears and hair of his horse's head, on the whip of a young man who accompanied him, on the hat of a gentleman, who, in attempting to brush it off, saw it extend over the greater part of his hat, and that at Williamstown, Vermont, it was observed by a company of fourteen persons, as they were returning from a religious meeting, on horses, bushes, fences, logs, and on each other. In one instance a quantity of logs and brush appeared perfectly luminous. The preacher broke off several boughs, on the ends of which the fluid rested. In several cases when he presented his hand, the fluid hissed with the appearance and noise of the electric spark. At this time, it snowed very fast, at the rate, as was supposed, of 6 inches an hour; the lightning was frequent, and the thunder heavy.

—John Farrar, *Memoirs of the Academy of Arts and Sciences,*
University of Cambridge, 1821

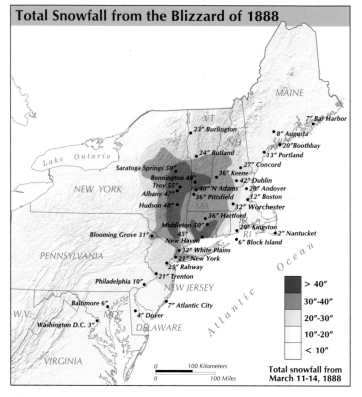

Total Snowfall from the Blizzard of 1888

7" Bar Harbor
MAINE
23" Burlington
8" Augusta
20" Boothbay
24" Rutland
13" Portland
27" Concord
Lake Ontario
Saratoga Springs 50"
36" Keene
Bennington 48"
42" Dublin
Troy 55"
40" N Adams
20" Andover
NEW YORK
Albany 47"
36" Pittsfield
12" Boston
Hudson 48"
32" Worcester
36" Hartford
Middleton 50"
20" Kingston
45"
Blooming Grove 31"
2" Nantucket
New Haven
6" Block Island
32" White Plains
21" New York
PENNSYLVANIA
25" Rahway
21" Trenton
Philadelphia 10"
NEW JERSEY
Baltimore 6"
7" Atlantic City
4" Dover
W.V.
Washington D.C. 3"
DELAWARE
VIRGINIA

> 40"
30"-40"
20"-30"
10"-20"
< 10"

0   100 Kilometers
0   100 Miles

Total snowfall from March 11-14, 1888

Blizzards of 1888

No blizzard in American folklore rivals the fame of the Blizzard of '88. In fact, there were two great blizzards in 1888. The first occurred between January 12 and 14 and was a classic Great Plains howler. It affected the Middle West from Texas to the Dakotas and then roared east to Wisconsin. Some 237 people lost their lives, a very high number considering how sparsely populated the region was at that time. So much livestock was lost that many historians point to the blizzard as the seminal event in the downfall of the Plains free-range cattle industry.

Stories of this great storm were widely reported in the media back East and it seemed so mythically huge and "Western" that everyone was surprised when a storm of equal ferocity struck civilized New York City and the Northeast on March 11–13. This blizzard remains the most severe on record to strike New York and Southern New England. Snowfall totals of up to 55" buried the region. Winds gusting to hurricane force and temperatures in the single digits combined to wreak havoc, especially in New York City, where 200 people perished, some literally frozen in their tracks in the city center.

Great Storm of 1899

A monstrous storm tracked up the eastern seaboard February 12–13, 1899, accompanied by blinding snow and unprecedented cold.

Snow was reported as far south as Fort Meyers, Florida, the farthest south snow has ever been seen on Florida's Gulf Coast (as opposed to the more southerly 1977 snowfall along Florida's Atlantic coast). The 22" of snow that fell in 24 hours on Atlantic City, New Jersey, remains that city's most intense snowstorm on record. An amazing 34" fell on Cape May. Depths over three feet were common in Maryland and Virginia. A monthly snow total of 50.3" was measured at Easton, Maryland, most of which fell during the storm on February 12–13. Washington, D.C.

received 20.5" of snow on top of an already existing snow cover of 15", leading to a total depth of 35", the deepest snow ever experienced in the nation's capital. The snow was followed by some of the coldest temperatures on record. Washington, D.C. fell to -15° and Chase, Maryland, on Chesapeake Bay, dropped to -25°, the coldest temperature ever in the Bay region. *See p. 63 for temperature details.*

Armistice Day Blizzard of 1940

Minnesota's worst blizzard occurred on Armistice Day of 1940, November 11. (This day of remembrance is now known as Veterans Day.) A sudden drop of temperature—from 60° to zero in 12 hours—took most people by surprise. Of the 49 fatalities in Minnesota attributed to the blizzard, 20 were duck hunters caught in the open countryside. Minneapolis received 16.8" of snowfall and amounts over 26" fell elsewhere. Drifts topped 25 feet in height across much of the state. The barometer at Duluth fell to 28.66", the equivalent of a Category 2 hurricane. Winds over 80 mph raked Lake Superior, sinking several ore ships with the loss of 59 crewmen. Ultimately, this storm was the deadliest blizzard in the Midwest during the 20th century.

The Armistice Day Blizzard of November 1940 was Minnesota's worst such event ever. This photo captures the storm's intensity in Minneapolis, where 17 inches of snow was whipped by 60-mph winds into huge drifts. (Minnesota Historical Society)

West's Worst: 1949

A series of blizzards and cold waves over the course of six weeks in January and February of 1949 were the severest combined storms ever to affect a large region of the West. In this case the storms extended from Nevada to the Dakotas and from Arizona to Montana. Snow drifts up to 50 feet formed, perhaps the deepest on record anywhere in the United States. The first blizzard struck on January 2–3 depositing 41" of snow on Chadron, Nebraska, an astonishing amount for a non-mountainous location so far from a moisture source such as the Gulf of Mexico. Even Las Vegas, Nevada received several snowstorms that totaled 16.7" by the end of January.

So severe was the situation that a military-style operation, Operation Haylift, had to be undertaken by the government to help the thousands of stranded people. The army airlifted supplies to stranded ranchers, and helped to extricate 7,500 passengers trapped on 50 trains stalled between Nevada and Nebraska. Wyoming was especially hard hit, the entire state being virtually snowbound for the month of January.

New York was paralyzed by the January 1996 blizzard which dumped 20–25 inches of snow on the city, almost double the amount that the 1993 Super Storm dropped on the town. The lesser snowfall during the 1993 Super Storm was occasioned by warm air wrapping around the storm and turning the snow to rain for several hours. (Michael Yamashita)

Super Storm of 1993

One of the largest and most impressive storm systems ever to affect the United States was the so-called Super Storm of March 12–15, 1993. No snowstorm has ever covered such a vast swath of territory and broken such a variety of records. Barometric pressure records were broken from Asheville, North Carolina, to White Plains, New York. Single-snowstorm state records were broken in Tennessee (56″ on Mt. LeConte), North Carolina (50″ on Mt. Mitchell), and Maryland (47″ in Grantsville). From Birmingham, Alabama (13″) to Syracuse, New York (43″), single-snowstorm city records were also shattered. Fifteen tornadoes connected to this storm system struck Florida, killing 44. Another 274 died from storm-related accidents, including 48 lost at sea. Never has a single storm disrupted the lives of so many Americans.

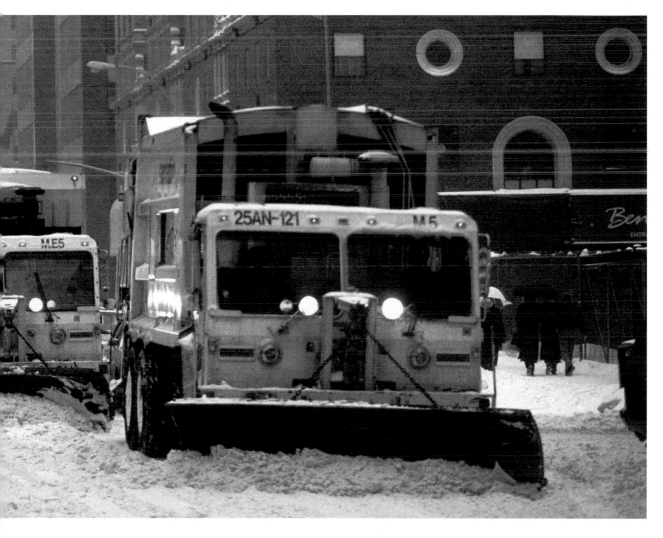

Classification of Icing

Class	General Description	Close Observation on tree branches
None	No observable indication.	
Very Light	Not observable by front light but can be detected by back lighting the glaze on the branches.	Lower side of branches are not covered by ice.
Light	Observable by front lighting but no appreciable bending of branches.	Branches enclosed in ice 2mm on top side.
Light-Moderate	Small branches start to bend.	Thickness on top side 3mm and considerably thicker than bottom.
Moderate	Trunks of young birch trees start to bend. Icicles sometimes odserved.	Maximum thickness about 4mm on top.
Moderate-Heavy	Trunks of young birch trees bend to ground. Icicles sometimes odserved. Some branches are broken.	Maximum thickness about 5mm on top.
Heavy	Considerable damage on old trees. Continuous icicles on power lines.	Maximum thickness about 6mm on top.
Very Heavy	Telephone lines slack.	Maximum thickness about 10mm on top.
Extremely Heavy	Telephone lines break.	Maximum thickness about 12mm on top.

ICE STORMS

The most unpleasant of all winter storms are ice storms. These occur when rain falls at the same time ground temperatures remain below freezing. The damage from ice storms is usually more severe than that caused by snowstorms, creating havoc with electric lines, highways, and forests. The regions of the United States most often affected by ice storms range from the south-central Great Plains, across the Ohio River Valley, and into the Mid-Atlantic and New England. Ice storms usually occur in a narrow band some 50 miles wide tucked between an area of snowfall to the north and rainfall to the south. Certain elevations in mountainous areas and their associated valleys are also prone to icing. One of the regions most prone to ice storms is the Columbia River Valley between Oregon and Washington. Warm Pacific storms sometimes move quickly inland, overrunning cold air at the surface that has been trapped in the valley.

Between November 18 and 22 in 1921, a massive Pacific storm dropped 8.9" of precipitation on The Dalles, a town on the Oregon side of the Columbia River Gorge. The precipitation was a mixture of snow, sleet, and freezing rain that built into a 54"-deep mass of snow and ice.

The worst ice storm in New England history happened just a week later on November 26–29, 1921, when up to 4" of ice accumulated at Worcester, Massachusetts, and other locations in eastern and central Massachusetts. *The Bulletin of the American Meteorological Society* computed that "ice on the side of any dense, unbroken, evergreen tree 50 feet high and on average 20 feet wide would have weighed 5 tons."

(opposite) A two-inch coating of ice almost snapped these telephone lines in central Oklahoma following a severe ice storm in 1997. (Peter Arnold, Inc.)

More recently, the historic ice storm of January 5–10, 1998, caused over $340 million dollars of damage in Maine. It was even worse in Quebec and became Canada's costliest natural disaster on record. Some communities were left without power for up to two weeks following the storm, and accumulations of ice 4″ thick were reported.

The deepest ice accumulations ever reported in the United States were measured in northern Idaho, where deposits up to 8″ thick coated exposed surfaces during a prolonged ice storm on January 1–3, 1961. Ice depths of 6 inches were reported in northwestern Texas on January 22–24, 1940, and in upstate New York during a storm on December 29–30, 1942.

Telephone lines are at great risk if they become iced over. To study how best to deal with this situation, the Bell Telephone Company commissioned several icing studies (in co-operation with electrical and railroad companies) in the late 1940s. Load factors on 12″-long sections of telephone wire were measured during severe ice storms. The worst icing condition mentioned in the report was of a 12″ wire section weighing 4.5 pounds during a particularly severe storm in Illinois on January 1, 1948. Weather historian David Ludlum cites an ice coating on a 12″ length of #14 telephone wire weighed 11 pounds, as a result of Michigan's famous ice storm on February 21–23, 1922.

STRANGE SNOWFALLS

COLORED SNOW

In Dr. Seuss's *The Cat in the Hat Comes Back,* the Cat and his Hat's mischievous inhabitants, Little Cats A, B, and C make a terrible pink mess in the home of two children. The Cat and the Little Cats manage to get the pink stains out of the house—and into the snow, which then becomes bright pink.

Preposterous as it may seem, it has, on very rare occasions, snowed pink. Not only pink but other unnatural shades as well. Pink snow was reported to have fallen in Durango, Colorado on January 9, 1932. Red snow coated the Alps on October 14, 1775, and again on February 3–4, 1852, over a wide area from Bergamo, Italy, to Zurich, Switzerland. Brown snow was reported on Mt. Hotham in the Snowy Mountains of Victoria, Australia, in July of 1935.

In the above cases, the coloring factor in the snow was dust which had risen into the atmosphere during desert dust storms. The dust that mixed with the snow in Europe was carried there by winds from the Sahara; the coloring in Australia originated in its interior desert.

Black and blue snowfalls have been reported on several occasions in New York State. The N.Y. State Weather Service report for April 1889 records black snow falling over Lewis, Herkimer, Franklin, and Essex Counties. Upon examination, the snow was found to contain a "sediment consisting principally of finely divided earth or vegetable mold."

A blue snowfall was reported in many towns of western New York State during January of 1955. Yellow snow that fell on South Bethlehem, Pennsylvania, on March 16, 1879, was found to contain pollen from pine trees that were in bloom throughout states further to the south. Unfortunately, there are no known photographs documenting these rare and unusual events.

GIANT SNOWFLAKES

Giant snowflakes are the result of hundreds of individual snowflakes coalescing as they fall. Wind, humidity, and temperature all play a part in causing these conglomerates to form. The largest flakes ever reported fell at Fort Keogh, Montana, on January 28, 1887. These flakes were "the size of milk pans" (15") in diameter! A snowstorm in Nashville, Tennessee produced flakes up to 5.5" wide on January 24, 1894.

In Berlin, Germany, a snowstorm yielded flakes some 4" in diameter. An observer noted that the large flakes

> *F*ell with both a greater speed and more definite paths than did the smaller flakes. They did not have the complicated, fluttering flight of the latter. In form the great flakes resembled a round or oval dish with its edges bent upward. During flight, they rocked to this side or that, but none were observed to turn quite over so that the concave side became directed downward.

Giant snowflakes, some two or more inches in diameter, blanket the city of Kashgar, China, in a rare, heavy snowfall during the winter of 2001. (Michael Yamashita)

FREAK SNOWFALLS

Snowfall in unusual places, or at strange times of the year, are what might be called freak snowfalls. Snow has fallen on several occasions in Florida, mostly in the Panhandle, where as much as 4″ accumulated at the Milton Experimental Station near the Alabama border on March 6, 1954. Flakes were observed as far south as Fort Myers (26.35°N) on the Gulf Coast of peninsular Florida during the great arctic outbreak of February, 1899, and, on the Atlantic side of Florida, as far south as Homestead (25.29°N), below Miami, on January 19, 1977.

On the western coast of the Gulf of Mexico, Brownsville, Texas, (25.54°N) had a freak snowfall of 3–6″ during the night of February 14–15, 1895. Snow flurries were reported as far south as Tampico, Mexico (22.18°N), during this storm, the most southerly sea-level fall of snow (below the Tropic of Cancer) ever observed in the Western Hemisphere.

Farther north, a hard-to-believe 20″ fell on Houston, Texas, and even coastal Galveston accumulated up to a foot of the white stuff. During the El Nino winter of 1997–1998 snow fell on Guadalajara (20.40°N) in the Jelasco State of Mexico on December 12. This is not altogether surprising since the city rests at 5,000 feet, and snow has on occasion even whitened the hills around Mexico City above the 8,000-foot level.

California has seen snowfall along the coast as far south as downtown San Diego, where a trace has been reported on three occasions—January 14, 1882, January 10, 1949, and again on December 13, 1967.

The Civic Center in Los Angeles had its greatest snowfall on January 15, 1932, when 2″ accumulated. Sea-level San Francisco rarely sees snow, although a spate of measurable snows fell during the winters of the 1880s with its greatest fall of 3.7″ occurring on the night of February 4–5, 1887. Twin Peaks, a hill 500 feet high in the western part of the city, reported depths over 7″ at this time. Snow is relatively common above 1,500 feet on the hills that rim San Francisco Bay.

The highest peaks of the Hawaiian volcanoes receive snow almost every winter, and snow has even been observed as low as the 8,000-foot level on Mauna Kea and Mauna Loa. At the highest elevations of these peaks, people ski.

Summer snowfalls have, on occasion, taken place in various parts of the United States. The most famous of such occurrences was during the summer of 1816, when snow fell during June as far south as the Appalachians of Pennsylvania. Snowflakes were reported along New York's Hudson River just ten miles from tidewater. The highlands of Vermont and New Hampshire received accumulations of over one foot of snow on June 6–8, as did Quebec City in Canada.

Widespread frosts in Ireland resulted in a failure of the potato crop that summer and the ensuing famine precipitated one of the early Irish emmigrations to America.

The cause for the worldwide unseasonable weather in 1816 was the massive eruption of Mount Tambora on Sumba Island, Indonesia, in 1815. The greatest of modern volcanic eruptions, some 36 cubic miles of matter were ejected into the earth's atmosphere in a blast some 50 times more powerful than the better-known eruption of Krakatoa which took place some 68 years later.

A bizarre snowfall occurred on July 2, 1927, in Wabash County, Indiana. A severe thunderstorm produced hail up to 2″ in diameter, but in a small area some 8″ of real snow accumulated within the larger hail streak. The report was investigated by the state director of the Weather Bureau and found to be correct. Most likely, an intense downdraft of cold air from within the thunderstorm, momentarily brought the snowflakes to the surface. This may also explain the freak snowfall of 5″ of slush which coated the decks of the Great Lake's steamer *Menominee* while she crossed Lake Michigan on the night of August 8, 1882.

Snow has been observed falling from a clear sky on many occasions, although with little or no accumulation. This happens when the air becomes supersaturated and ice fog condenses into precipitation.

Giant snow rollers formed in a field in the Lamoille River Valley of Vermont during a windy day in February 1973. (Ronald L. Hagerman)

SNOW ROLLERS

Snow rollers are a strange sight. When the surface snow conditions are just right (light, wet, and sticky) strong, gusty winds can pick the snow up and begin to push it along the surface, forming large barrels of snow, just as a child might do to make a snowman. The result is fields or frozen lakes covered with hundreds, if not thousands, of these barrels all evenly spaced, as if made by the hands of mischievous elves. If the snow is deep and moist enough, and the winds strong enough, these barrels may reach three or more feet in diameter.

SNOW TORNADOES

Although there are no confirmed reports of a tornado actually forming in a snowstorm, occasionally they have traveled over snow covered fields and forests and sucked the snow into their vortices, giving them the appearance of the proverbial white tornado of Top Job fame. Such was the case in Utah on December 2, 1970, when a twister moved across the Timpanogas Divide, where a snow cover of some 38″ had accumulated. The tornado was powerful enough to snap trees one foot in diameter and suck snow over a thousand feet high into its funnel. This gave it a solid white appearance.

chapter 4
RAIN & FLOODS

RAINIEST CITIES AND PLACES IN THE UNITED STATES

How does one determine which places in the U.S. are the rainiest? Should it be done by measuring the average annual rainfall or by counting the number of rainy days per year? Using the first criteria, cities such as Mobile, Alabama, and New Orleans would rank near the top of the list. Employing the second criteria we would find that Seattle would rank above them.

Although New Orleans normally receives about 64" of rain a year and Seattle only about 38", New Orleans averages just 114 rainy days a year compared to Seattle's 158. Thus New Orleans has far fewer rainy days than Seattle, but when it does rain, it rains a lot harder. In fact, 20% of the rainy days in New Orleans net an inch or more of precipitation, compared to just 2% of such days in Seattle. The bottom line is that Seattle has more dreary, drizzly weather than New Orleans.

Those regions of the U.S. that have the most rainy (and/or snowy) days a year include the extreme Pacific Northwest, southeast Alaska, the Great Lakes region, northern New England, and the northern half of the Appalachian highlands. The areas with the fewest rainy days are: the desert Southwest, southwestern Texas, and Central and Southern California. The charts on this and the following pages rank the ten wettest and driest cities in the U.S.

10 Wettest Cities by Average Annual Precipitation

	Location	Rainfall
1.	Aberdeen, WA	83.7"
2.	Astoria, OR	67.1"
3.	Mobile, AL	66.3"
4.	Miami (Hialeah), FL	66.0"
5.	Baton Rouge, LA	65.1"
6.	North Bend, OR	64.4"
7.	Pensacola, FL	64.3"
8.	New Orleans, LA	64.2"
9.	Fort Lauderdale, FL	64.2"
10.	Tallahassee, FL	63.2"

source: NCDC

10 Wettest Cities by Average Number of Days of Precipitation

	Location	Days
1.	Astoria, OR	196
2.	Marquette, MI	175
3.	Buffalo & Syracuse, NY Elkins, WV	169
4.	Sault Ste. Marie, MI	165
5.	Olympia, WA	164
6.	Erie, PA & Binghampton, NY	162
7.	Caribou, ME & Youngstown, OH	160
8.	Seattle, WA	158
9.	Rochester, NY	157
10.	Cleveland, OH	155

source: NCDC

A devastating flood overwhelmed Florence, Italy, in 1966, when the Arno River went on a rampage following days of torrential rainfall. Many museums and historical sites were terribly damaged. (Balthazar Korab)

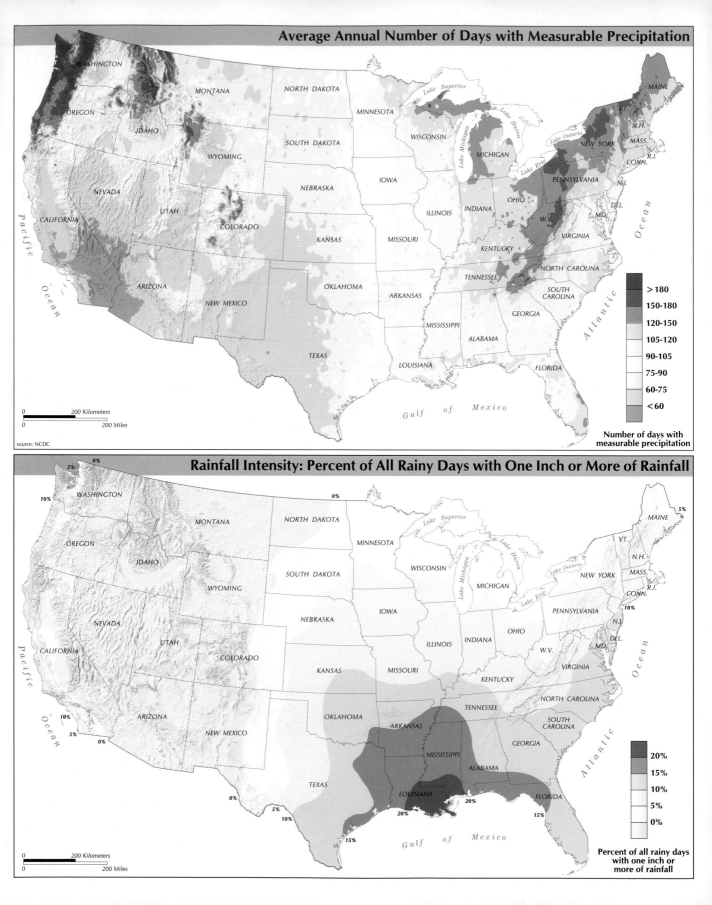

Average Annual Number of Days with Measurable Precipitation

WASHINGTON
OREGON
MONTANA
NORTH DAKOTA
Lake Superior
MINNESOTA
MAINE
IDAHO
WISCONSIN
Lake Michigan
Lake Huron
MICHIGAN
Lake Ontario
NEW YORK
V.T.
N.H.
MASS.
R.I.
CONN.
PENNSYLVANIA
N.J.
WYOMING
SOUTH DAKOTA
NEVADA
NEBRASKA
IOWA
ILLINOIS
INDIANA
OHIO
W.V.
DEL.
MD.
VIRGINIA
CALIFORNIA
UTAH
COLORADO
KANSAS
MISSOURI
KENTUCKY
Pacific Ocean
ARIZONA
NEW MEXICO
OKLAHOMA
ARKANSAS
TENNESSEE
NORTH CAROLINA
SOUTH CAROLINA
GEORGIA
Atlantic Ocean
TEXAS
MISSISSIPPI
ALABAMA
LOUISIANA
FLORIDA
Gulf of Mexico

0   200 Kilometers
0   200 Miles

source: NCDC

> 180
150-180
120-150
105-120
90-105
75-90
60-75
< 60

Number of days with
measurable precipitation

Rainfall Intensity: Percent of All Rainy Days with One Inch or More of Rainfall

5%   0%
10%   WASHINGTON
0%
MONTANA
NORTH DAKOTA
Lake Superior
OREGON
MINNESOTA
MAINE   5%
IDAHO
WISCONSIN
Lake Michigan
Lake Huron
MICHIGAN
Lake Ontario
NEW YORK
VT.
N.H.
MASS.
WYOMING
SOUTH DAKOTA
PENNSYLVANIA
CONN.   R.I.
10%
N.J.
NEVADA
NEBRASKA
IOWA
ILLINOIS
INDIANA
OHIO
W.V.
DEL.
MD.
UTAH
COLORADO
KANSAS
MISSOURI
KENTUCKY
VIRGINIA
CALIFORNIA
Pacific Ocean
10%
5%
0%
ARIZONA
NEW MEXICO
OKLAHOMA
ARKANSAS
TENNESSEE
NORTH CAROLINA
SOUTH CAROLINA
GEORGIA
Atlantic Ocean
TEXAS
MISSISSIPPI
ALABAMA
0%
5%
10%
LOUISIANA
20%
FLORIDA
15%
20%
15%
Gulf of Mexico

0   200 Kilometers
0   200 Miles

20%
15%
10%
5%
0%

Percent of all rainy days
with one inch or
more of rainfall

RAINFALL INTENSITY

By dividing the average number of days with a rainfall of .01" or more by the average number of days receiving 1.00" or more, we determine the rainfall-intensity ratio. For instance, Santa Barbara, California, has few rainy days, but when it does rain it usually rains hard. More than 10.3% of its rainy days result in a fall of one inch or more. This is a higher ratio than that of any other city in the western United States. Seattle's ratio is just 1.9% and San Francisco's 7.9%. The reason for this is Santa Barbara's southwestern exposure at the base of the Santa Ynez Mountains. Winter storms off the Pacific usually blow in from the southwest and slam headfirst into the high mountains looming above the city. The ensuing orographic lift intensifies the rainfall.

Rainfall-intensity ratios along the Gulf Coast are the highest in the country. New Orleans and Baton Rouge top the list with over 20% of their rainy days producing an inch or more of precipitation. Most of their precipitation originates from the great pool of tropical, moist air that sweeps north from the Gulf of Mexico.

The Great Basin, from the Rocky Mountains to the eastern flanks of the Sierra Nevada and Cascades, is a region that rarely sees more than an inch of rain in a 24-hour period. The exceptions would be the highest mountain ranges, where a few snowfalls every winter melt to more than an inch of water.

10 Driest Cities by Average Annual Precipitation

Location	Rainfall
1. Yuma, AZ	3.0"
2. Las Vegas, NV	4.5"
3. Bishop, CA	5.0"
4. Palm Springs, CA	5.2"
5. Bakersfield, CA	6.5"
6. Reno, NV	7.5"
7. Winslow, AZ	8.0"
8. Phoenix, AZ	8.3"
9. Yakima, WA	8.3"
10. Winnemuca, NV	8.3"

source: NCDC

10 Driest Cities by Average Number of Days of Precipitation

Location	Days
1. Yuma, AZ	17
2. Las Vegas, NV	26
3. Bishop, CA	31
4. Phoenix, AZ	35
5. Los Angeles, CA	36
6. Bakersfield, CA	37
7. Santa Barbara, CA	39
8. San Diego, CA	43
9. Fresno, CA	45
10. El Paso, TX	46

source: NCDC

IT RAINS SOMETIMES IN SEATTLE

It is the first word out of the mouths of arriving visitors. It is the symbol and mascot of the region. A joke. A myth. The inspiration of poems and shimmering prose effusions and muttered curses. The trigger of countless cases of depression. As precious as gold. As common as dirt. Hard to predict. Impossible to summon, control, or stop.

Rain. No matter that Seattle and Portland receive less of it than New York City or Washington, D.C. No matter that much of the territory east of the Cascades is starved for it. No matter that it usually nearly ceases almost everywhere in the region for two or three months a year. No matter. Rain is indelibly imprinted on the climatic reputation of the Pacific Northwest.

—from David Laskin's book, *Rains all the Time*, Sasquatch Books, Seattle, 1997

RAINY SPOTS

Hawaii

One of the rainiest spots in the world is far up the eastern slope of Mount Waialeale on the Hawaiian Island of Kauai (the gauge is located at 5,075 feet on the 5,148-foot mountain). An average of about 460″ of rain falls here every year due to the prevailing easterly trade wind, which is abruptly lifted by the steep eastern slopes of the mountains.

Other mountainous islands in the Hawaiian chain also receive enormous rainfalls—Pu'u Kukui in the mountains of West Maui catches around 400″ a year on average, and in its wettest year on record, 1982, picked up an astonishing 705″. The rainfall is usually spread evenly throughout the year with 25 to 40″ falling each month. In March 1942, however, 107″ fell here, the most rainfall in a single month ever recorded anywhere in the United States.

Mt. Waialeale, on Hawaii's Kauai Island, is the wettest location in the United States and one of the wettest in the world. An average year will see 320 days and 460 inches of rainfall. (Paul Chesley)

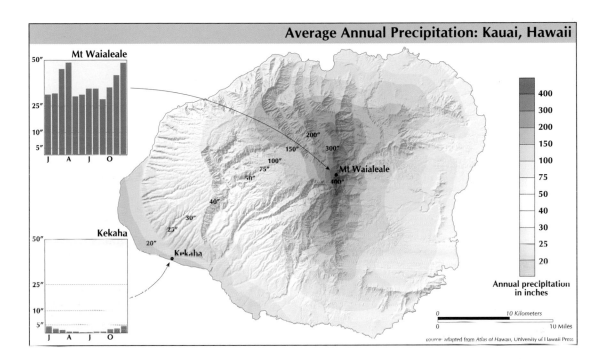

Average Annual Precipitation: Kauai, Hawaii

Mt Waialeale

Kekaha

200"
150"
300"
100"
75"
50"
40"
30"
25"
20"

Mt Waialeale
400"

Kekaha

400
300
200
150
100
75
50
40
30
25
20

Annual precipitation in inches

0   10 Kilometers
0   10 Miles

Source: adapted from Atlas of Hawaii, University of Hawaii Press

A curious feature of Hawaii's rainfall is its localization. On the leeward side of the islands some spots receive as little as 10″ to 15″ annually, the same as in typically dry, desert locations. A place called Grandview once received only .19″ of rain in an entire year, similar to the driest years on record in Death Valley.

The rainfall gradient (the increase/decrease in rainfall amounts over a specific distance) can be dramatic. On Kauai the gradient is 118″ per mile along a line 2.5 miles long from the Hanalei Tunnel to Mt. Waialeale. In other words, you only have to travel a few miles to see an increase in annual precipitation of almost 300″! Variability from year to year is also enormous. Mana, also on Kauai, has had as much as 48″ and as little as 5″ of rainfall in a calendar year.

Alaska

Alaska follows Hawaii as having some of the rainiest spots in the United States. Little Port Arthur on the southern tip of Sitka Island receives an annual average of 226″ of rainfall. This is less than Canada's wettest location, which is not so far away at Henderson Lake in British Columbia, with 256″ in the typical year. MacLeod Harbor on Alaska's Montague Island saw 332″ of rainfall in 1976. Of this total, 71″ fell in the single month of November.

Pacific Northwest

In the lower 48 states, the windward slopes of the Olympic Peninsula in Washington and the coast ranges of Oregon collect the greatest rainfalls. Wynoochee Lake, about 20 miles northeast of Aberdeen in Washington, and nearby Quinalt Lake, vie for the honor of highest annual average precipitation in the lower 48 states with 135–137" falling in a typical year. Even higher amounts likely fall at the 2,000–3,000-foot level of the Olympic Mountains north of here, but no rain gauges are there to measure them.

The town of Aberdeen itself receives an average of 84" annually, and ranks as the wettest town in America with a population of 10,000 or more. In Oregon, a spot on the map called Valsetz, deep in the Coast Range about 20 miles northeast of Newport, receives 130" a year, the second rainiest location in the lower 48. Laurel Mountain, just a few miles from Valsetz but 2,400 feet higher (Laurel Mountain is 3,589 feet) has the record for the most rain ever to fall in a single year in the United States with 204.12" recorded in 1996.

Locations beyond the reach of rain gauges probably receive much greater amounts. It is estimated that 180–200" of rain a year fall on the rainiest locations of Oregon's Coast Range at the 2,000- to 4,000-foot levels.

Northern California's Coast Range also receives tremendous winter rainfalls of near or over 100", virtually all of which comes down between October and April, with January normally being the wettest month of the year (as is true throughout California). The tiny hamlet of Honeydew in Humboldt County is generally credited as the state's wettest town with an average of 104" per season.

Rocky Mountains

From the Rocky Mountains to the Appalachians, precipitation uniformly increases eastward and southward towards the shores of the central Gulf Coast. However, the Rocky Mountains have several isolated wet spots, foremost being the windward slopes and summits of the Bitterroot Mountains along the border of Montana and Idaho. Although there are no weather stations in the wettest locations of these mountains, it is estimated that 60–70" of precipitation, much of it snowfall, falls here on an annual basis. The Pipestone Plateau in the southern portion of Yellowstone National Park also receives an estimated 60" annually.

In general, other wettest (and snowiest) locations in the central and southern Rockies include the highlands of the Unita and Wasatch Ranges in Utah, and the higher mountain plateaus of Colorado's San Juan and Sawatch Mountains. These areas receive an estimated 40–50" of precipitation, including some of the heaviest snowfalls in the United States.

The temperate rain forest of Washington's Olympic Peninsula receives an average of 100–200 inches of rain annually, most of it falling during the wet season of October through April. (Greg Vaughn)

WETTEST AND DRIEST LOCATIONS BY STATE

Location	Driest	State	Wettest	Location
Montgomery	48.36″	**ALABAMA**	67.31″	Robertsdale
Kuparuk (near Prudoe Bay)	3.61″	**ALASKA**	225.35″	Little Port Walter
Yuma Valley	2.63″	**ARIZONA**	39.62″	Hawley Lake
Omaha	40.40″	**ARKANSAS**	62.65″	Big Fork
Cow Creek	1.60″	**CALIFORNIA**	104.18″	Honeydew
Center	6.98″	**COLORADO**	47.46″	Wolf Creek Pass
Bridgeport	44.15″	**CONNECTICUT**	56.93″	West Hartford
Middletown	41.58″	**DELAWARE**	49.02″	Wilmington Resevoir
Key West	38.94″	**FLORIDA**	69.48″	Milton Experiment St.
Mount Vernon	40.49″	**GEORGIA**	73.36″	Clayton
Waikoloa Beach Resort	6.81″	**HAWAII**	460.00″	Mt. Waialeale, Kauai
Grand View	7.11″	**IDAHO**	48.87″	Burke
Waukegan	32.32″	**ILLINOIS**	48.70″	Rosiclare
Monroeville	33.74″	**INDIANA**	49.72″	English
Akron	25.81″	**IOWA**	39.37″	Lamoni
Big Bow	14.60″	**KANSAS**	46.01″	Pittsburg
Wheelersburg	38.38″	**KENTUCKY**	58.92″	Closplint
Red River Lock #1	46.46″	**LOUISIANA**	68.95″	Thibodaux
Presque Isle	35.27″	**MAINE**	57.30″	Acadia NP
Cumberland	37.21″	**MARYLAND**	49.77″	Benson
Nantucket	37.43″	**MASSACHUSETTS**	61.65″	Chester
Mio	26.73″	**MICHIGAN**	40. 06″	Dowagiac
Karlstad	18.56″	**MINNESOTA**	34.99″	Caledonia
Cleveland 3 N	50.97″	**MISSISSIPPI**	72.10″	Vancleave
Tarkio	33.52″	**MISSOURI**	50.75″	Puxico
Belfry	6.85″	**MONTANA**	40.80″	Many Glacier
Mitchell	13.22″	**NEBRASKA**	35.08″	Falls City
Indian Springs	2.91″	**NEVADA**	30.60″	Mt. Rose Bowl
Monroe	36.12″	**NEW HAMPSHIRE**	101.91″	Mt. Washington
Atlantic City (Marina)	38.37″	**NEW JERSEY**	53.67″	Morris Plains
Newcomb	5.96″	**NEW MEXICO**	28.19″	Cloudcroft
Chazy	30.25″	**NEW YORK**	63.61″	Slide Mountain
Asheville	37.32″	**NORTH CAROLINA**	91.72″	Lake Toxaway
Sherwood	13.13″	**NORTH DAKOTA**	21.87″	Wahpeton
Middlebourne	30.44″	**OHIO**	47.54″	Cincinnati Fernbank
Regnier	15.62″	**OKLAHOMA**	55.71″	Smithville
Fields	6.55″	**OREGON**	127.71″	Valsetz
Wellsboro	33.18″	**PENNSYLVANIA**	54.04″	Chalk Hill
Block Island	39.79″	**RHODE ISLAND**	53.17″	North Foster
McColl	39.04″	**SOUTH CAROLINA**	84.38″	Jocasee 8mi WNW
Belle Fourche 22mi NNW	13.88″	**SOUTH DAKOTA**	29.85″	Lead
Bristol	41.33″	**TENNESSEE**	81.28″	Mt. Leconte
Tornillo	8.02″	**TEXAS**	62.75″	Orange 9ni N
Wendover	4.77″	**UTAH**	52.26″	Alta
South Hero	32.57″	**VERMONT**	78.80″	Mt. Mansfield
Timberville	35.18″	**VIRGINIA**	54.88″	Big Meadows
Priest Rapids Dam	6.84″	**WASHINGTON**	137.21″	Quinault Ranger St.
Moorefield	32.88″	**WEST VIRGINIA**	66.26″	Pickens
Washington Island	28.54″	**WISCONSIN**	37.05″	Lake Geneva
Lysite	5.11″	**WYOMING**	31.54″	Snake River

source: NCDC

AVERAGE ANNUAL PRECIPITATION

HAWAII

460″ ▲ Kaua'i
Niihau
O'ahu
Moloka'i
Lanai Maui
HAWAIIAN ISLANDS 6.81″ ▲ Hawaii

| 100″ + |
| 70″-100″ |
| 50″-60″ |
| 40″-50″ |
| 30″-40″ |
| 20″-30″ |
| 12″-20″ |
| 5″-12″ |
| 0″-5″ |

Average annual precipitation

▲ *46.01″* - Wettest location in each state

▼ *46.01″* - Driest location in each state

Scale for all areas except, Alaska and Hawaii.
Scale in those areas vary.

Eastern United States

In the eastern United States, the topography of the southern Appalachians conspires to create the rainiest area outside of the Pacific Northwest. The mountainous region of the Nantahala National Forest in southwestern North Carolina receives the full force of moisture-laden winds from the Gulf of Mexico. The village of Highlands, just north of the Georgia border, catches an average of 88″ of rain in a typical year. Nearby Rossman had a record 130″ in the year of 1964. Adjoining areas of South Carolina and Georgia have also received over 100″ in a single year.

Other wet locations in the eastern U.S. include the high ridges of West Virginia and the Catskills of New York State. Annual rainfalls in the 60–65″ range are typical in the wettest locations, 20″ more than nearby low-lying areas such as New York City and Washington, D.C. Pickens, West Virginia, with an average of 65″ and Slide Mountain, New York, with 64″, are the wettest locations in each area.

AROUND THE WORLD

Wettest Locations in the World (Annual Average Precipitation)

CONTINENT	LOCATION	ANN. PRECIPITATION
SOUTH AMERICA	Lloro, Colombia	523.6″
	Quibdo, Colombia	354.0″

The precipitation at Lloro is an estimated amount, the location being 14 miles southeast of Quibdo and at a higher elevation. Quibdo once recorded 781.06″ in a single year, 1936. If the estimate for Lloro's wetter location is correct, one might assume that during 1936 over 1,100″ would have fallen at Lloro, which would have constituted the most rain to ever fall in a single year anywhere in the world. Bahia Felix in Chile's Tierra del Fuego has an average of 325 days a year with rain, the most in the world.

ASIA	Mawsynram, Meghalaya State, India	467.4″
	Cherrapunji, Meghalaya State, India	463.0″
	Emei Shan, Sichuan Province, China	321.6″

Mawsynram and Cherrapunji are both located at about 3,000 feet in the Khasi Hills of Assam, India. These hills catch the full brunt of the southwest monsoon blowing off the Bay of Bengal between May and October. About 90% of their rain falls during this time of the year. June alone averages over 100″, the highest monthly average rainfall in the world. Variability from one monsoon to another can be tremendous. Cherrapunji has had as much as 905″ and as little as 282″ in a single monsoon season (May–October).

CONTINENT	LOCATION	ANN. PRECIPITATION
OCEAN ISLANDS	Mt. Waialeale, Kauai, Hawaii	460.0"

The location on Mt. Waialeale is uninhabited and measurements are made about once every one to three months. The 460" cited here is for the period of record from 1931–1960. A more recent period of record, 1941–1970, yielded an average of 486"; and 472" was the average for the period 1912–1949. It is not clear to the author why the middle period of record is the standard most often cited.

AFRICA	Ureca, Bioko (Fernando Poo Island)	411.4"
	Debundscha, Cameroon	405.5"
AUSTRALIA	Bellenden Ker, Queensland	340.0"

Bellenden Ker is located at an altitude of 5,102 feet on the slope of Mt. Bartlefrere, some 40 miles south of Cairns in Queensland. This is the highest mountain in the state. February is the wettest month with an average of 56" of rainfall. A tropical storm in January 1979, produced 100" of rain in just three days here.

CARIBBEAN	Bowden Pen, Jamaica	307.9"

Annual rainfall is estimated to be as high as 400" on the highest mountain peaks of the Caribbean Islands of Guadeloupe and Dominica.

NORTH AMERICA	Henderson Lake, B.C., Canada	256.0"
	Little Port Arthur, Alaska	226.8"

Henderson Lake is on the west coast of Vancouver Island and the rainfall measurements were made at a fish hatchery that is no longer in operation.

NEW ZEALAND	Milford Sound	245.2"
EUROPE	Crkvica, Bosnia-Herzegovina	183.0"

Driest Locations in the World (Annual Average Precipitation)

CONTINENT	LOCATION	ANN. PRECIPITATION
SOUTH AMERICA	Arica, Chile	0.03"

Arica, the driest place in the world, is located in Chile's Atacama Desert about 18° south of the equator. Years often pass without any rainfall at all, the longest such stretch being from October

The Shillong Plateau in the Meghalaya State of India has received as much as 1,000 inches of rain in a single year as measured in the town of Cherrapunji. This is the greatest such total ever measured on earth. (Nick Haslam, Hutchison Agency)

1903 to January 1918, an amazing 14 consecutive years. Rain usually occurs only during El Niño events, which occasionally even provide flooding rains. The .03" figure at Arica is based on a 60-year period of record.

CONTINENT	LOCATION	ANN. PRECIPITATION
ANTARCTICA	South Pole Station	0.08"

(estimate based upon liquid content of snow accumulation)

The arctic region is extremely dry. Arctic Bay, in Canada's Northwest Territories once received only .05" of precipitation in a calendar year (1949) and Rea Point, also in the Northwest Territories, averages only eight days a year with measurable precipitation.

AFRICA	Wadi Halfa, Sudan	0.10"

Much of the northern and eastern sections of the Sahara Desert go for several years at a time with no measurable rainfall, as is the case at Wadi Halfa. The .10" annual average is based on 40-year period of record.

NORTH AMERICA	Batagues, Mexico	1.20"

Death Valley is the driest location in the United States, with an average annual rainfall of just 1.60″ at Cow Creek. The longest rainless stretch of time on record in North America occurred at Baghdad, California, when no rain fell for 767 consecutive days from October 3, 1912 to November 8, 1914.

CONTINENT	LOCATION	ANN. PRECIPITATION
ASIA	Aden, Yemen	1.80″

Parts of the Lut Desert of Iran are surely drier than Aden but good records do not exist.

OCEAN ISLANDS	Mindelo, Cape Verde Islands	3.90″
AUSTRALIA	Mulka, South Australia	4.05″
EUROPE	Astrakhan, Russia	6.40″

Probably the driest region on the planet is the Atacama Desert of Chile. Decades often go by with no measurable precipitation, and the village of Arica averages only .03″ of rain a year. (Ed Darack)

EXTREME RAINSTORMS

Few people have experienced a truly extreme rainfall of more than five inches in an hour. In the U.S., such rainstorms are most likely to be encountered in the states along the Gulf Coast or in the Midwest during intense thunderstorms. However, many of the most intense rainfalls on record in the U.S. have occurred outside those regions in such states as Virginia and Pennsylvania.

This is due in part to the effect that topography has on rainfall. Mountains create what is known as an orographic event. These occur when moisture-laden winds are lifted by an elevated land mass and condensed, thus enhancing the intensity of the rainfall and occasionally resulting in phenomenal amounts of precipitation. Often, however, topography plays no role, or only a minor one, in determining the amount of rain that falls. A confluence of atmospheric conditions such as the training of thunderstorm cells (when a series of storms pass in a line over the same place), low-level moisture inflows, and other factors will be enough to trigger a super rainstorm. On the following pages are the world and U.S. record rainfalls.

U.S. Record Point Rainfalls

Time	Rainfall	Location	Date
1 minute	1.23"	Unionville, MD	7/4/1956
5 minutes	2.03"	Alamogordo Creek, NM	6/5/1960
12 minutes	2.30"	Embarrass, WI	5/28/1881
15 minutes	3.95"	Galveston, TX	6/4/1871
30 minutes	7.00"	Cambridge, OH	7/16/1914
40 minutes	9.25"	Guinea, VA	8/24/1906
42 minutes	12.00"	Holt, MO	6/22/1947*
1 hour	13.80"	Central WV	5/4-5/1943
1 hour 30 minutes	14.60"	Central WV	5/4-5/1943
2 hours	15.00"	Woodward Ranch, (D'Hanis) TX	5/31/1935
2 hours 30 minutes	19.00"	Rockport, WV	7/18/1889
2 hours 45 minutes	22.00"	Woodward Ranch, (D'Hanis) TX	5/31/1935*
3 hours	28.50"est.	Smethport, PA	7/18/42*
4 hours 30 minutes	30.70"	Smethport, PA	7/18/42*
12 hours	34.30"	Smethport, PA	7/17-18/1942
18 hours	36.40"	Thrall, TX	9/9/1921
24 hours	43.00"	Alvin, TX	7/25-26/1979
4 days	62.00"	Kukaiau, Hamakua, HI	2/27-3/2/1902
8 days	82.00"	Kukaiau, Hamakua, HI	2/27-3/6/1902
1 month	107.00"	Kukui, Maui, HI	3/1942
1 month (mainland)	71.54"	Helen Mine, CA	1/1909
1 year	704.83"	Kukui, Kauai, HI	1982
1 year	332.29"	MacLeeod Harbor, AK	1976
1 year (mainland)	204.12"	Laurel Mountain, OR	1996

*constitutes a world record

The map to the right depicts what hydrologists have estimated to be the maximum possible amount of rain that can fall in a 24-hour period given the physics of the atmosphere. It does not take into account the effect mountain terrain may have on the rainfall intensity, hence the entire West is excluded and the data for the Appalachian and Ozark Mountains is considered unreliable. The effect mountains can have on enhancing these amounts can be seen by the 34" of rain which fell in less than 24 hours on Smethport, Pennsylvania, in 1942.

Estimate of Maximum Possible 24-hour Rainfall

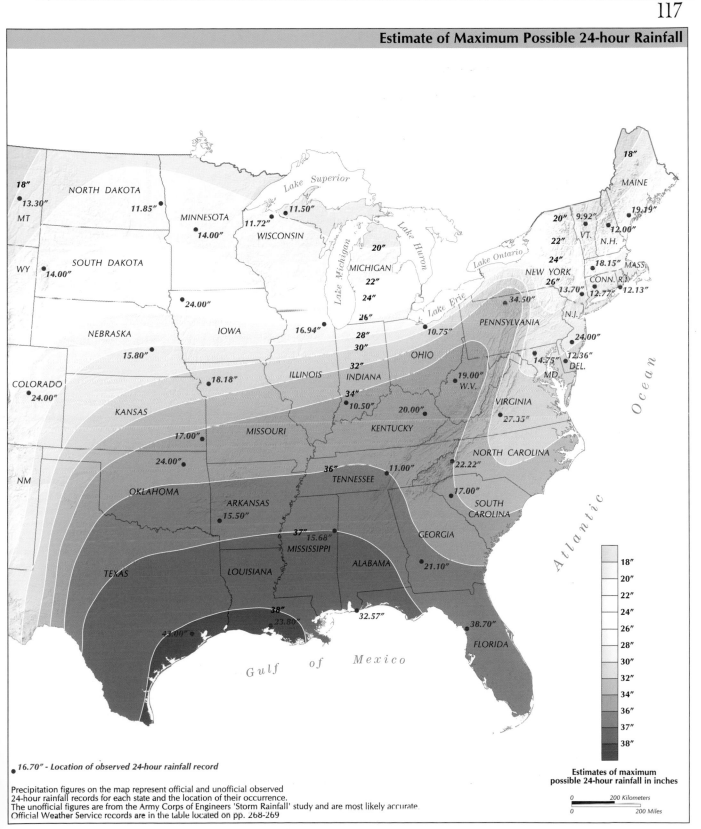

18" MAINE

18"
● **13.30"**
MT

NORTH DAKOTA

● **11.85"** MINNESOTA

11.72"

● **14.00"** WISCONSIN

11.50"

Lake Superior

20" 9.92"
VT. **12.00"**
22" N.H.

24" ● **18.15"** MASS.
CONN. R.I.
● **13.70"** ● **12.13"**
● **12.77"**

WY
● **14.00"** SOUTH DAKOTA

● **24.00"**

IOWA

16.94"

MICHIGAN **20"**

22"

24"

26"

Lake Huron

Lake Ontario

NEW YORK
26"

34.50"

PENNSYLVANIA **10.75"**

Lake Erie

N.J.

● **24.00"**

OHIO

28"

30"

32"

14.75" **12.36"**
DEL.
MD.

NEBRASKA
● **15.80"**

COLORADO
● **24.00"**

KANSAS

ILLINOIS INDIANA

34"
10.50"

● **18.18"**

19.00"
W.V.

VIRGINIA
● **27.35"**

● **20.00"**

KENTUCKY

● **17.00"** MISSOURI

● **24.00"**

NM

OKLAHOMA

36"
TENNESSEE

11.00"

NORTH CAROLINA
● **22.22"**

● **17.00"**

SOUTH
CAROLINA

ARKANSAS
● **15.50"**

37"
● **15.68"**
MISSISSIPPI

GEORGIA

● **21.10"**

TEXAS

LOUISIANA

ALABAMA

38"
● **23.80"**

● **38.70"**

32.57"

FLORIDA

● **41.00"**

Gulf of Mexico

Atlantic Ocean

Scale legend
	18"
	20"
	22"
	24"
	26"
	28"
	30"
	32"
	34"
	36"
	37"
	38"

**Estimates of maximum
possible 24-hour rainfall in inches**

● **16.70"** - Location of observed 24-hour rainfall record

Precipitation figures on the map represent official and unofficial observed
24-hour rainfall records for each state and the location of their occurrence.
The unofficial figures are from the Army Corps of Engineers 'Storm Rainfall' study and are most likely accurate.
Official Weather Service records are in the table located on pp. 268-269

0 _____ 200 Kilometers
0 _____ 200 Miles

PHENOMENAL CLOUDBURSTS: COLORADO & TEXAS, MAY 30–31, 1935

In an event that strains credulity, an astonishing 24" of rain fell in six hours (22.80" of which fell in just four hours) at two locations in eastern Colorado on the afternoon and evening of May 30, 1935. The amounts were recorded by two remote rain gauges which were located about a hundred miles apart, Gauge #Sec. 34, T9S, R564W was located about 25 miles northeast of Colorado Springs, and Gauge #AB Sec. 26, T5S, R55W, just north of Burlington, near the Kansas border. The amounts of rainfall were recorded five hours apart, at Gauge #34 between noon and 6 p.m., and at #26 between 7 p.m. and 3 a.m. A third gauge halfway between the two picked up 11" in three hours, between 6 p.m. and 9 p.m.

Two feet of rainfall in less than six hours would be close to the most intense such fall ever recorded anywhere in the world. Furthermore, unlike most other rainfall records of this intensity, the rain occurred over flat terrain in a non-tropical environment. In fact, this region of Colorado normally receives only 16–18" of rain annually. The figures, however, appear to be accurate since there were three separate readings over a 100-mile area. In addition, an official weather observer in Seibert recorded 9" of rain in two hours.

Naturally, extreme flooding ensued. The Weather Bureau's *Climatological Data*, Colorado Section, May, 1935, reported the following in the Monthly Review:

> On the 30th, excessive local downpours occurred in the vicinity of Colorado Springs and along the northern slope of the Arkansas-Platte Divide. Four lives were lost and a total estimated property damage of $1,800,000 [about $20 million adjusted for current dollars] occurred along Monument Creek and Fountain Creek in Colorado Springs and vicinity. At the height of the flood, skies over extreme eastern counties (where the phenomenal rainfalls were recorded) were a chocolate brown. This was due to a most unusual situation. Along the Colorado-Kansas border there was a heavy dust storm. Clouds of dust could be seen for miles, while to the west torrents of floodwater roared, and at Bovina, hailstones, some as large as baseballs, were reported to have fallen. The coppery-hued sky cast a brown shadow, giving the scene a weird appearance.

In the end, 21 people were killed in floods statewide, and property damage totaled between $8 and $10 million (about $100 million in current dollars), ranking this event as one of the costliest disasters in Colorado history.

Just hours after the phenomenal Colorado rainstorm, another storm of even greater intensity occurred at the Woodward Ranch in south-central Texas (17 miles northwest of D'Hanis, which is about 40 miles west of San Antonio). Reportedly, 22" fell in two hours and 45 minutes between 3 a.m. and 6 a.m. on May 31, a world record for rainfall intensity.

It was at 3 a.m. that gauge #26 in Colorado was just finishing its measurement of 24". Of the Woodward Ranch total, 15" fell in just two hours and about 10" in one hour.

DELUGE AT SMETHPORT, PENNSYLVANIA, JULY 18, 1942

Perhaps the most extraordinary rainfall ever measured in the world occurred in the unlikely location of McKean County, Pennsylvania, in and around the town of Smethport, during the morning of July 18, 1942. Over 600 sheets of data were compiled to substantiate this record. The rainfall, as measured by Mr. E. Hultz located between Smethport and Port Alleghany, began at midnight July 17. Six inches of rain fell between then and about 9 a.m. the following day, ordinarily a flooding rainfall in its own right. Then between 9 a.m. and noon another 28.50″ fell. All told 34.50″ fell in an 18-hour period, of which 34.30″ fell in 12 hours and 30.70″ in just 6 hours.

No such rain intensity has ever been recorded anywhere else in the world. Other station records in the area included 20.40″ in Emporium over a 12-hour period and 18.50″ at Mt. Jewett in 18 hours. Salamanca, New York reported 6.70″ in 3 hours. Hillsides in the Smethport area were reported stripped of vegetation to the bedrock. Eleven people drowned as a result of flooding during this storm.

Heaviest 24-hour Rainfalls in U.S. History			World 24-hour Rainfalls of 40″or More		
Rainfall	Location	Date	Rainfall	Location	Date
43.00″	Alvin, TX	7/25-26/1979	73.62″	Cilaos, Reunion Island	3/15-16/1952
38.70″	Yankeetown, FL	9/5/1950	71.85″	Foc-Foc, Reuinion Island	1/7-8/1966
38.20″	Thrall, TX	9/9 10/1921	66.49″	Belouve, Reunion Island	2/27-28/1964
38.00″	Kilauea Plantation, HI	1/24-25/1956	62.33″	Aurere, Reunion Island	4/7-8/1958
34.50″	Smethport, PA	7/17/1942	55.12″	Muduocaicang, Nei Monggol, China	8/1-2/1977
33.60″	State Fish Hatchery, TX	7/1-2/1932	49.13″	Paishih, Taiwan	9/10-11/1963
32.52″	Dauphin Island Sea Lab, AL	7/19-20/1997	46.98″	Halaho, Taiwan	9/9-10/1963
30.00″	Trenton, FL	10/21/1941	45.99″	Baguio, Philippines	7/14-15/1911
29.20″	Vic Pierce, TX	6/26-27/1954	44.92″	Belleden Ker, QLD, Australia	1/4/1979
27.00″	Massies Mill, VA	8/19-20/1969	44.80″	Hiso, Tokushima	9/11-12/1976
26.12″	Hoegees Camp, CA	1/22-23/1943	43.70″	Saigo, Japan	7/25-26/1957
26.00″	Broome, TX	9/16/1936	43.00″	Alvin, TX, U.S.	7/25-26/1979
25.83″	Santa Anita Canyon, CA	1/21-22/1943	41.76″	Linzhuang, Henan, China	8/5-6/1975
24.00″	Ewan, NJ	8/31-/9/1/1940	41.70″	Kadena AFB, Okinawa	9/8/1956
24.00″	Hallett, OK	9/4/1940	41.10″	Paling, Taiwan	9/10/1963
24.00″	Cherry Creek, CO	5/30-31/1935	41.06″	Bowden Pen, Jamaica	1/22-23/1960
24.00″	Boyden, IA	9/17-18/1926	40.80″	Cherrapunji, India	6/14/1876
23.80″	Miller Island, LA	8/7-8/1940	40.71″	Funkiko, Japan	7/20/1913
23.22″	New Smyrna, FL	10/9-10/1924	40.10″	Jowai, India	9/11/1877

Reunion Island Rainstorms

Rainfall	Time	Date
Storm of March 11-19, 1952 at Cilaos		
73.62"	24 hours	3/15-16[1]
98.42"	2 days	3/15-17[2]
127.56"	3 days	3/15-18[2]
137.95"	4 days	3/14-18
151.73"	5 days	3/13-18
159.65"	6 days	3/13-19
161.81"	7 days	3/12-19
162.59"	8 days	3/11-19
Storm of April 4-9, 1958 at Aurere		
62.33"	24 hours	4/7-8
97.12"	2 days	4/7-9
123.21"	3 days	4/6-9
134.74"	4 days	4/5-9
136.83"	5 days	4/4-9
Storm of February 26-March 5, 1964*		
at Belouve (recording gage)		
15.18"	2 hours	2/28
30.16"	6 hours	2/28
42.79"	9 hours	2/28[2]
52.76"	12 hours	2/28-29
66.49"	18 hours 30min	2/28-29[2]
95.09"	2 days	2/27-29
105.90"	3 days	2/26-29
109.50"	4 days	2/26-3/1
110.58"	5 days	2/25-3/1
110.58"	6 days	2/24-3/1
112.04"	7 days	2/26-3/4
118.54"	8 days	2/26-3/5
Storm of January 7-8, 1966 at Foc-Foc		
(Tropical Cyclone Denise)		
45.00"	12 hours	1/7-8
62.56"	18 hours	1/7-8[2]
66.81"	20 hours	1/7-8[2]
70.08"	22 hours	1/7-8[2]
71.85"	24 hours	1/7-8
Storm of January 14-28, 1980 at Commerson		
(Tropical Cyclone Hyacinthe)		
155.55"	5 days	1/23-27[2]
169.41"	6 days	1/22-27[2]
183.19"	7 days	1/21-27[2]
194.33"	8 days	1/20-27[2]
210.31"	9 days	1/19-27[2]
223.53"	10 days	1/18-27[2]
234.21"	11 days	1/17-27[2]
238.23"	12 days	1/16-27[2]
239.06"	13 days	1/15-27[2]
239.45"	14 days	1/15-28[2]
239.49"	15 days	1/14-28[2]

GREATEST RAINSTORMS IN THE WORLD: REUNION ISLAND

The most extraordinary rainstorms ever recorded on earth have occurred on the small and mountainous Indian Ocean Island of Reunion. Reunion is located 21° south of the equator, some 500 miles east of the island of Madagascar.

The small island has a French-speaking population of 755,000. It lies directly in the heart of the Southwest Indian Ocean cyclone belt and consequently is struck by tropical storms on average three or four times between the months of December and April. The island is only about 50 miles in diameter at its widest but has several very tall volcanic mountain peaks, the highest of which reaches an elevation of 10,070 feet (3,069 meters). Consequently, when tropical storms strike, the mountainous terrain creates phenomenal orographic lift for the storm moisture, and mountain canyons funnel the rainfall upslope, further enhancing the rain's intensity.

The rainfall stations (maintained by the French Meteorological Service) recording the most impressive deluges are located at elevations between 3,000 feet and 8,000 feet on the slopes of the volcanic mountains. These consist of Aurere (3,084 feet), Cilaos (3,937 feet), Belouve (4,921 feet), Foc-Foc (7,511 feet), and Commerson (7,610 feet).

The greatest storms, as far as rainfall is concerned, hit the island in March 1952, April 1958, February 1964, January 1966, and January 1980. It was at these stations and during these storms that the records in the table at left were established.

Many of these rainfalls are world records as can be seen in the table on the opposite page.

[1] World's greatest 24-hour rainfall   source: Paulhus, J.H. and NWS
[2] World Record
* These figures questioned by the French Meteorological Service

World Record Point Rainfalls

Time	Rainfall	Location	Date
1 minute*	1.50″	Barot, Guadeloupe, West Indies	11/26/1970
3 minutes	1.75″	Haughton Grove, Jamaica	9/30/1925
5 minutes	2.48″	Porto Bello, Panama	11/29/1911
8 minutes	4.96″	Fussen, Bavaria, Germany	5/25/1920
15 minutes	7.80″	Plumb Point, Jamaica	5/12/1916
20 minutes	8.10″	Curtea-de-Arges, Roumania	7/7/1947
30 minutes	11.02″	Sikeshugou, Hebei, China	7/3/1974
42 minutes	12.00″	Holt, Missouri, USA	6/22/1947
1 hour*	15.78″	Shangdi, Inner Monggol, China	7/3/1975
1 hour 12 minutes	17.32″	Gaoj, Gansu, China	8/12/1985
2 hours	19.25″	Yujiawanzi, Inner Mongolia, China	7/19/1975
2 hours 30 minutes	21.65″	Bainaobao, Hebei, China	6/25/1972
2 hours 45 minutes	22.00″	D'Hannis, Texas (17 m. NW)	5/31/1935
3 hours	23.62″	Duan Jiazhuang, Hebei, China	6/28/1973
4 hours 30 minutes	30.80″	Smethport, Pennsylvania	7/18/1942
6 hours*	33.07″	Muduocaidang, Nei Monggol, China	8/1-2/1977
8 hours*	41.34″	Muduocaidang, Nei Monggol, China	8/1-2/1977
9 hours	42.79″	Belouve, La Reunion Island, Indian Ocean	2/28/1964
10 hours*	55.12″	Muduocaidang, Nei Monggol, China	8/1-2/1977
12 hours	52.76″	Belouve, Reunion Island, Indian Ocean	2/28/1964
18 hours	62.56″	Foc-Foc, Reunion Island	1/7-8/1966
18 hours 30 minutes	66.49″	Belouve, Reunion Island	2/28/1964
20 hours	66.81″	Foc-Foc, Reunion Island	1/7-8/1966
22 hours	70.08″	Foc-Foc, Reunion Island	1/7-8/1966
24 hours	73.62″	Cilaos, Reunion Island	3/15-16/1952
2 days	98.42″	Cilaos, Reunion Island	3/15-17/1952
3 days	127.56″	Cilaos, Reunion Island	3/15-18/1952
4 days	146.50″	Cherrapungi, Assam, India	9/12-15/1974
5 days	155.55″	Commerson, Reunion Island	1/23-27/1980
6 days	169.41″	Commerson, Reunion Island	1/22-27/1980
1 week	183.19″	Commerson, Reunion Island	1/21-27/1980
8 days	194.33″	Commerson, Reunion Island	1/20-27/1980
9 days	210.31″	Commerson, Reunion Island	1/19-27/1980
10 days	223.53″	Commerson, Reunion Island	1/18-27/1980
11 days	234.21″	Commerson, Reunion Island	1/17-27/1980
12 days	234.21″	Commerson, Reunion Island	1/16-27/1980
13 days	239.06″	Commerson, Reunion Island	1/15-27/1980
2 weeks	239.45″	Commerson, Reunion Island	1/15-28/1980
15 days	239.49″	Commerson, Reunion Island	1/14-28/1980
1 month	366.14″	Cherrapungi, Assam, India	7/1861
2 months	502.63″	Cherrapungi, Assam, India	6-7/1861
3 months	644.44″	Cherrapungi, Assam, India	5-7/1861
4 months	737.70″	Cherrapungi, Assam, India	4-7/1861
5 months	803.62″	Cherrapungi, Assam, India	4-8/1861
6 months	884.03″	Cherrapungi, Assam, India	4-9/1861
11 months	905.12″	Cherrapungi, Assam, India	1-11/1861
1 year	1041.78″	Cherrapungi, Assam, India	8/1860-7/1861
2 years	1605.05″	Cherrapungi, Assam, India	1860-1861

* questionable veracity

Conversion Chart

Inches	Millimeters
50	1270
	1200
	1150
45	1100
	1050
40	1000
	950
	900
35	850
	800
30	750
	700
	650
25	600
	550
20	500
	450
	400
15	350
	300
10	250
	200
	150
5	100
	50
0	0

FLOODS

Flooding is divided into six categories by the United States Geological Survey (USGS). These categories include:

1 **Regional floods**—where a river overflows its banks on a large scale, flooding entire regions.

2 **Flash floods**—when an intense local rainfall causes a stream or river to suddenly flood a small area, usually the local watershed of that river.

3 **Ice-jam floods**—when melting ice-floes dam a portion of a river, normally in the spring, causing it to flood upstream from the ice blockade.

(above) The Little Colorado River's Grand Falls in flood following heavy monsoon rains. These cliffs are 200 feet high and located in the Painted Desert north of Winona, Arizona. (Tom Brownold)

(opposite) One of the worst regional floods in U.S. history took place during the summer of 1993. The photograph at left shows the state capitol of Missouri in Jefferson City surrounded by the floodwaters of the swollen Missouri River. (Missouri Highway and Transportation Department)

4 Storm-surge floods—when a sudden rise in the sea level inundates coastal areas, usually during intense hurricanes prior to the landfall of the eye. Also happens on a lesser scale from intense storms of a non-tropical nature, like the "northeasters" of the Atlantic Coast.

5 Dam failure flood—more often than not the result of faulty engineering, and consequently not covered in this book.

6 Mudflow flood—when the ash from a volcanic eruption washes into rivers and creates mudflow-like conditions.

The deadliest varieties of floods are those caused by storm surges during hurricanes *(covered in HURRICANES on p. 217)* and flash floods caused by intense local rainfalls. Because flash floods take the local population by surprise, many people die from them every year. Often no rain at all has fallen in the area where the flooding occurs, so no one is alert to possible flooding. However, at some distance away, desert lands are being drenched. The hard, dry ground has no time to absorb the moisture, and a wall of water forms, suddenly washing down narrow canyons or arroyos and drowning unsuspecting people.

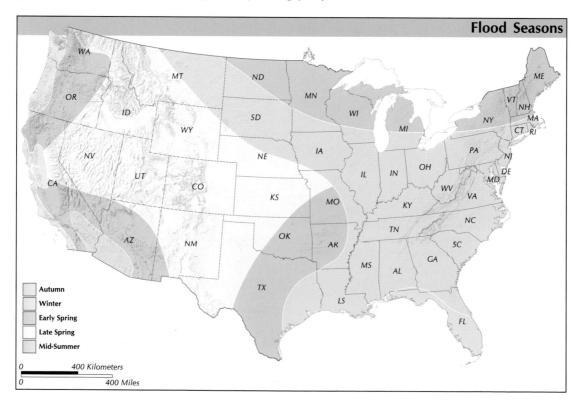

Flood Seasons

- Autumn
- Winter
- Early Spring
- Late Spring
- Mid-Summer

0   400 Kilometers

0   400 Miles

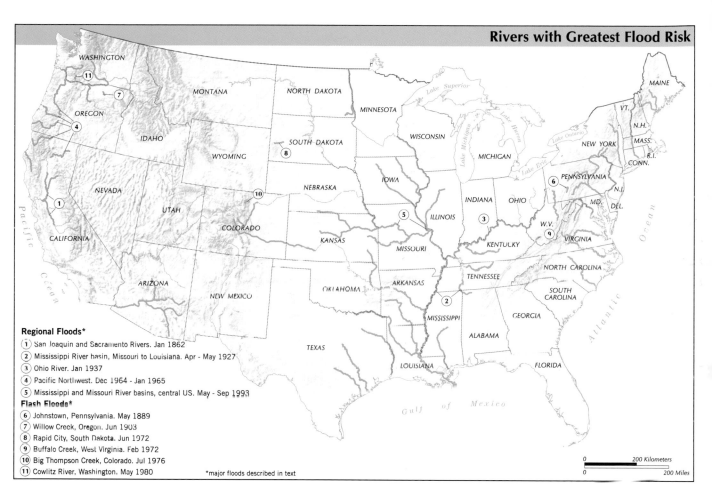

Rivers with Greatest Flood Risk

Regional Floods*
1 San Joaquin and Sacramento Rivers. Jan 1862
2 Mississippi River basin, Missouri to Louisiana. Apr - May 1927
3 Ohio River. Jan 1937
4 Pacific Northwest. Dec 1964 - Jan 1965
5 Mississippi and Missouri River basins, central US. May - Sep 1993

Flash Floods*
6 Johnstown, Pennsylvania. May 1889
7 Willow Creek, Oregon. Jun 1903
8 Rapid City, South Dakota. Jun 1972
9 Buffalo Creek, West Virginia. Feb 1972
10 Big Thompson Creek, Colorado. Jul 1976
11 Cowlitz River, Washington. May 1980

*major floods described in text

0   200 Kilometers
0   200 Miles

Regional floods are more destructive than, but usually not as deadly as, flash floods. In the course of regional floods, thousands of square miles of territory become submerged for weeks on end. The map above shows the rivers of the United States that have been most prone to regional flooding. The locations of the worst flash floods are also marked. The USGS points out the following "flood facts":

◈ Most flood-related deaths are due to flash floods.

◈ 50% of all flash-flood fatalities are vehicle related.

◈ 90% of those who die in hurricanes drown as a result of the storm surge.

◈ Most homeowners' insurance policies do not cover floodwater damage. The Federal Emergency Management Agency (FEMA) provides insurance through its National Flood Insurance Program subsidized by the government.

FAMOUS REGIONAL FLOODS

January 1862 — California

The worst flooding in California history occurred during the winter of 1861–1862, and culminated in mid-January, when the entire San Joaquin and Sacramento Valleys lay under water. Sacramento recorded 20.65″ of rain between December 21 and January 22, or 121% of the entire seasonal norm.

San Francisco had 24.36″ in January alone, or 110% of what the city normally receives in an entire rainy season. This was its wettest month in history, surpassing its second rainiest month on record by over 8″! From Redding in the north to Merced in the south and from the coastal hillsides in the west to the Sierra foothills to the east, a vast lake some 250-miles long and 20-miles wide formed. The Sacramento River rose 24 feet above its low-water mark, exceeding all flood stages on record. In the southern part of the state, Anaheim was under four feet of water that stretched unbroken to Coyote Hills seven miles from the overflowing Santa Ana River. A reoccurrence of such a flood today probably would result in the costliest flood in U.S. history.

March 1913 — Ohio River

The deadliest regional flood in U.S. history took place in Ohio the last week of March 1913. Four days of heavy rain soaked the entire state, with 11.16″ falling in Bellefontaine. Extreme

Fifth Street in Dayton, Ohio, submerged under 10 feet of water during the city's most catastrophic flood ever, in March 1913. (Montgomery County Historical Society)

The Mississippi River flood of 1927 displaced some 750,000 people. Here a barge loaded with refugees is guided down the flooding Sunflower River in western Mississippi. (Library of Congress)

flooding began along all of the state's rivers with record flood stages along the Miami River in Dayton cresting some 34 feet above flood level.

Four hundred sixty-seven people lost their lives, including 123 in Dayton alone. A report described the situation: "A brown wave of water, six feet high, rolled its foaming crest westward on the streets and meeting at each corner a similar wave from the north, piled the water into a raging torrent which filled the streets with foam and wreckage." The water reached a depth of 18 feet in Dayton's Union Railway Station, and 600 people were trapped there for three days before the water receded.

April 1927 — Mississippi River

The flooding of the Mississippi River in April of 1927 was the seminal event that led to the federal flood-control program and gave the Army Corps of Engineers the job of controlling the nation's rivers via the erection of dams, dikes, and other measures of flood abatement. The flood was a result of 12–24" of rain deluging Missouri, Oklahoma, Tennessee, and Louisiana. In just three hours, on the evening of April 15, 1927, 13" of rain fell on the Jefferson and Plaque drainage district just south of New Orleans. A total of 20.4" was recorded here in just an 18-hour period.

It was upriver from New Orleans that the worst flooding occurred. Eighteen million acres (several thousand square miles) of agricultural land went under water, along with all the cities and towns near them. When the river crested at Cairo, Illinois, it was at its highest level ever, and some portions of the river, along the border of Tennessee, rose an astonishing 56.5 feet above flood stage. The river actually became 80 miles wide at one point in Arkansas.

New Orleans was saved from destruction when President Hoover ordered the demolition of the Poydras Levee, thereby allowing the Mississippi to bypass the threatened city and flow directly into the Gulf of Mexico. However, the action destroyed virtually all of the homes south of the city and made more than 10,000 people homeless. At the height of the disaster there were 750,000 refugees under the care of the Red Cross. A total of 313 lives were lost.

January 1937 — Ohio River

The worst flooding to affect the Ohio River occurred during the last two weeks of January 1937. The cities of Cincinnati, Ohio, and Frankfort and Louisville, Kentucky, all suffered their worst flooding in history, as did countless smaller towns along the river between Pittsburgh, Pennsylvania, and Cairo, Illinois. One hundred thirty-seven lives were lost and 175,000 people were made homeless. In Cincinnati the Ohio River crested at 28 feet above flood stage, its highest ever, and 10% of the city was submerged.

Louisville fared even worse with 67% of its residential and 90% of its commercial districts under water after the river rose 10 feet higher than had ever been recorded.

December 1964 — Pacific Northwest

On December 18, 1964, a powerful Pacific storm roared ashore in northern California, Oregon, and Washington. Record snowfalls fell at low levels, with five to 10 feet accumulating at the higher elevations.

The snow changed over to a torrential warm rain, with two months' worth falling in just five days and melting the snow at even the highest elevations. Government Camp at 3,900 feet in the Oregon Cascades saw a 55" snowpack reduced to 6" in just 48 hours. Eugene received 10.3" of rainfall in four days, and totals in the coastal hills and Cascades were at least double that.

Massive flooding resulted as the melted snow and heavy rains swelled the rivers of southern Oregon to their highest levels ever. The Willamette and Umpqua Rivers and all their tributaries went into extreme flood, some of them at levels 50% higher than ever recorded. Forty-seven people died and damage of $430 million was reported in southern Washington, Idaho, Oregon, and northern California.

May–August 1993 — Mississippi and Missouri Rivers

Probably the costliest regional flood in U.S. history took place during the late spring and summer of 1993. Rainfall averaged almost 200% of normal over an enormous swath of the Midwest following an already wet fall and winter. For what may have been the first time in history, both the Missouri and Mississippi Rivers went into flood at the same time. More than 22,000 homes were damaged or destroyed and some entire towns, located on flood plains, were wiped out. St. Louis and Kansas City both reached their greatest flood stages on record with the river cresting 19.6 feet and 16.9 feet above flood stage in each location. The Mississippi flowed past St. Louis at a rate four times greater than normal. Only 48 deaths were attributed to the floods but the costs reached a staggering $23.1 billion, the second-most-expensive natural disaster (Hurricane Andrew being the first) in U.S. history.

A section of a railroad bridge over the Missouri River near Glasgow, Missouri, is washed away leaving the rails themselves dangling in midair during the devastating flood of 1993. (Army Corps of Engineers)

FLASH FLOODS

May 31, 1889 — Johnstown, Pennsylvania

The most devastating flash flood ever to occur in the United States was the famous Johnstown flood of 1889. Only three other peace-time events have killed so many Americans in a single day—the Galveston Hurricane of 1900; the attacks on the World Trade Center and Pentagon in 2001; and the attack on Pearl Harbor in 1941.

The flood was the result of torrential rains leading to the collapse of the South Fork Dam above the Little Conemaugh Valley in western Pennsylvania. The dam had been built to create a reservoir for a resort community catering to Pittsburgh's industrial barons that included Andrew Mellon and Andrew Carnegie. When the dam gave way, a wall of water at times

A contemporary lithograph depicts the destruction of Johnstown during the catastrophic 1889 flash flood. (Library of Congress)

A tree trunk impales a ruined Johnstown home and illustrates the incredible depth the floodwater attained when it overwhelmed the city. (Carnegie Library of Pittsburgh)

nearly 40 feet high, plunged down the valley and overwhelmed the working-class city of Johnstown, drowning it and 2,200 of its citizens under 20 feet of water.

The rescue operation that ensued brought Clara Barton and the newly founded Red Cross to national attention.

June 14, 1903 — Willow Creek, Oregon

The town of Heppner, located in the foothills of the Blue Mountains in western Oregon, was virtually wiped out by a flash flood on the afternoon of June 14, 1913. A heavy rain and hail-storm caused the flood, but it was the exposed location of the village that caused its destruction. The town was overwhelmed by a wave of 10 to 15 feet of water that suddenly flowed down the normally tranquil Willow Creek. Within 30 minutes, 220 of Heppner's 1,400 residents were drowned.

This little-known event remains the third deadliest flash flood in U.S. history (after the Johnstown and Rapid City floods).

November 3, 1927 — Vermont

The worst flash flood disaster to occur in the Northeast struck the Winooski and White River Valleys of Vermont on November 3 and 4, 1927. Unprecedented rainfalls of up to 9" soaked the entire state, following weeks of drenching rainfall.

Kinsman Notch in New Hampshire's White Mountains recorded 14" of rain in 48 hours, of which 7.8" fell in just six hours. Somerset, Vermont, recorded a total of 9.65" of which 8.77" fell in 24 hours, a state record for a 24-hour period at the time. Elsewhere in New England, Westerly, Rhode Island, had 9.4" in eight hours between 7 p.m. and 2 a.m. In Vermont, 84 lives were lost, most in the Winooski Valley. Montpelier, the capital, was especially hard hit.

June 9, 1973 — Rapid City, South Dakota

The second-worst flash flood in U.S. history struck the Black Hills of South Dakota on the afternoon and evening of June 9, 1973. A stationary thunderstorm lingered over the Box Elder and Rapid Creek watersheds high in the Black Hills some 30 miles northwest of Rapid City. A rain gauge at Nemo recorded 15" of rainfall in just six hours and another gauge near Sheridan Lake picked up 14.5" in five hours. Although Rapid City itself reported only 2–4" of rain, the massive rainfalls upstream caused the Canyon Lake Dam to collapse just as the natural flood crests of all the streams approached Rapid City. A torrent of water 10 feet high swept into the city causing over $100 million in damage. In Rapid City and along the flooding streams of the Black Hills, 236 lives were lost. Worst hit was the town of Keystone, where every single building was destroyed.

July 31, 1976 — Big Thompson River, Colorado

Colorado's Big Thompson Canyon flood was the result of a stationary thunderstorm dropping phenomenal amounts of rain over a watershed of many narrow stream valleys (much as had happened in the Black Hills flood of 1973). In this instance, an estimated 14" of rain fell in a four-hour period near Estes Park at the headwater's of the Big Thompson River. A dam collapsed and a wall of water perhaps 20-feet deep roared down the canyon, sweeping all before it. In places where the creek would normally be only 18" deep it became 34 feet deep as measured by the debris stream. Tragically, the event occurred on a Saturday afternoon in summer when many tourists and locals were camping along the river valley or staying in motels in the canyon. Deaths numbered 144 and many bodies were never recovered.

MUDFLOW FLOOD

May 18, 1980 — Eruption of Mount St. Helens

The deadliest flood as a result of a volcanic mudflow, or "lahar"—when an eruption melts a mountain's snow pack causing a flood of ash and water—was that associated with the eruption of Mount St. Helens in southwestern Washington on May 18, 1980. A significant number of the 60 deaths associated with the eruption occurred along the Toutle and Cowlitz Rivers, which inundated areas downstream under millions of tons of mud and debris.

MUD SHOWERS OVER NEW YORK

An intense dust storm over Illinois on April 11, 1902, lifted such a quantity of dirt into the air that it was transported all the way to the East Coast, where, on April 12, it mixed into storm clouds and fell as mud-rain over a large area of the Mid-Atlantic states. Mud showers were reported in Stroudsburg, Pennsylvania, the Delaware Water Gap, Elizabeth, New Jersey, New Haven, Connecticut, and along the Hudson River from Newburg to Poughkeepsie in New York. A Rev. George P. Sewall wrote from Aurora, New York the following:

Last Saturday, April 12, a strange phenomena [sic] was witnessed in a remarkable shower sweeping from the west and extending the entire length of Lake Cayuga, 40 miles. All Saturday forenoon the air was calm, but a strange ashy-looking, very dense mist hung over the lake. The atmosphere was heavily charged with moisture, but no rain fell until noon, when a sudden downpour dashed upon us from the clouds which were so dense as to throw a pall of darkness over everything. Some say the light over the lake was of a pinkish color. To me it was more the hue of the sand dunes of Michigan City. So threatening was the appearance of the sky upon the approach of the shower that many feared a veritable cyclone.

The shower once upon us, was discovered to be supercharged with sand, or mud, which discolored or soiled everything exposed to it. Our windows were besmeared with muddy water, ridges of light colored sand lay along our tin roofs when the moisture dried away. In Ithaca umbrellas caught in the shower looked as if they had been trailed in the mud, and farmers in the fields or on the roads in Aurelius, north of us, say they could feel the sand on their faces. Careful examination with a strong microscope showed the deposit to be composed of minute particles of various shapes, some like tiny grains of wheat, mustard seed, or tiny cylinders, while others were brilliant transparent crystals, no two of which were alike in shape, but looked like mica or flakes of quartz.

—Rev. George P. Sewall, Aurora, New York, *Monthly Weather Review*, May 1902

chapter 5
THUNDERSTORMS & HAIL

t any given time there are about 1,800 thunderstorms raging around the world. These most common manifestations of extreme weather are awesome and beautiful events with their billowing clouds, lightning, thunder, and hail. No wonder ancient Greeks made the highest of their gods, Zeus, the king of thunder and lightning. Cumulonimbus clouds, the harbingers of all thunderstorms, can rise into the atmosphere as high as 78,000 feet at their most exalted (as was recorded on June 15, 1992, during a tornado outbreak in Kansas), well into the stratosphere and twice as high as the cruising altitude of jetliners.

Most severe thunderstormheads top out in the 45,000–55,000-foot range, with a few monster 55,000–65,000-foot storms developing each year in the United States. An observer in Abilene, Texas, monitored a cumulus cloud as it grew 35,000 vertical feet in just "60 seconds" on June 10, 1938, during an outbreak of severe thunderstorms and tornadoes (this would indicate an updraft of around 400 mph and so is most likely an exaggeration). These were associated with a "dry line" moving through the area at the time. A dry line is the leading edge of a dry air mass that plows into a more humid pool of air located at the surface and causes many of the severe thunderstorm outbreaks in the southern plains. Most severe thunderstorms occur as "supercells," monstrous cumulonimbus clouds which grow rapidly into individual thunderstorm clusters. When these supercells begin to rotate they become mesocyclones, leading perhaps to the development of tornadoes.

The critical factor in the growth of a thunderstorm is the "lapse rate," or differential in temperature at ascending altitudes (the greater the temperature difference the more likely a thunderhead is to expand upwards). Equally important is the amount of moisture available in the lower levels of the atmosphere. Dew points should be 60–65 degrees or higher at the surface.

A typical evening thunderstorm rolls over the desert Southwest during the summer monsoon season. (Warren Faidley/Weatherstock, Inc.)

Other facilitators of rapid growth are mechanisms that uplift columns of air from the surface to the mid-levels of the atmosphere, such as a powerful overhead jet stream drawing the air upwards.

Computer modeling takes into account all the above factors, making forecasting severe thunderstorms much more accurate than in years past.

More difficult to predict are the rare supercells that develop over the mountains of the West and the Appalachian Mountains. These storms are responsible for some of the deadliest natural disasters to occur in the United States in the past decades: the Buffalo Creek flash flood in West Virginia on February 2, 1972; the Black Hills flash-flood disaster of June 10, 1972; the Big Thompson Canyon flash flood in Colorado on July 31, 1976; and the Conemaugh River flood in Pennsylvania on July 20, 1977. These four events together resulted in almost 600 deaths and were all caused by supercell thunderstorms.

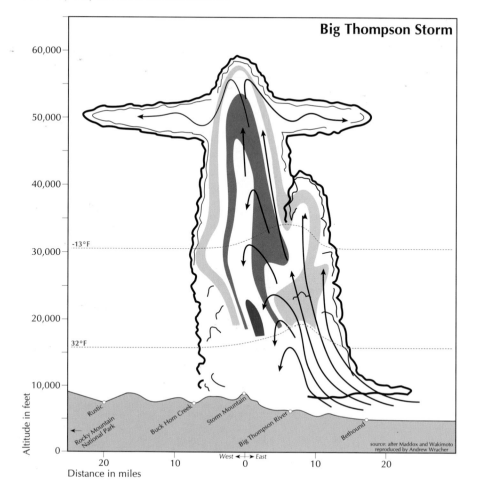

The tremendous thunderhead associated with the Big Thompson River flood looms some 60,000 feet above Estes Park in Colorado, July 31, 1976. (NOAA)

These super storms cause the most destruction when they become stationary over one area for an extended period of time and proceed to hurl phenomenal amounts of rain down upon the land. Forecasting these storms is exceptionally difficult—only broad warnings can be made well in advance, and they are usually too vague to be of much use.

The Big Thompson River thunderstorm was very well documented. The supercell developed rapidly and grew to a height of 65,000 feet in just three to four hours during the afternoon and evening of July 31, 1976. It remained almost completely stationary, dropping up to 14" of rain on a small area, most of which fell in just a four- to six-hour period. Doppler radar estimates indicated that at one point 6" of rain fell in less than one hour over Glen Comfort between 8 p.m. and 9 p.m. When the Big Thompson River flooded its canyon below Estes Park that night, 144 lives were lost

No other country in the world (with the possible exceptions of India and Bangladesh) experiences as many severe thunderstorms as does the United States. This is a result of North America's unique geography. This vast land mass stretches from the subtropical waters of the

Gulf of Mexico, source of abundant moisture and warmth, to the subarctic, source of abundant cold and dry air. Unlike the continents of Asia or Europe, there are no mountain barriers keeping the two air masses from colliding. It is this clash of warm, moist, unstable air with cool, dry, stable air that causes violent storms, such as supercells, to form. Supercells are most likely to develop during the spring months of March through May. Of course, severe thunderstorms (and tornadoes) are most frequent in the Midwest, where the air masses from the south and north most often clash.

A thunderstorm is classified as being severe when it produces large hail (greater than one inch in diameter), strong winds (gusts above 60 mph), frequent cloud-to-ground lightning, or tornadoes. Severe thunderstorms are most frequent (17 or more occurrences a year per 2,000-square-mile area) in central Oklahoma, northeastern Kansas, west-central Missouri, southeastern Nebraska, central Indiana, and northern Alabama. These regions invite tornadoes as well. Fortunately, the vast majority of thunderstorms are not severe. They may and do occur at any location in the United States.

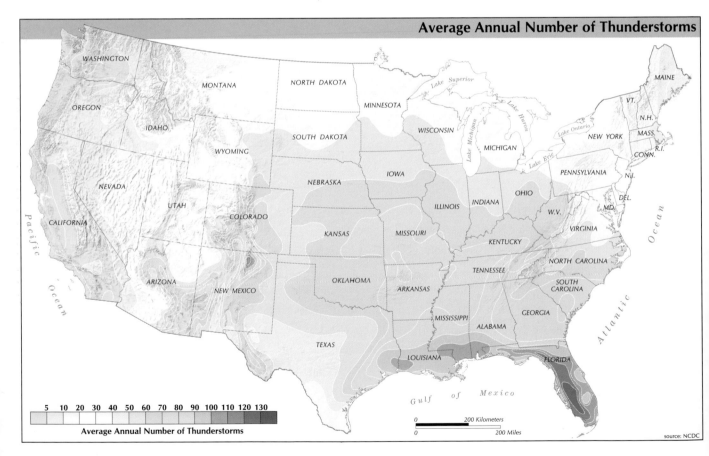

Average Annual Number of Thunderstorms

5 10 20 30 40 50 60 70 80 90 100 110 120 130

Average Annual Number of Thunderstorms

0   200 Kilometers
0   200 Miles

source: NCDC

STORMIEST PLACES IN THE U.S.

Two regions rank as the stormiest in the United States—at least as far as thunderstorms or "thunderstorm days" are concerned:

1 **Eastern Gulf Coast and peninsular Florida,** where 80 to 100 days a year with thunderstorms normally take place.

2 **Spine along the front ranges of the Rocky Mountains** of Colorado and New Mexico, averaging 50 to 100 thunderstorms a year.

There are two ways of counting thunderstorms. The first is by the number of thunderstorm days. A thunderstorm "day" means that thunder is heard at least once during that day by a weather station observer. If there are two or more storms in a single day, it is still considered just one "thunderstorm day."

Another way of counting thunderstorms is by counting the actual number of thunderstorms that are observed, regardless of how many might happen on any given day. This can lead to

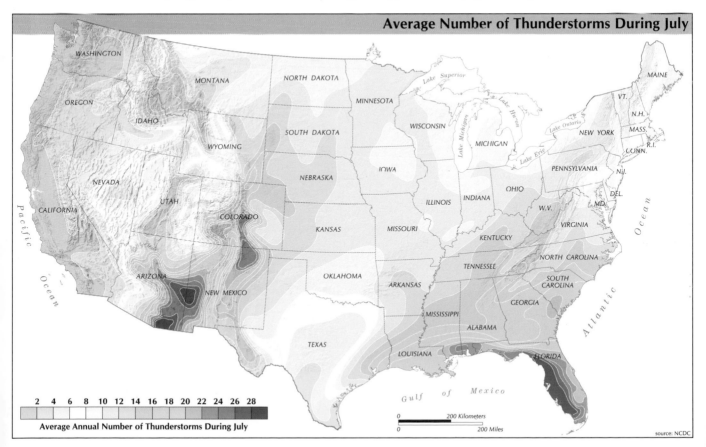

Average Number of Thunderstorms During July

2 4 6 8 10 12 14 16 18 20 22 24 26 28

Average Annual Number of Thunderstorms During July

0   200 Kilometers
0   200 Miles

source: NCDC

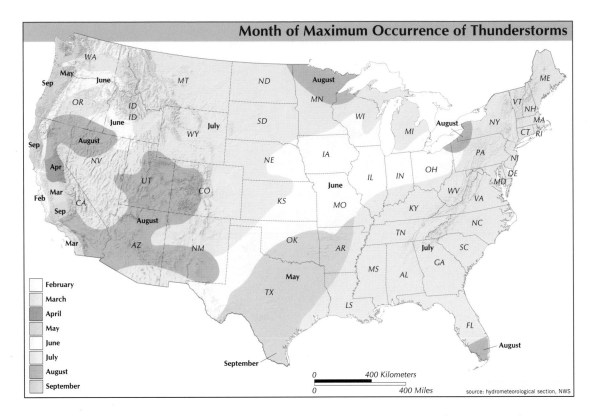

Month of Maximum Occurrence of Thunderstorms

	February
	March
	April
	May
	June
	July
	August
	September

0   400 Kilometers
0   400 Miles

source: hydrometeorological section, NWS

some confusion when climatologists attempt to compare one stormy place to another.

Lakeland, Florida, averages 100 thunderstorm days a year, but receives closer to 130 actual thunderstorms, the most anywhere in the United States. An analysis of thunderstorm activity over Cape Canaveral, Florida, indicates that as many as six thunderstorms have occurred on a single day.

Outside of Florida, perhaps the single most-thunderstorm-prone location is the Cimarron Range of northeast New Mexico. The town of Cimarron (elev. 6,400 feet) reports an estimated 110 thunderstorms a year. The month of July alone accounts for 30 of these storms, so on average, there is a thunderstorm every afternoon. Not surprisingly, the government has a lightning research facility near here.

In spite of Florida's edge in thunderstorm activity, the best places in the U.S. to observe storms are in the West. Visibility here is usually excellent because the air is dry and the spaces

(previous pages) A thunderstorm skirts the city of Tucson, Arizona. In a typical July, more thunderstorms visit the mountains of southeastern Arizona than anywhere else in the country.
A popular vantage point for viewing the pyrotechnics is on Mt. Lemmon, just north of Tucson.
(A. T. Willett/lightningsmiths.com)

wide open. Furthermore, the mountains that ring those wide spaces make perfect perches for thunderstorm watching. The thunderstorms of the Southwest are the result of the summer Southwest monsoon, whose winds draw moisture across the area from Mexico's Gulf of California. The winds are normally active only during the months of June, July, and August.

The most active Southwestern hot spot for thunderstorms is the Huachuca Mountains in southeastern Arizona near the Mexican border. In this area the month of July averages an incredible 34 thunderstorms—more than one a day.

Tucson may expect about 40 thunderstorms during the summer, with more forming over the

---

EXTREME SPORT: PARACHUTING INTO A THUNDERSTORM

U.S. Marine Corps pilot William Rankin had the unfortunate experience of almost becoming a human hailstone when the engine in his fighter jet, a supersonic F8U, failed at 47,000 in the upper levels of a severe thunderstorm over Norfolk, Virginia, in 1959. He bailed out and free fell 37,000 feet through the storm cloud before his parachute engaged. Instead of floating gently to earth, he was caught in the violent updrafts of the storm and actually rose 6,000 feet. For 45 minutes he was lifted and dropped through the thunderstorm's cloud.

I was blown up and down as much as 6,000 feet at a time. It went on for a long time, like being on a very fast elevator, with strong blasts of compressed air hitting you. Once when a violent blast of air sent me careering up into the chute and I could feel the cold, wet nylon collapsing about me, I was sure the chute would never blossom again. But, by some miracle, I fell back and the chute did recover its billow.

The wind had savage allies. The first clap of thunder came as a deafening explosion that literally shook my teeth. I didn't hear the thunder, I actually felt it—an almost unbearable physical experience. If it had not been for my closely fitted helmet, the explosions might have shattered my eardrums.

I saw lightning all around me in every shape imaginable. When very close, it appeared mainly as a huge, bluish sheet several feet thick, sometimes sticking close to me in pairs, like the blades of a scissors, and I had the distinct feeling that I was being sliced in two. It was raining so torrentially that I thought I would drown in midair. Several times I had held my breath, fearing that otherwise I might inhale quarts of water. How silly, I thought, they're going to find you hanging from some tree, in your parachute harness, your lungs filled with water, wondering how on earth you drowned.

In fact, Rankin did eventually land in a tree, unharmed if a bit shaken. He manged to walk away and was eventually picked up in Rich Square, North Carolina, 65 miles southwest of, and more than nine miles below, where he first bailed out.

surrounding mountains. Mt. Lemmon, which rises to 9,100 feet just north of town, is one of the most popular lightning observation spots in the world. The clarity of the air and panoramic vista of the surrounding valley and mountains make it the perfect weather-watching spot.

The Front Range of the Rocky Mountains in Colorado also has almost daily thunderstorm development in July and August, especially on and around Pikes Peak just west of Colorado Springs.

Many thunderstorms in the West are "dry" storms producing little rainfall but lots of lightning. These are the cause of the majority of wildfires that rage here every summer.

In the both the West and the East most thunderstorms occur between noon and 6 p.m. when solar heating is most effective. However, in the Central Plains there is a peak of thunderstorm activity between 6 p.m. and midnight. This is because storms that have formed over the Rocky Mountain Front Range during the afternoon move eastward and are fueled by the latent heat

A wall cloud forms over northern Kansas as a dry line presses into the state from the west. This storm will likely continue to intensify into the nighttime hours as it moves east into more humid air. (Warren Faidley/Weatherstock, Inc.)

and moisture built up over the prairie following a hot summer day. When these afternoon storms originating from the Rocky Mountains reach the plains after sunset, they are fueled by this energy and continue to develop into the nighttime hours.

Almost any place in the U.S., except for the West Coast, is likely to experience thunderstorms during the spring and summer. Along the West Coast, thunderstorms are very rare (severe ones exceptionally rare), despite their widespread appearance in Hollywood films. San Francisco and Los Angeles average no more than one or two thunderstorms a year, and these normally spark only a handful of lightning strokes. When they do occur, it is usually during the wintertime, when strong Pacific storms slam into the coast.

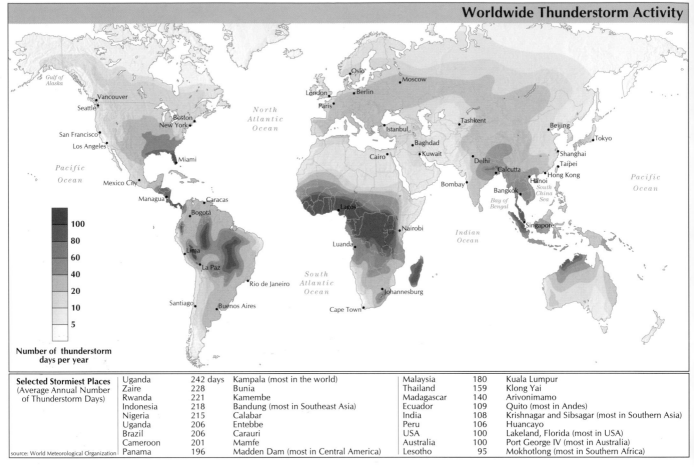

Worldwide Thunderstorm Activity

Number of thunderstorm days per year

| 100 |
| 80 |
| 60 |
| 40 |
| 20 |
| 10 |
| 5 |

Selected Stormiest Places (Average Annual Number of Thunderstorm Days)	Uganda	242 days	Kampala (most in the world)	Malaysia	180	Kuala Lumpur
	Zaire	228	Bunia	Thailand	159	Klong Yai
	Rwanda	221	Kamembe	Madagascar	140	Arivonimamo
	Indonesia	218	Bandung (most in Southeast Asia)	Ecuador	109	Quito (most in Andes)
	Nigeria	215	Calabar	India	108	Krishnagar and Sibsagar (most in Southern Asia)
	Uganda	206	Entebbe	Peru	106	Huancayo
	Brazil	206	Carauri	USA	100	Lakeland, Florida (most in USA)
	Cameroon	201	Mamfe	Australia	100	Port George IV (most in Australia)
source: World Meteorological Organization	Panama	196	Madden Dam (most in Central America)	Lesotho	95	Mokhotlong (most in Southern Africa)

STORMIEST PLACES IN THE WORLD

Although the United States may suffer from the greatest frequency of *severe* thunderstorms, tropical regions of the world account for the vast majority of typical thunderstorms. At one time the large Indonesian city of Bogor, on the island of Java, was reported to experience 322 thunderstorm days per year, the most in the world. This has since been proven not to be the case, and the actual number to strike the city is much fewer, although storms build up almost daily on the volcanic peaks to the south of the city. Most likely it is Kampala, Uganda, that holds the record for the most thunderstorm days in the world, with an average of 242. As is true everywhere, local influences play a major role in thunderstorm formation, and Kampala, near the north shore of Lake Victoria, seems to be in a particularly prime location for thunderstorm development. F. E. Lumb in the journal *Weather* explains:

> *L*and-breeze convergence over the lake during the night releases the latent instability of the moist lower layers of air over the lake which participate in the land breeze circulation, resulting in the development of cumulonimbus clouds and thunderstorms over the lake most nights of the year.

However, in Kampala, as with Bogor, most of the 242 storm days do not result in a thunderstorm hit on the city. Instead, storms remain centered over the lake and within earshot. The town of Mbarara, 80 miles from the western shore of the lake and 120 miles southwest of Kampala, receives only seven thunderstorm days a year. The only other cities in the world to report more than 200 thunderstorm days per year are listed in the map opposite (along with the world's other stormiest locations).

LIGHTNING

One of nature's most spectacular and beautiful phenomena, lightning, is also the deadliest, or second deadliest, weather event to plague America (depending upon whose statistics you follow). Approximately 100 Americans are killed by lightning strikes every year. The singular nature of the lightning bolt is such that people even miles away from a thunderstorm can be directly struck, the proverbial bolt out of the blue. This happens because electrical discharges are capable of traveling as far as 20 miles through the atmosphere.

Moments after this photograph was taken by Mary McQuilken, her younger brother Sean, on the left, was struck by lightning and was seriously injured. Another hiker nearby was killed. Mary, Sean, and Michael (right) were hiking in California's Sequoia National Park in August 1975, when they were caught in an afternoon thunderstorm atop Moro Rock. (photo courtesy of Michael McQuilken)

The majority of lightning victims (54%) are struck down in open fields such as ballparks or golf courses, 23% are killed while taking cover under trees that are struck, 12% on beaches or in boats, 7% while operating farm equipment, and the remaining 4% in miscellaneous ways—for example while riding bicycles, or standing near open windows. Not everyone struck by lightning is killed, and most victims survive, albeit with injuries.

Roy C. Sullivan, a one-time park ranger in the Appalachians of Virginia, was struck by lightning seven times between 1942 and 1977 without suffering any serious, long-term injury. On the other hand, 21 people were killed by a single bolt of lightning while taking refuge in a hut near Mutari, Zimbabwe, during a thunderstorm on December 23, 1975. In the U.S., Florida reports the most lightning deaths each year.

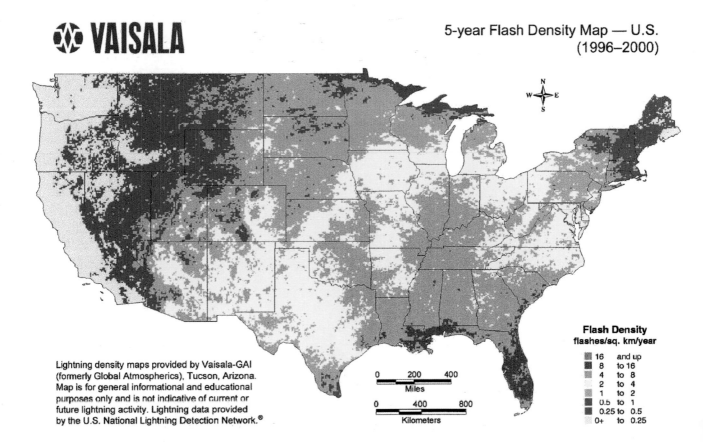

VAISALA

5-year Flash Density Map — U.S.
(1996–2000)

Lightning density maps provided by Vaisala-GAI (formerly Global Atmospherics), Tucson, Arizona. Map is for general informational and educational purposes only and is not indicative of current or future lightning activity. Lightning data provided by the U.S. National Lightning Detection Network.®

0   200   400
Miles

0   400   800
Kilometers

Flash Density
flashes/sq. km/year

16	and up
8	to 16
4	to 8
2	to 4
1	to 2
0.5	to 1
0.25	to 0.5
0+	to 0.25

This map shows the average number of lightning flashes detected per year per square kilometer for the period of 1996–2000. it gives a fairly good idea of where lightning is most likely to strike, although there is considerable variability from one year to the next.

Superbolts often originate from the top of thunderstorm clouds and are positively charged rather than negative as are normal lightning bolts. (Alan Moller)

SUPERBOLTS

On rare occasions a single bolt of lightning is three to maybe a hundred times more powerful in brightness and voltage than an ordinary lightning bolt. For some reason these superbolts are most common during winter storms over Japan and the northeast Pacific. These superbolts can create tremendous damage to the objects or sites they strike.

Top mast hit: The 800-pound top-gallant mast of the HMS *Rodney* was struck by a superbolt in 1838 and "instantly converted to shavings. The sea looked as if the carpenters had swept their shavings overboard ," wrote Frank Lane in *The Elements Rage.*

Church belfry: Frank Lane mentions another report concerning a church belfry in England that was struck by a tremendous lightning bolt and "...a stone weighing 350 pounds was hurled 60 yards over the roof, and another stone was flung nearly a quarter mile away."

Hole in the cornfield: On April 2, 1959, the residents of a rural area near Leland, Illinois, reported a tremendous flash of lightning, and thunder so loud it shook many houses, breaking windows, knocking flowerpots off sills, and scattering loose objects. Many homeowners assumed that their own house had been directly hit, but the bolt had actually hit an open field three-quarters of a mile from the nearest home. A hole seven to 12 feet in diameter and more than a foot deep had been carved in a cornfield.

Apocryphal Lightning Pranks

Some very strange phenomena are attributed to lightning. Perhaps the strangest of all is the effect taking place when an image is literally etched into glass or on human flesh as the result of a nearby lightning flash.

A 19th-century French scientist, Camille Flammarion, reported a case of a woman who received the likeness of a flower imprinted on her skin following a nearby lightning strike, the flower in question having stood in the route of the flash. Likewise, a sheep killed by a lightning strike was reported to have had the surrounding landscape etched on the side of its stomach by the lightning.

Images appearing to be etched on glass panes following lightning flashes have been reported as well. In all cases, the image fades away after several minutes. Scientists claim the etchings seen on human skin are the result of a branching of electrical discharges through tissues, and it is only coincidence or wishful thinking that they might resemble a flower or landscape. Furthermore, lightning has no known photographic properties.

Perhaps the most outrageous story concerning lightning ever to appear in the popular press was a report from New Salem, Vermont, in July 1891. A collector of silver-plated, Revolutionary-era swords, Mr. Arent S. Vandyck submitted the following story to a Boston newspaper following a strike on his home by a terrific bolt of lightning.

> Suddenly the younger Mr. Vandyck [his son] pointed to an old-fashioned sofa. Upon it lay what was apparently the silver image of a cat curled up in an exceedingly comfortable position. Each glittering hair was separate and distinct, and each silvery bristle of the whiskers described a graceful curve as in life. Father and son turned towards the swords which hung upon the wall just above the sofa and there saw that the sword had been stripped of all its silver. The hilt was gone and the scabbard was but a strip of blackened steel. The family cat had been electroplated by lightning.

Lightning Frequency

How intense can lightning storms become? During the Barneveld, Wisconsin, tornado disaster in June 1984, lightning flashes visible from Madison, some 20 miles to the east, became so frequent that it was impossible to differentiate one stroke from another. The effect was that of a strobe light in the darkness. It would have been impossible to count the flashes, but one could estimate that about 200 strokes a minute were visible over a 15-minute period at the storm's height.

Flash rates over 600 per minute reportedly occurred during the Worcester, Massachusetts, tornado in 1953. These reports were made visually, and lightning-detection technology has

(previous pages) A lightning bolt strikes an oil-and-gas tank farm in Arizona. In 1926, a lightning strike triggered a tremendous explosion which completely destroyed an ammunition depot in New Jersey. (Warren Faidley/Weatherstock, Inc.)

since made it possible to actually count individual strikes. Severe storms in Florida have produced flash rates as high as 425 a minute as recorded by the LDAR (Lightning Detection and Ranging) system at Kennedy Space Center.

According to the U.S. National Lightning Detection Network, the Gulf Coast regions of Louisiana, Alabama, and Mississippi, along with much of the peninsula of Florida, receive 20 to 40 flashes of lightning per square mile each year (*see Five-Year Flash Density Map on p. 148*). At the very top of the list is a small area just north and east of Tampa Bay, specifically Pasco County, which is the only place in the United States averaging more than 40 flashes of lightning per square mile on an annual basis. That translates into an average of one or two lightning strikes every thousand square yards, each and every month!

Some extraordinary phenomena have been attributed to lightning and, stranger still, ball lightning.

BALL LIGHTNING

One of the last great mysteries of the meteorological world is the existence of ball lightning. Although there have been thousands of eye-witness accounts, the phenomenon has never been videotaped, positively photographed, or otherwise recorded scientifically. Attempts at reproducing ball lightning in laboratory simulations have all failed. A detailed analysis of some 5,000 eyewitness accounts compiled by electrical engineers Vladimir Rakov and Martin Uman of the University of Florida came to the conclusions that follow.

Lightning balls range in size from less than one inch to over six feet in diameter but average in the four-inch to two-foot range, with most reports referring to "basketball" size. The duration of the observations ranged from less than one second to over two minutes but average just one to four seconds. The color of the lightning balls has been variously described as white, red, orange, yellow, green (very rarely), blue, violet, or a mixture of several colors. However, the most common colors reported are either white, yellow, or orange.

What may be one of the very few photographs of ball lightning taken in Japan some years ago. (Weatherstock, Inc.)

Ball lightning has fascinated scientists and terrified witnesses for centuries, as this 19th-century lithograph illustrates. It remains one of meteorology's great mysteries.

The balls dissipate with an explosive bang, or they may silently fade away. They have been observed to rotate, roll, or even bounce off the ground, but they usually move slowly and in a horizontal direction and can penetrate metal screens, glass windows, and small holes. Some observations have reported the balls as making a hissing, buzzing, or fluttering sound and leaving an acrid burning odor.

The intensity of lightning balls is not great, about the brightness of a 60-watt light bulb, writes Rakov. In almost all cases, however, the balls maintain the same shape, size, color, brightness, and motion during the course of their occurrence. Although balls of lightning have sometimes struck humans and burned or singed them, there are no known instances of human fatalities as a result of such encounters. Cases of fatalities dating from the 18th and 19th centuries were most likely deaths caused by ordinary lightning or meteors, since the word "fireball" referred to both meteors and ball lightning.

The balls are almost always observed seconds after a nearby cloud-to-ground strike of ordinary lightning, although there are quite a few cases of the balls appearing inside aircraft in flight.

A harrowing account of such was reported in a 1952 issue of *Nature* magazine, which described what happened when a BOAC flying boat encountered a thunderstorm at 8,500 feet near Toulouse, France. A lightning ball entered the cockpit through the captain's open window,

REMARKABLE STORM OF BALL LIGHTNING IN MT. DESERT, MAINE

On the evening of February 16, 1853, a blizzard raged across Down East Maine. Not an unusual occurrence for this region at this time of year, but what happened over the course of the night at Bar Harbor was unusual in the extreme and seems like something right out of a Stephen King novel.

The wind during the day had been from the N.E., accompanied with snow, with a temperature of from 15 to 20 degrees above zero. At 6 p.m. the wind had increased to a heavy gale, and at 7 p.m. ceased to blow, and flashes of vivid lightning commenced. In a few minutes more thunder was heard in the N.W., and at 8:30 p.m. the scene was grand and awful beyond description. The lightning was of a purple color, and sometimes appeared like balls of fire, coming through windows and doors and down chimneys, while the houses trembled and shook to their very foundations.

Mrs. E. Holden was near a window, winding up a clock; a ball of fire came in through the window and struck her hand, which benumbed her hand and arm. She then, with all in the house, retreated into the entry. Another flash succeeded, and, in the room from which they had retired, resembled [sic] a volume of fire, whirling around and producing a cracking noise. A similar appearance of fire was seen, and cracking noises were heard in a large number of houses. Some who heard the noise say it sounded like breaking glass.

Capt. Maurice Rich had his light extinguished, and his wife was injured. He got his wife onto a bed and found a match; at that instant another flash came and ignited the match and threw him several feet backwards. John L. Martin received such a shock that he could not speak for a long time.

A great many people were slightly injured. Some were struck in the feet, some in the eye while others were electrized [sic] , some powerfully and some slightly. But what was very singular, not a person was killed or seriously injured, nor a building damaged; but a cluster of trees within a few rods of two dwelling houses was not thus fortunate. The electric fluid came down among them, taking them out by the roots, with stones and earth, and throwing all in every direction. Some were left hanging by their roots from the tops of adjacent standing trees—roots up, tops down.

The lightning, after entering the earth to a depth of several feet and for a space some 8 to 10 feet in diameter, diverged into four different directions. One course which it took led thro open land, making a chasm to the depth of several feet, and continued its march unobstructed by solid, frozen ground or any other substance, to the distance of 370 feet; lifting, overturning and throwing out chunks of frozen earth, some of which were 10 or 11 feet long by 4 feet wide; and hurling at a distance, rocks, stones and roots. It really seemed that God's mercy is manifested in sparing our lives amidst such danger and destruction.

—-*Ellsworth Herald,* Friday, March 4, 1853

Another witness was later quoted in the New York Times *as saying, "I don't believe there ever was a worse frightened lot of people in the world than the inhabitants of Bar Harbor were that night. That purple ball [of] lightning flashed about and obtruded itself everywhere. There was scarsely [sic] a house that was not visited by it."*

singed his eyebrows and hair, and then burned holes in his safety belt and briefcase before moving on into the main passenger cabin where it exploded loudly, assuredly to the consternation of passengers on board.

An Eastern Airlines plane was struck by a ball of lightning while flying through a storm over Atlanta, Georgia, on November 1, 1963. The pilot reported smoke in the cockpit and the smell of ozone. Upon landing, investigators found an 18" section of the rudder lost, a hole in the radome, and burn spots on various parts of the fuselage. Just another reason not to fly.

Theories as to what causes the formation of ball lightning include one that postulates the balls have an internal energy source, probably particles ejected from the earth when a normal lightning stroke hits then somehow becomes electrified and airborne. In *Nature* magazine, J. Abrahamson and J. Dinnis theorized that "...silicon nanoparticles evaporated from sand by the lightning form a ball and by their oxidation illuminate the ball...."

Rakov and Uman (of the University of Florida study) conclude, however, that:

> ...given a wide range of observed characteristics, conditions under which it
> occurs, and locations where it occurs, there may well be more than one type of
> ball lightning and more than one mechanism that creates it. Supporting this point
> of view is the fact that similar phenomena are generated accidentally from high-
> power electrical apparatus in absence of nearby thunderstorms.

HAIL IN THE UNITED STATES

Hailstorms are closely associated with thunderstorms, and the more severe the storm the larger the hail is likely to be. In the United States there are two regions of maximum occurrence—the High Plains of southeastern Wyoming and northeastern Colorado, and the immediate coastline of the Pacific Ocean from northern California to Washington. Both these areas report an average of six or more days with hail each year. The single location reporting the most hailstorms anywhere in the United States is Cheyenne, Wyoming, where an average of nine to 10 days of hail occur each year.

There is a big difference between the hailstorms that affect the West Coast and those of the High Plains. The West Coast hailstorms occur in showers that form in the cold unstable air that follows the passage of cold fronts coming off the ocean, and they are not usually associated with thunderstorms. The cloud tops of these showers are usually no more than 15,000 feet high, and consequently the hail they produce is almost always very small—no more than a quarter inch in diameter. Hailstorms in the rest of the country are almost always associated with powerful thunderstorms.

Hail falls from a thunderstorm over western Texas. (Warren Faidley/Weatherstock Inc.)

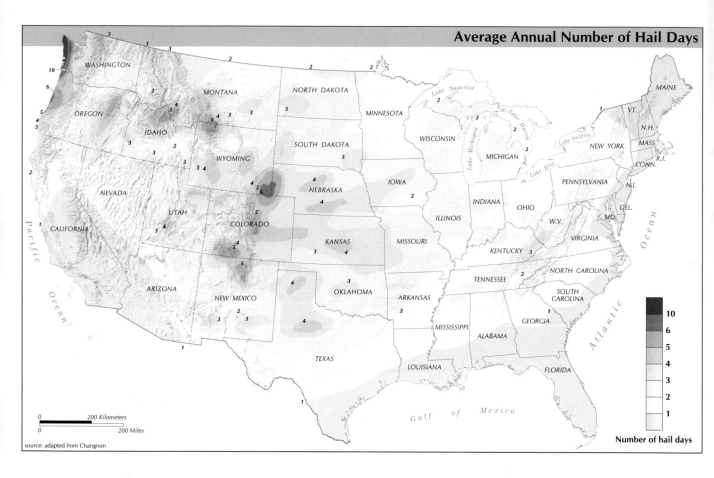

Average Annual Number of Hail Days

Number of hail days

source: adapted from Changnon

The higher a storm's cloud tops, the larger the hail is likely to become. In the High Plains giant cumulonimbus reaching over 50,000 feet are capable of producing monstrous hailstones the size of softballs.

Elevation also plays a key role in the frequency of hail: the higher the elevation the more likely is hail to reach the ground. This is because the air will be cooler, and there is less chance of the ice melting before making it to the surface. Because elevation is so important in the formation of hail, isolated mountainous areas frequently experience hail. These include the Yellowstone Plateau in northwest Wyoming, the San Juan Mountains of Colorado, the Selway-Bitterroot region of Idaho and Montana, and the Blue Mountains of eastern Oregon. These areas all receive an average of five to six hail days per year. The hail that falls in these mountain areas, however, is usually small—no more than one inch in diameter. Large and damaging hail is most likely to fall from western Texas north to eastern Wyoming and east across the states of South Dakota, Nebraska, Kansas, and Oklahoma.

The largest hailstones ever measured in North America fell during violent, tornadic thunderstorms that swept across southeast Nebraska on June 22, 2003. In the town of Aurora, hailstones the size of volleyballs fell, puncturing roofs and leaving craters a foot in diameter in the ground. The largest stones retrieved measured 7" in diameter and 18.75" in circumference, and weighed 1.7 pounds. The thunderstorms producing these monster hailstones had tops over 70,000 feet high.

The largest hailstone ever to be officially measured was this "volleyball" seven inches in diameter that fell during a severe thunderstorm at Aurora, Nebraska, in June 2003. (NOAA)

Hailstorm Intensity Scale

Intensity	General Description	Range of Hailstone Size During Hailstorm	
H0	None	.2"—.4"	pea-sized
H1	Makes holes in leaves and flower plants	.2"—.8"	marble-sized
H2	Strips leaves from plants, damages vegetables	.2"—1.2"	penny-sized
H3	Breaks glass panes, scrapes paint, marks wookwork, dents trailers, tears tents	.4"—1.8"	nickel-sized
H4	Breaks windows, cracks windscreens, scrapes off paint, kills chickens and small birds	.6"—2.4"	golf ball-sized
H5	Breaks some roof tiles and slates, dents cars, strips bark off trees, cuts branches from trees, kills small animals	.8"—3.0"	tennis ball-sized
H6	Breaks many roof tiles and slates, cuts through roof shingles and thatch, makes some holes in corrugated iron, breaks wooden window frames	1.2"—3.9"	baseball-sized
H7	Shatters many roofs, breaks metel window frames, seriously damages car bodies	1.8"—4.9"	grapefruit-sized
H8	Cracks concrete roofs, destroys other roofs, marks pavement, splits trees, can seriously injure people	2.4"—5.0"	softball-sized
H9	Marks concrete walls, makes holes in walls of wooden houses, fells trees, can kill people	3.2"—5.0"+	softball-sized
H10	Destroys wooden houses, seriously damages brick houses, can kill people	4.0"—7.0"+	volleyball-sized

Hail of similar size was recorded in Coffeyville, Kansas, on September 3, 1970 (6.5" diameter and 1.3 pounds), and near Potter, Nebraska (6" diameter and 1.5 pounds), in July of 1928.

Giant hail is not confined to the High Plains. Hailstones four inches in diameter fell across Cumberland County, Maine, on June 1, 1986, and the hail accumulated to a depth of six inches in the town of Naples, Maine, during the same storm. Incredible accumulations can occur when a hail-producing thunderstorm stalls or redevelops over a particular location. Hail accumulated four feet deep in Dix, Nebraska, on August 4, 1987. More recently, a freak storm in Los Angeles resulted in a hail accumulation of one foot on the evening of November 12, 2003, over an area of a few city blocks on the south side of the city.

Hail causes an average of $1 billion worth of damage to crops and property in the U.S. every year. The single costliest hailstorm in U.S. history occurred along the Front Range of the Rocky Mountains from Colorado Springs to Fort Collins on July 11, 1990. Damage to crops and property, especially automobiles, reached $625 million.

In spite of the frequency of hailstorms in the U.S., only two people have died from being hit by large hailstones. The figure is low because very large hail usually falls in sparsely populated regions, and when it does fall the largest stones are usually spaced about ten feet or so apart from one another.

Hail Around the World

The worst hailstorms in the world occur in Bangladesh and on the Deccan Plateau of India. In these regions fatalities resulting from hail strikes are relatively common because so many people live here and because their dwellings are poorly constructed. In one hailstorm in the Moradabad and Beheri districts of India, 246 people were killed by hail on April 30, 1888, the deadliest such event in modern history. This area also reports the largest incidence in the world of large hail, where fully 25% of the hailstorms produce stones one inch in diameter or larger. This compares to a similar ratio of only 4% to 7% in the High Plains of the United States. Southwest France is another region where hailstorms occur frequently (as often as three to six days annually), and it has the second highest ratio of large hail after India. On average 14% of the storms produce stones one inch in diameter or larger.

Probably the costliest hailstorm in world history struck Sydney, Australia, on April 14, 1999. Damage exceeded $1 billion US.

Hail can accumulate to remarkable depths when a storm becomes stationary over one place for a period of time. The hail in this photograph, however, drifted this deep after floodwaters washed it into these giant heaps in a low-lying area. (Michael Mee, FEMA)

The heaviest authenticated hailstone in the world, 2.25 pounds, fell in the Gopalganj district of Bangladesh on April 14, 1986. Its diameter was not reported, although anecdotal evidence points to stones the size of "pumpkins." This region also has the highest annual number of hail days in the world, with an average of 10 to 15.

Other reports of huge hailstones have come from Kazakhstan, where a single stone weighing 4.18 pounds was said to have fallen in 1959.

A hailstone weighing 2.14 pounds was measured and photographed in Strasbourg, France, on August 11, 1958. China has reported some catastrophic hailstorms. A hailstorm in Guangxi Province, China, on May 1, 1986, killed 16 people and injured another 125. The stones were reported to have weighed up to 11 pounds, but it is likely these were actually masses of hailstones that had melted together on the ground.

Other regions of the world with a high frequency of hail fall (six or more days per year) are listed on the map below.

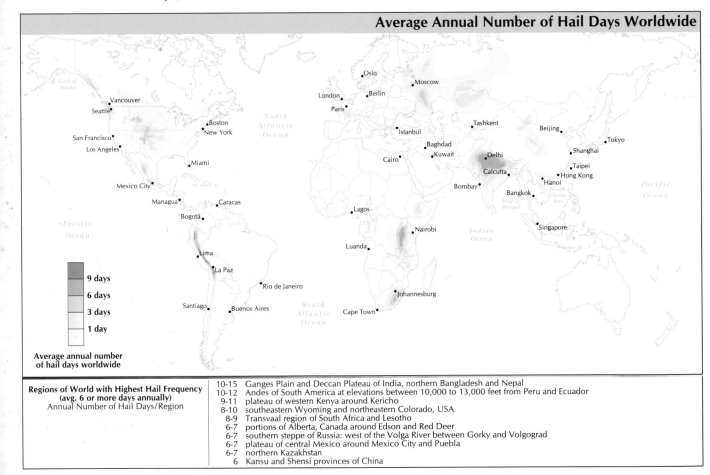

Average Annual Number of Hail Days Worldwide

| 9 days |
| 6 days |
| 3 days |
| 1 day |

Average annual number
of hail days worldwide

Regions of World with Highest Hail Frequency (avg. 6 or more days annually) Annual Number of Hail Days/Region	
10-15	Ganges Plain and Deccan Plateau of India, northern Bangladesh and Nepal
10-12	Andes of South America at elevations between 10,000 to 13,000 feet from Peru and Ecuador
9-11	plateau of western Kenya around Kericho
8-10	southeastern Wyoming and northeastern Colorado, USA
8-9	Transvaal region of South Africa and Lesotho
6-7	portions of Alberta, Canada around Edson and Red Deer
6-7	southern steppe of Russia: west of the Volga River between Gorky and Volgograd
6-7	plateau of central Mexico around Mexico City and Puebla
6-7	northern Kazakhstan
6	Kansu and Shensi provinces of China

(opposite) Towering cumulonimbus clouds often produce large hail. The higher the cloud top the larger the hail is likely to be. (A. T. Willett/lightningsmiths.com)

chapter 6
TORNADOES

ornadoes are without question the ultimate manifestation of extreme weather. No other meteorological event is as violent or awe inspiring. Their occurrence is relatively rare even in the most tornado-prone regions of the United States, and few people have ever witnessed one. Fewer still have survived a direct hit by a tornado, and only a handful have seen the very funnel that passed over them *(see p. 170)*.

The United States reports more twisters than any other country in the world and, as with severe thunderstorms, the reason is the result of its unique topography. Only in North America does a solid land mass stretch from the sub tropics to the arctic with no mountain barriers to inhibit the mixture of air masses originating from these two regions. This is also why most tornadoes, especially the more violent ones, occur in the Great Plains and Midwest.

Approximately one thousand tornadoes are counted on an annual basis in the United States, but the number varies considerably from year to year, with recent years averaging more than a thousand. Some climatologists estimate that more than two thousand occur nationwide, but many go by unseen and unrecorded. So the increase in the number of tornadoes being reported in recent years is probably due to better detection, thanks to Doppler radar and increased public awareness, rather than increased frequency.

Tornadoes can be deadly, but with increased awareness the annual death toll has been dropping over the years. An average of about 56 Americans were killed by tornadoes during the decades of the 1980s and 1990s, down from an average of 97 during the 1960s and 1970s. That in turn was a dramatic improvement compared to the 160 deaths per year on average during the 1940s and 1950s, when efforts to understand and forecast tornadoes began. In the single decade of the 1920s, 3,169 Americans were killed by tornadoes, an average of 317 a year.

A tornado swirls across the Black Rock Desert in Nevada only a few hundred yards from the lens of the photographer. It is very unusual for tornadoes to form anywhere in Nevada. (Doug Keister)

Fujita Tornado Scale

Rating	Wind Speed	Damage
F-0	40-72 mph 64-116 kmh	*Slight.* Some damage to chimneys, breaks branches off trees, pushes over shallow-rooted trees, damages sign boards.
F-1	73-112 mph 117-180 kmh	*Moderate.* The lower limit is the beginning of hurricane wind speed, peels surface off roofs, mobile homes pushed off foundations or overturned, moving autos pushed off the roads, attached garages may be destroyed.
F-2	113-157 mph 181-253 kmh	*Considerable.* Roofs torn off frame houses, mobile homes demolished, boxcars pushed over, large trees snapped or uprooted, light object missiles generated.
F-3	158-206 mph 254-331 kmh	*Severe.* Roof and some walls torn off well constructed houses, trains overturned, most trees in forest uprooted.
F-4	207-260 mph 332-418 kmh	*Devastating.* Well-constructed houses leveled, structures with weak foundations blown off some distance, cars thrown and large missiles generated.
F-5	261-318 mph 419-512 kmh	*Incredible.* Strong frame houses lifted off foundations and carried considerable distances to disintegrate, automobile sized missiles fly through the air in excess of 100 meters, steel re-inforced concrete structures badly damaged.

Fujita Scale

The strength of a tornado is measured using the Fujita Scale, named after Tetsuya Theodore Fujita, a Japanese-born University of Chicago physicist who, in the 1950s, became one of the world's first and foremost tornado scientists. He developed his eponymous scale ranking tornadoes from F0 to F5 after surveying hundreds of damage sites in the early 1950s. The vast majority of tornadoes, about 70%, are of the F0 and F1 intensity, indicating wind speeds in the 40- to 110-mph range—relatively weak, but still capable of causing significant, even extreme damage.

At the top of the scale are the true monsters, the F5's, sometimes a mile or more in width with winds having been measured as high as 318 mph, that completely devastate everything in their path. It would appear that tornadoes of F5 strength are unique to the United States. Fortunately, these are extremely rare and only a handful occur each decade (*see the list of all F5 tornadoes recorded in the U.S. since 1900 on p.169*).

"Typical" Tornado

Tornadoes may form at any time of the day and happen at any time of the year but are most frequent during the spring months of April, May, and June. At this time warm, moist air from the Gulf of Mexico is likely to collide with cooler, drier air coming down from Canada or blowing in off the Rocky Mountains.

Almost 60% of tornadoes occur between noon and sunset when the sun best heats the atmosphere, creating unstable conditions at various altitudes. The jet stream plays an important role in tornado production when it shears through the axis of moist air and dry air rotating into a surface low-pressure system.

Other factors that produce conditions conducive to tornado development include vorticity, low-level jets, and atmospheric "caps." To understand more about these terms and tornado development, it is best to read one of the excellent texts devoted to tornado research written by experts such as Howard B. Bluestein or Thomas P. Grazulis (see bibliography).

Statistically, the most likely time, date, and place for a tornado to form would be at 5 p.m. during the first or second week of May in central Oklahoma. April and May are the months of maximum tornado activity for the states of Texas, Oklahoma, and Kansas; the Gulf States have their peak tornado season in March. In the Upper Midwest and Northern Plains it is June, and in New England and the Mid-Atlantic it is July.

The time of day a tornado is most likely to form also varies from region to region. In the Gulf States, tornadoes are most likely to occur around 3 p.m. The "typical" tornado has

◆ The intensity rating of F1 (winds from 73–112 mph)

◆ Path length of about one mile

◆ Width of 150 feet

◆ Forward speed of 40 mph

◆ Movement from the southwest to northeast

In general, a tornado's path length and width increase with its strength. An F5 tornado has a typical path length of almost 25 miles and funnel width of 1,600 feet.

Month of Maximum Occurrence of Tornadoes

source: NOAA

Extreme Tornadoes

The longest continuous path a tornado has traveled is probably the 217 miles covered by the great Tri-State Tornado of March 18, 1925 (*see p.194*). Some tornadoes don't travel far at all, but simply kiss the earth before ascending like malevolent angels back into the sky.

Tornado expert Thomas Grazulis suggests that the widest tornado to have been measured (by damage path) was probably the 2.2-mile-wide monster that cut through Pennsylvania's Moshannon State Forest on May 31, 1985. This F4 tornado had a path length of 69 miles and, of course, was not two miles wide for its entire length but only at its widest. Fortuitously, this extremely violent storm passed through a largely uninhabited portion of what is otherwise a heavily populated state. Another contender for widest-ever tornado is the storm that struck the Red Springs–McColl area of North Carolina on March 28, 1984. This tornado may have been 2.5 miles wide at its greatest.

In their dissipation stage, some tornadoes become extremely narrow and rope-like.

The fastest tornado on record was, once again, the disastrous Tri-State storm of 1925, which attained a forward motion of an estimated 73 mph near the end of its course in Indiana.

On the other hand, tornadoes have been known to become stationary for a brief period of time, obviously not a comfortable place to be.

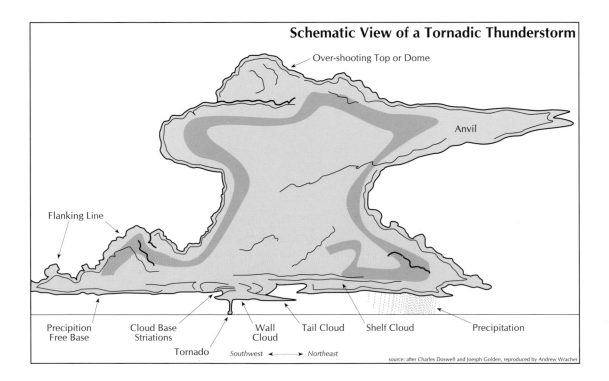

Schematic View of a Tornadic Thunderstorm

Over-shooting Top or Dome

Anvil

Flanking Line

Precipition Free Base

Cloud Base Striations

Tornado

Wall Cloud

Tail Cloud

Shelf Cloud

Precipitation

Southwest ◄——► Northeast

source: after Charles Doswell and Joesph Golden, reproduced by Andrew Wracher

F5 Tornadoes 1900-1959

Date	Location	Deaths
1900-1909		
May 10, 1905	OK	97
June 5, 1905	MI	5
April 23, 1908	NE	3
June 5, 1908	NE	12
1910-1919		
June 15, 1912	MO	15
June 11, 1915	KS	0
May 25, 1917	KS	23
May 21, 1918	IA	4
June 22, 1919	MN	57
1920-1929		
March 11, 1923	TN	20
May 14, 1923	TX	23
March 18, 1925	MO, IL, IN	695
April 12, 1927	TX	74
May 7, 1927	KS	10
April 10, 1929	AR	23
1930-1939		
May 22, 1933	NE	8
April 5, 1936	MS	216
April 26, 1938	NE	14
June 10, 1938	TX	14
April 14, 1939	OK, KS	7
1940-1949		
March 16, 1942	IL	7
April 29, 1942	KS	15
June 17, 1944	SD, MN	8
April 12, 1945	OK	69
April 9, 1947	TX, OK, KS	181
May 31, 1947	OK	7
1950-1959		
March 21, 1952	TN	55
May 11, 1953	TX	114
May 29, 1953	ND	2
June 8, 1953	MI	115
June 27, 1953	IA	1
December 5, 1953	MS	38
May 25, 1955	OK, KS	20
May 25, 1955	OK, KS	80
April 3, 1956	MI	18
May 20, 1957	KS, MO	44
June 20, 1957	ND, MN	10
December 18, 1957	IL	1
June 4, 1958	WI	20

source: NCDC, Thomas P Grazulis

F5 Tornadoes 1960-2003

Date	Location	Deaths
1960-1969		
May 5, 1960	OK	5
May 19, 1960	KS	0
May 30, 1961	NE	0
April 3, 1964	TX	7
May 5, 1964	NE	2
April 11, 1965	IN	36
April 11, 1965	OH	18
May 8, 1965	SD	0
March 3, 1966	MS	57
June 8, 1966	KS	12
October 14, 1966	IA	6
April 23, 1968	OH	6
May 15, 1968	IA	13
May 15, 1968	IA	5
June 13, 1968	MN	9
1970-1979		
May 11, 1970	TX	28
February 21, 1971	MS	46
May 6, 1973	TX	0
April 3, 1974	IN	6
April 3, 1974	KY, IN	31
April 3, 1974	OH	34
April 3, 1974	IN, KY, OH	3
April 3, 1974	AL	28
April 3, 1974	MS, AL	30
March 26, 1976	OK	2
April 19, 1976	TX	0
June 13, 1976	IA	0
April 4, 1977	AL	22
1980-1989		
April 2, 1982	OK	0
June 7, 1984	WI	9
May 31, 1985	OH, PA	18
1990-1999		
March 13, 1990	KS	1
March 13, 1990	KS	1
August 28, 1990	IL	29
April 26, 1991	KS	17
June 16, 1992	MN	1
July 18, 1996	WI	0
May 27, 1997	TX	27
April 8, 1998	AL	32
April 16, 1998	TN	3
May 3, 1999	OK	35
TOTALS:	**94 F5 TORNADOES**	**3202 DEATHS**
2000-2004	No F5 tornadoes reported as of April 2004	

EYEWITNESS TO THE CENTER OF A TORNADO

Very few people have actually seen the interior of a tornado funnel, for obvious reasons. Probably the best and most reliable eyewitness account given was that of Will Keller, a farmer in Kansas, who was caught in an open field on the afternoon of June 22, 1928 when a tornado headed in his direction. He ran for his cyclone cellar and made it just moments before the funnel was upon him. As he paused momentarily prior to closing the door, he looked up one last time and noticed that the funnel was beginning to lift.

As I paused to look I saw the lower end [of the tornado funnel] which had been sweeping the ground was beginning to rise. I knew what that meant, so I kept my position. I knew that I was comparatively safe and I knew that if the tornado again dipped I could drop down and close the door before any harm could be done.

Steadily the tornado came on, the end gradually rising above the ground. I could have stood there only a few seconds, but so impressed was I with what was going on that it seemed a long time. At last the great shaggy end of the funnel hung directly overhead. Everything was still as death. There was a strong gassy odor and it seemed that I couldn't breathe. There was a screaming, hissing sound coming directly from the end of the funnel. I looked up and to my astonishment I saw right up into the heart of the tornado. There was a circular opening in the center of the funnel, about 50 or 100 feet in diameter, and extending straight upward for a distance of at least one half mile, as best I could judge under the circumstances. The walls of this opening were of rotating clouds and the whole was made brilliantly visible by constant flashes of lightning which zigzagged from side to side. Had it not been for the lightning, I could not have seen the opening, not any distance into it anyway.

Around the lower rim of the great vortex small tornadoes were constantly forming and breaking away. These looked like tails as they writhed their way around the end of the funnel. It was these that made the hissing noise.

I noticed that the direction of rotation of the great whirl was anti-clockwise, but the small twisters rotated both ways—some one way and some another.

The opening was entirely hollow except for something which I could not exactly make out, but suppose that it was detached wind cloud. This thing was in the center and was moving up and down.

—from *The Elements Rage* by Frank W. Lane, 1965

Although tornadoes can move in any direction, most move from the southwest to northeast, perpendicular to the center of a surface low-pressure system. Now and then tornadoes get quirky, making, say, a U-turn; or as was the case on June 21, 1949, in Caddo County, Oklahoma, an almost perfect circle, two miles in diameter.

One of the first photographs ever taken of a tornado was this one near Howard, South Dakota, on August 28, 1884. It is possible that this photograph inspired L. Frank Baum to include a tornado in his book The Wizard of Oz, *as he was a reporter for a newspaper in nearby Aberdeen at the time. (Warren Faidley/Weatherstock, Inc.)*

TORNADOES IN THE UNITED STATES

Where are you most likely to witness or, heaven forbid, be killed by a tornado? There's probably no greater expert on the subject than climatologist Thomas P. Grazulis, whose exhaustive survey of tornado phenomena is reported in his 1,400-page book, *Significant Tornadoes: 1680-1991* (a must-read for any tornado aficionado, and one of the most remarkable reference works ever assembled).

(following pages) An eerily beautiful F1 tornado swirls through the plains outside of Littlefield, Texas. The vast majority of tornadoes are relatively weak ones such as this. (Warren Faidley)

Grazulis defines a "significant" tornado as one of F2 strength or stronger, capable of causing extreme damage. He lists a multitude of ways to rank tornado risk and occurrence (by state) based on official tornado statistics between 1953 and 1995. The most pertinent of these rankings are the following:

TORNADO OCCURRENCES

HIGHEST ABSOLUTE NUMBER		HIGHEST NUMBER PER YEAR PER 10,000 SQ. MI.	
1	Texas	1	Florida
2	Oklahoma	2	Oklahoma
3	Florida	3	Indiana

KILLER TORNADO OCCURRENCES

HIGHEST ABSOLUTE NUMBER		HIGHEST NUMBER PER YEAR PER 10,000 SQ. MI.	
1	Texas	1	Arkansas
2	Oklahoma	2	Indiana
3	Arkansas	3	Mississippi

TORNADO FATALITIES

HIGHEST ABSOLUTE NUMBER		HIGHEST NUMBER PER YEAR PER 10,000 SQ. MI.	
1	Texas	1	Massachusetts
2	Mississippi	2	Mississippi
3	Alabama	3	Indiana

The eastern plains of Colorado experience a significant amount of tornado activity every year. On rare occasions, a tornado forms over the state's rugged mountainous region, which lies to the west. Here, a tornado funnel spins over 14,000-foot-high peaks in the Colorado Rockies. (David Smyth/Curtis Martin Photography)

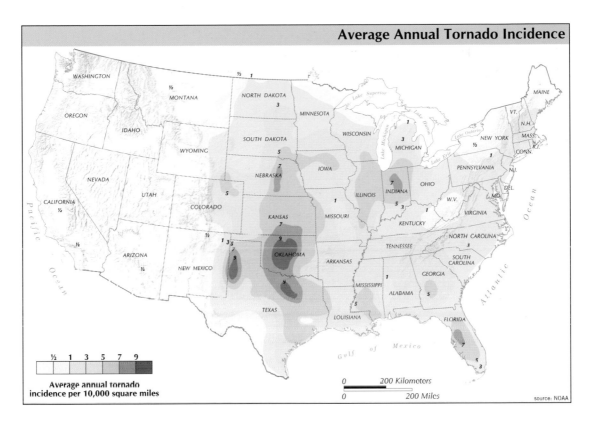

Average Annual Tornado Incidence

Average annual tornado incidence per 10,000 square miles

½ 1 3 5 7 9

0 200 Kilometers
0 200 Miles

source: NOAA

Yes, statistically, a resident of Massachusetts is at the greatest risk of dying from a tornado. But this is a statistical anomaly.

As geographer Mark Monmonier put it in his book *Cartographies of Danger; Mapping Hazards in America*, "...threat ratings computed for whole states ignore intrastate differences in population..." And so, in no way are the residents of Massachusetts at greater risk from tornadoes than, say, people who live in central Oklahoma.

The residents of Oklahoma City, probably the largest, most tornado-prone, urban area in the world, know this only too well. The Massachusetts statistic is largely due to the occurrence of a single deadly event, the Worcester tornado of June 1953, when 94 people were killed in and around the city. This has skewed (for the time being) the statistical hazard of death by tornado in Massachusetts.

Nevertheless, a very real and anomalous risk of tornadoes exists in Massachusetts. The western and central districts of the state are unusually tornado-prone. And tornadoes in this region are exceptionally dangerous because it is hard to see them approaching, due to hilly terrain and dense vegetation. By "anomalous" I mean that there seems to be a higher risk of tornado occurrence in some regions relative to their surrounding areas, but keep in mind that virtually

20 Deadliest Tornado Outbreaks in U.S. History

Deaths	Locations	Date	No. F2+ Tornadoes
747	MO, IL, IN, AL, TN, KY	Mar 18, 1925	9
454	AR, TN, AL, MS, GA, SC	Apr 5-6, 1936	12
330	AL,TN,KY, GA, SC	Mar 21-22, 1932	36
324	AR, NE, TX, AL, LA, MS	Apr 23-24, 1908	30
317	LA, MS	May 7, 1840	1
315	12 states and Canada	Apr 3-4, 1974	95
305	MO, IL	May 27, 1896	10
256	IA, WI, IL, IN, MI, OH	Apr 11, 1965	38
236	MI, MA	Jun 8-9, 1953	14
224	MS, AL, TN	Apr 20, 1920	31
217	TX, KS, AR, MO, IL	May 8-9, 1927	32
199	NE, IA, AR,IL, IN, MO, LA	Mar 23-24, 1913	14
177	AR, TN, MS, AL	Mar 21, 1952	17
167	11 states	Feb 19-20, 1884	37
165	8 states	Apr 18, 1880	22
159	8 states	May 26-28, 1917	20
163	WI, Il, PA, WV, OH, MD	Jun 22-23, 1944	9
153	8 states	Mar 28, 1920	31
148	IL,MS, TN, KY, IN, AL, TN	Mar 16, 1942	26
132	8 states	Apr 29-30, 1909	38

10 Greatest Tornado Swarms

No. of Tornadoes	Date	Deaths
148	Apr 3-4, 1974	315
111	Sep 19-23, 1967	5
99	May 26-27, 1973	22
95	Nov 21-23, 1992	26
94	May 4-5, 2003	37
85	May 22-23, 2004	1
80	May 18-19, 1995	4
78	May 3-4, 1999	46
70	May 11-12, 1982	2
67	Apr 26-27, 1994	3

20 Deadliest Single Tornadoes

	Deaths	Injured	Location	Date	Length in miles	Width in yards
1	695	2027	MO, IL, IN	Mar 18, 1925	219	1200
2	317	109	Natchez,MS	May 7, 1840	35	1000
3	255	1000	St. Louis, MO	May 27, 1896	12	800
4	203	1600	Gainesville, GA	Apr 6, 1936	7	400
5	216	700	Tupelo, MS	Apr 5, 1936	15	1000
6	181	970	Woodward, OK	Apr 9, 1947	170	1500
7	143	770	Amite, LA — Purvis, MS	Apr 24, 1908	135	1000
8	117	200	New Richmond, WI	Jun 12, 1899	30	300
9	115	115	Flint, MI	Jun 8, 1953	27	800
10	114	597	Waco, TX	May 11, 1953	23	600
11	114	250	Goliad, TX	May 18, 1902	15	250
12	103	350	Omaha, NE	Mar 23, 1913	40	400
13	101	638	Matoon, IL	May 26, 1917	155	800
14	100	381	Shihstoan, WV	Jun 23, 1944	60	300
15	99	200	Marshfield, MO	Apr 18, 1880	64	800
16	98	300	Poplar Bluff, MO	May 9, 1927	60	600
17	98	180	Gainesville, GA	Jun 1, 1903	4	300
18	97	150	Snyder, OK	May 10, 1905	40	800
19	94	1288	Worcester, MA	Jun 9, 1953	46	1000
20	91	400	Natchez, MS	Apr 24, 1908	105	700

Tornadoes keyed to map

Significant Tornado Paths: 1889–1990

Significant Tornadoes = F2 or stronger

0   200 Kilometers

0   200 Miles

source: Thomas P. Grazulis

anywhere in the Midwest from southeastern South Dakota, south to most of Texas, west to north-eastern Colorado, and east to western Ohio, has a considerably higher risk of tornadoes than the Northeast or Mid-Atlantic *(see map p. 175)*.

Most of these "anomalous" regions are populated areas with numerous weather stations and storm spotters, so it is reasonable to assign some of this "elevated risk" to record keeping. Nevertheless, if you live in central Oklahoma for more than a few years you are almost guaranteed to see a tornado sooner or later. In fact, of all the nation's metropolitan areas, Oklahoma City is at the greatest risk from tornadoes *(see below)*. The most at-risk, large metropolitan areas—250,000 people or more—include the following (in descending order of risk).

MOST AT-RISK LARGE U.S. CITIES

1 Oklahoma City, Oklahoma

Sitting in what is practically the epicenter of North America's most active "tornado alley," Oklahoma City has been struck by tornadoes 112 times since the city's first recorded twister in 1893. Residents may expect a damaging tornado to hit somewhere in the metro area about once every two years.

Oklahoma City was devasted by several tonadoes on May 3, 1999, including one of F5 intensity. The damage portrayed in this photograph illustrates the kind of destruction an F5 tornado can cause. This tornado outbreak was the costliest in U.S. history. (Peter Arnold, Inc.)

2 Dallas–Fort Worth, Texas

As more and more people move into this sprawling metropolitan area, more and more residents find themselves at risk from destructive tornadoes. Downtown Fort Worth was devastated by a tornado as recently as March 28, 2000. Dallas County was hit by no less than 54 tornadoes between 1950 and 1991. For a detailed analysis of tornado risk in the Dallas–Fort Worth area, visit:*www.dfwinfo.com/weather/study/index.asp*

3 Lubbock, Texas

Lubbock is in the center of a particularly active tornado zone. The city has been struck by several devastating tornadoes in the past, including its worst disaster ever on May 11, 1970, when an F5 tornado ripped through the heart of the city killing 28.

4 Kansas City, Missouri

Kansas City has been struck by tornadoes many times, most recently in May 2003, when four tornadoes passed through the metropolitan area. Fortunately, the city has been spared a truly devastating strike.

(opposite) A violent F5 tornado swept away the village of Glazier, Texas, on April 9, 1947. Only one damaged structure remained standing. The town was never rebuilt. (Amarillo Globe-News)

A "mothership" supercell rotates across the prairie near Childress, Texas. Storm clouds of this nature are precursors of tornado formation, as was the case here. (Roger Hill)

5 Indianapolis, Indiana

This sprawling city is in the middle of the Indiana "tornado alley" and has been hit by tornadoes a dozen times in its history. In April and May the city is most at risk, although a tornado could strike here at any time of the year.

6 St. Louis, Missouri

Although not in the middle of a particularly active tornado zone, the St. Louis metropolitan area has had several devastating incidents, including one of the nation's worst tornado disasters, when, on May 27, 1896, 306 people were killed. Another 79 were killed in and around the city by a tornado on September 29, 1927.

7 Jackson, Mississippi, and Birmingham, Alabama

In the Southeast, it is hard to say whether Birmingham, Alabama, or Jackson, Mississippi, better deserves a spot on this list. Birmingham has had several terrible tornadoes, including one on April 8, 1998, when 33 people died. Jackson, on the other hand, has been spared bad tornado strikes even though it is in an area that is prone to their occurrence and devastating tornadoes have struck at nearby locations.

8 Little Rock, Arkansas

Little Rock has been hit at least a dozen times in the past by destructive tornadoes and probably will continue to be so again in the future. So far, it has not had a truly devastating event.

9 Omaha, Nebraska

Omaha is a bit to the north and east of the heart of the Plains "tornado alley" but has a history of being struck by dangerous storms. A major tornado rolled through the city on March 23, 1913, killing 103 people in one of the nation's worst tornado disasters. It seems that almost every spring or summer it has some close calls.

10 Chicago, Illinois

Surprisingly, Chicago has never been directly hit by a major tornado, although its suburbs have had a few near misses. Given the vast size of the city and its location in an active tornado region, it seems inevitable that a big tornado will make a direct strike one day.

Probably the strongest tornado ever to traverse the Rocky Mountains was an F4 that mowed down miles of forest, some above 10,000 feet, in the Teton Wilderness area of Wyoming. (James W. Partacz)

Although tornadoes are most common in the American Midwest they have occurred in every state and in places that would normally be considered "risk-free." Tornadoes have been reported in Alaska near the arctic circle and near Yellowknife in the Northwest Territory of Canada. They occur every winter in California's Central Valley, especially in the Fresno area.

One of the most powerful tornadoes ever to strike the West happened in the high-mountain area of the Teton Wilderness just northeast of Moran Junction on July 21, 1987. A mile-wide, F4 tornado cut a destructive path for 24 miles across mountainous, densely forested terrain and even crossed the Continental Divide at an elevation of 10,170 feet.

Tornadoes Around the World

Almost every country in the world has reported tornadoes at one time or the other, although they are common to only a handful. Canada has had over a hundred killer tornadoes in the past century. Most strike the southern and central portions of Alberta, southern Saskatchewan, and southern Manitoba, but a smaller tornado alley exists in southern Ontario from Windsor to Orangeville.

Canada's deadliest tornado, ever, hit the heart of Regina, Saskatchewan, on June 30, 1912, killing 28 people. More recently, Edmonton, Alberta, was devastated by a monstrous F4 tornado on July 31, 1987, resulting in 27 deaths. The funnel was almost a mile wide at one time and traveled on the ground for 22 miles causing US$528 million in damage.

One of the most powerful tornadoes ever to strike Canada was this F4 monster, which roared through Edmonton on July 31, 1987. Twenty-seven lives were lost. (Robert Carlton, University of Alberta)

Bangladesh

After the United States and Canada, the country receiving the most violent, if not frequent, tornadoes is Bangladesh. These storms are deadly because the area is densely populated and buildings are poorly constructed. Tremendous thunderstorms develop here at the beginning of their monsoon season, usually April and May, when deep tropical air masses collide with cool dry air spilling over the Himalayas. Like the U.S. Midwest, the flat terrain of Bangladesh enhances the development of tornadoes and allows them to travel considerable distances. A tornado on April 26, 1989, killed a reported 1,300 people north of Dacca, probably the single deadliest tornado in world history. Another 700 lives were lost in a storm that struck Tangail, Bangladesh, on May 13, 1995. The intensity of these storms is not fully known, since detailed engineering studies of the damage were not recorded, but F4 storms probably occur, the most powerful outside of North America.

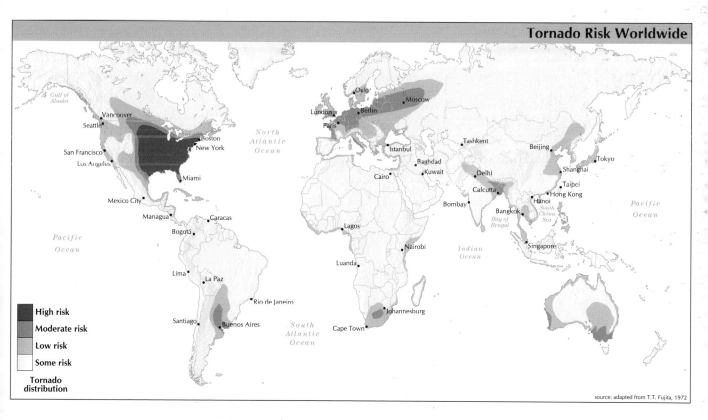

Tornado Risk Worldwide

High risk
Moderate risk
Low risk
Some risk

Tornado
distribution

source: adapted from T.T. Fujita, 1972

(opposite) Audra Thomas poses in front of an F1 tornado that narrowly missed striking her family's farm (in the background) south of Beaver City, Nebraska. In Bangladesh, even relatively weak tornadoes, such as the one shown opposite, can cause widespread death and destruction where the rural population density is about a thousand times that of central Nebraska. (Merrillee Thomas/Tom Stack Associates)

Russia

Tornado expert Thomas Grazulis speculates that Russia probably receives the largest absolute number of tornadoes a year after the United States, due to its vast size and "potential for small tornadoes." Violent tornadoes, however, have been known to occur in Russia, including an outbreak that killed 400 people in towns 150 to 200 miles north of Moscow on June 9, 1984.

Western Europe

In Europe, France and Great Britain experience damaging tornadoes annually though they rarely, if ever, reach the viciousness of the North American variety. Great Britain averages 30 to 35 weak tornadoes per year. Most strike an area that runs from the north Midlands southeastwards to Kent. The last known instance of a fatality was on October 27, 1913, when two persons were killed in southern Wales. An outbreak on November 23, 1981, produced 105 tornadoes in a single day from Anglesey to Essex. Little damage was caused by any of these mini-twisters, which attests to the relative moderation of the British variety of tornado.

A weak tornado whirls through Algeciras, Spain. Powerful tornadoes in the F4 or F5 range are almost unheard of anywhere in the world outside the United States. (Royal Geographical Society, Alamy Images)

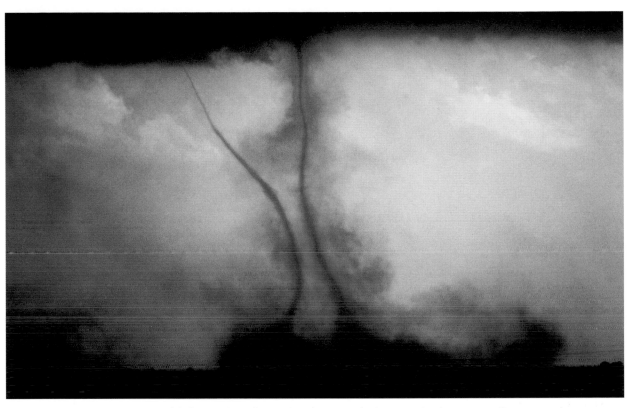

On rare occasions, multiple vortices form simultaneously from a single supercell, as was the case here, when two weak tornadoes spun from a storm near Dimmitt, Texas in June 1995. (Warren Faidley/Weatherstock, Inc.)

Perhaps Western Europe's worst tornado disaster ever occurred near Monville, France on August 19, 1845, when a large tornado struck three textile and paper mills killing at least 70 people and perhaps as many as 200.

South Africa

South Africa is often hit by violent thunderstorms during the summer months of November through February, and these storms sometimes produce devastating hailstorms and tornadoes. An outbreak between November 30 and December 2, 1952, killed 31 people in the towns of Albertynesville and Paynesville near Johannesburg.

(following pages) This classic photograph of an entire supercell with its attendant tornado was photographed by tornado expert Howard B. Bluestein while tornado chasing near Spearman, Texas, on May 31, 1990. Large hail was falling from the clouds in the illuminated area to the right of the tornado. (Howard B. Bluestein)

Australia and New Zealand

Tornadoes are relatively common in Australia, with an average of about 25 a year. Occasionally they become strong enough to warrant a rating as high as F3. The areas affected are usually those in the southwestern part of Western Australia around Perth and in the Highlands of Victoria state in southeastern Australia.

New Zealand can expect 25 or so weak tornadoes each year. These usually occur on the North Island around Auckland and the Bay of Plenty and in the vicinity of Mount Taranaki (Egmont).

Japan

In Japan about 20 tornadoes, or "tatsumaki" (dragon whirls), are reported annually. Most occur along the Pacific coastline in the summer months and are associated with supercells. Ishikawa Prefecture on the Sea of Japan is also tornado prone, but only during the winter season. These winter tornadoes are known as "tornadoes with snowcloud" and form over the sea as waterspouts before moving ashore to become weak tornadoes. Violent winter tornadoes sometimes

form on Japan's Pacific coast as well, and one hit the coastal city of Mobara, 20 miles southeast of Tokyo, on December 1, 1990. This violent, F4 tornado damaged or destroyed over 1,000 homes and injured 100 people.

Argentina

Tornadoes in Argentina are rare. Only a handful occur every year, but sometimes they are exceedingly violent. Most occur in the states of Cordoba and Santa Fe to the northwest of Buenos Aires. A powerful tornado struck this region on January 10, 1973, causing the death of at least 54 people and the destruction of over 500 homes in the town of San Justo.

An ominous squall line approaches the beach at Cariló, Argentina about 200 miles south of Buenos Aires. The storm struck during the middle of Argentina's summer in February 2003 and caused no damage. These mean-looking gust front's bark usually being worse than their bite. (Rafael Ketelhohn)

LUMINOUS TORNADOES

The luminous glow sometimes witnessed during tornadoes at nighttime has intrigued scientists for years. Although frequent and intense lightning is common when tornadoes occur, especially in the minutes preceding their formation, tornadoes themselves are not known to have any electrical properties, save perhaps static electricity that develops as a result of debris in motion. Several very well documented cases that include photographic evidence are on record among the dozens of historical observations. *Monthly Weather Review* Vol. 83, 1955, reprinted the following account made by the Weather Service observer in Blackwell, Oklahoma, who witnessed a luminous tornado on the evening of May 25, 1955.

> *A*s the storm was directly east of me (at a distance of 3,000 feet), the fire up near the top of the funnel looked like a child's Fourth of July pinwheel. There were rapidly rotating clouds passing in front of the top of the funnel. These clouds were illuminated only by luminous bands of light. The light would grow dim when these clouds were in front, and then it would glow bright again as I could see between the clouds. As near as I can explain, I would say that the light was the same as an electric arc welder but very much brighter. The light was so intense that I had to look away when there were no clouds in front of it. The light and clouds seemed to be turning to the right like a beacon in a lighthouse.

Other witnesses reported that eventually the entire tube of the funnel from the ground to the cloud became suffused with a steady, deep-blue light with the center of the tube having an orange, fire-like color.

> *I*t looked like a giant neon tube in the air. As it swung along the ground level, the orange fire or electricity would gush out from the bottom of the funnel, and the updraft would take it up in the air causing a terrific light and it was gone! As it swung to the other side, the orange fire would flare up and do the same.

Even stranger, scientist H. Jones noticed a large, blinking, circular, blue spot rotating through the thunderstorm cloud about an hour prior to the tornado's formation. Sferic storm-tracking equipment in Jones's laboratory recorded electromagnetic radiation emanating from this storm cloud. The blue spot would blink on and off at two-second intervals.

The best documented of all luminous tornado events was that of the F4 tornado that roared through Toledo, Ohio, on the night of April 11, 1965, killing 16 people. This tornado was part of the historic Palm Sunday outbreak which produced 48 funnels killing 256 people from Wisconsin to Michigan and Ohio. There were dozens of eyewitness accounts of luminous phenomena. Mrs. Highiet of Toledo remembers,

> *T*he beautiful electric blue light that was around the tornado was something to see, and balls of orange and lightning came from the cone point of the tornado. The cone or tail of the tornado reminded me of an elephant trunk. It would dip down as if to get food then rise again as if the trunk of an elephant put food in his mouth. My son and I watched the orange balls of fire roll down the Race Way Park.

Another witness of ball lightning during the storm reported a ball...

...*A*bout the size of a basketball and six feet away from me and about five feet off the ground. It was white, blue and yellow in color and coming slowly toward me at less than the speed a person could walk and when it hit the door it made the door sound like it was singing.

Other residents reported huge, bluish-white searchlights darting around in the clouds. A professional photographer, James R. Weyer, had the presence of mind to photograph some of these mysterious lights (unfortunately only with black-and-white film). The original negatives of these photographs have been extensively analyzed and proven to be not only genuine but positive evidence that the sources of light were from the tornado and not from any other optical or photographic effect.

It would seem that ball lightning often accompanies these luminous tornadoes, and some scientists have speculated that there may be an enticing connection between these light phenomena and the existence of a natural form of cold fusion.

The two shafts of light in this photograph are illuminated tornadoes. This is the only photograph ever authenticated by experts. The image was captured by photographer James R. Weyer near Toledo, Ohio, on the night of April 11, 1965. (courtesy of James R. Weyer)

Some of the horrendous damage caused by the Tri-state Tornado can be seen in this photograph taken of Griffin, Indiana. Seventy-one people were killed in Indiana, many in rural areas outside Griffin. (National Archives)

HISTORIC TORNADO DISASTERS

Deadliest: U.S. Tri-State Tornado of March 18, 1925

The worst tornado disaster in North American history occurred on the afternoon of March 1, 1925, when a single, monstrous tornado developed around 1 p.m. outside Ellington, Missouri. Four hours later the storm had raced 219 miles across Illinois and into Indiana, destroying nine towns (two of them completely wiped out) and hundreds of farms. It killed 695 people and injured over 2,000.

The tornado may well have had a forward speed of 73 mph and was so wide that no discernable funnel could be seen. Those who saw it reported that the tornado was a roiling black mass of cloud one mile in diameter. A witness in Illinois observed that the tornado resembled "a fog rolling" toward him. In fact, the tornado may have consisted of two or more funnels that merged in and out with one another, making it a multi-vortex tornado.

In the town of Murphysboro, Illinois, 234 people lost their lives, probably the single worst death toll for one town or city in United States history. The nearby village of Gorham lost every single structure and half its residents. In Indiana, along one 10-mile stretch of the tornado's path, 85 farms were wiped out. The rural death toll of 65 in Illinois was unprecedented. By any measurement—the destruction it caused (of property, loss of life, injuries) or size (length of path, forward speed)—this tornado has never been matched in climate history. On the same day, eight other tornadoes in Alabama, Tennessee, and Kentucky took another 52 lives, bringing the day's toll to 747.

Most Extensive Tornado Outbreak: Jumbo Outbreak of April 3–4, 1974

If the Tri-State Tornado disaster of 1925 was the nation's deadliest, the "jumbo outbreak" of April 3–4, 1974, was its most prolific. An incredible 148 individual tornadoes developed across

a vast stretch of the United States from Illinois to Alabama and east to Virginia during a single 18-hour period. It is possible that at one point between 7 p.m. and 8 p.m., as many as 20 tornadoes were on the ground at the same time. Three hundred and fifteen people were killed during this outbreak.

Worst hit was the town of Xenia, Ohio, where a F5 funnel destroyed 300 homes and killed 34 people. The damage to Xenia alone was calculated at $100 million ($350 million in inflation-adjusted dollars), a huge sum for such a small city.

Other devastated towns included Brandenburg, Indiana, with 28 deaths, Guin, Alabama, where 20 people died (and the entire town was wiped out), and Moulton, Alabama, where another 28 lives were lost in and around town.

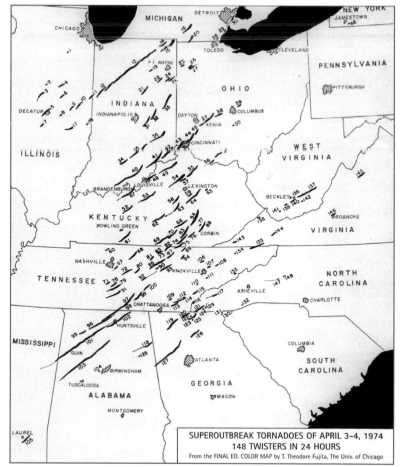

SUPEROUTBREAK TORNADOES OF APRIL 3–4, 1974
148 TWISTERS IN 24 HOURS
From the FINAL ED. COLOR MAP by T. Theodore Fujita, The Univ. of Chicago

The most remarkable aspect of this outbreak was the intensity of the tornadoes. Ninety-five of the tornadoes were "significant" (having winds of 113 mph or higher); another 30 were rated F4 or stronger with winds of 207 mph and up; and an unprecedented six of the storms were rated at F5—capable of producing extraordinary damage with their winds of 261 to 318 mph. No tornado outbreak, before or since, has produced more than two F5 tornadoes. Normally, only one in two thousand tornadoes reaches F5 intensity in the United States, and often years go by without a single F5 tornado.

Most Expensive Tornado: Oklahoma City, May 3, 1999

It shouldn't be too surprising that a recent F5 tornado was the most costly in terms of property damage—property values having soared nationwide in recent years. The fact that the tornado struck a major metropolitan area compounded the cost.

On the afternoon and evening of May 3, 1999, supercell thunderstorms produced an outbreak of over 70 tornadoes across the state of Oklahoma, the most severe such outbreak in the state's tornado-prone history. The worst storm of all gave birth to a rare F5 tornado just to the south of downtown Oklahoma City. The tornado cut through the suburbs of Moore, Del City, and Midwest City killing 36 and injuring 675. The death toll was miraculously low in relation to the incredible damage caused by the tornado. State and local meteorologists, the media, and law enforcement agencies can take credit for this, as the low number of fatalities was a direct result of their excellent forecasting and advance warnings. An amazing 2,343 homes, apartments, and businesses were completely destroyed and another 7,160 badly damaged in the space of an hour. The total damage has been estimated at $1.2 billion, making this one of the costliest natural disasters in United States history and considerably more expensive than the tornado that rates second. That one struck Wichita Falls, Texas, on April 10, 1979, causing $902 million in damage (inflation-adjusted dollars).

WATERSPOUTS

Waterspouts are tornadoes that form, or pass over, bodies of water. Usually far weaker than tornadoes, they are also more common. Most occur in tropical or subtropical waters, but they have also been noted off the coasts of New England and California and over the Great Lakes. The Lower Florida Keys Waterspout Project, conducted between May and September of 1969, counted 390 waterspouts within 50 miles of Key West. The survey established that waterspouts are much more common than previously thought. Indeed, anyone who lives along the coast of Florida will be sure to witness a waterspout sooner or later.

Other waters of the world that favor the formation of waterspouts include parts of the Gulf of Mexico, as well as the Caribbean, the Bay of Bengal, the southeast coast of Brazil, the South China Sea, and the North Atlantic equatorial convergence zone (see map).

In the rare event that a waterspout comes ashore, it becomes a tornado and can cause the same kind of damage that a FO or weak F1 tornado might cause.

This was the case in Bermuda on April 5, 1953, when a waterspout-cum-tornado roared ashore killing one person, injuring nine, and damaging more than 50 homes. In December of 1969, the Mediterranean island of Cyprus was also hit by deadly waterspouts. Four people were killed by a swarm of six waterspouts that all came ashore in the space of a few hours.

It is not unusual to witness swarms of waterspouts forming at about the same time. As many as nine have been counted churning simultaneously across the ocean like so many sea devils. They come in all shapes and sizes, although they usually are quite thin and long compared to their terrestrial sisters. An unusually slender waterspout was spotted off the Australian coast at Eden, New South Wales: measuring 5,000 feet in height, it was only 10 feet wide.

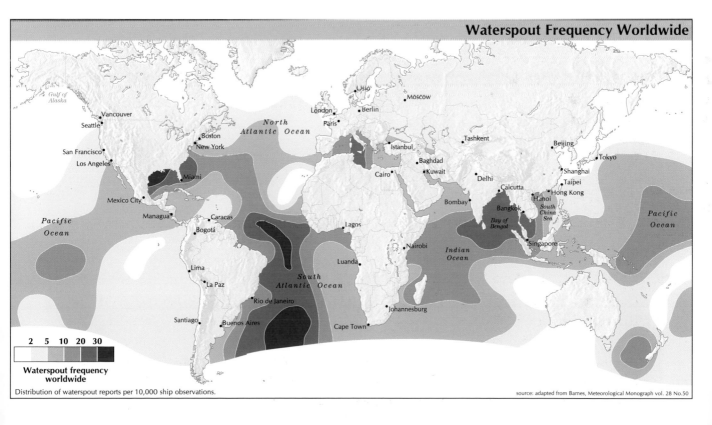

Waterspout Frequency Worldwide

2 5 10 20 30

Waterspout frequency worldwide

Distribution of waterspout reports per 10,000 ship observations.

source: adapted from Barnes, Meteorological Monograph vol. 28 No.50

Another observation from Rabat, Morocco, made on December 18, 1917, claimed a funnel 1,000 feet high but only three feet in diameter dancing off the Atlantic shoreline. But as said above, not all waterspouts fit the tall-and-thin description. Some are enormous. The captain of the British steamship *River Avon*, sailing in the vicinity of Bermuda in January of 1888, reported seeing a waterspout a mile wide. And a large and powerful waterspout struck the White Star ocean liner *Pittsburgh* while she was sailing in the mid-Atlantic on March 30, 1923. The hit badly damaged her bridge and chart room and brought the ship to a stop for more than an hour.

An enormous waterspout churns in the Atlantic perilously close to a passing freighter. (Royal Air Force)

ANIMAL FALLS

Popular topics for *Ripley's Believe it or Not* and its ilk, are the many well-documented cases of fish, frogs, periwinkles, and other such objects falling from the sky during the course of a rainstorm. Outrageous as it may sound, these are perfectly explainable phenomena. Tornadoes, and especially waterspouts, tend to suck up whatever lies in their path and, if the objects are light enough, lift them, via the funnel's updraft, right into the storm's parent cloud. For a time following the dissipation of the funnel cloud, these objects are often carried miles before falling back to earth. Most common of these animal falls are fish falls.

A particularly appetizing account is related in William R. Corliss's *Catalog of Geophysical Anomalies*. It describes a British military regiment near Pondicherry, India, that was inundated with fish sometime in 1809.

> While our army was on the march, a short distance from Pondicherry, a quantity of small fish fell with the rain, to the astonishment of all. Many of them lodged in the men's hats; when General Smith, who commanded, desired them collected, and afterwards, when we came to our [camping] ground, they were dressed, making a small dish that was served up and eaten at the general's table.

Sometimes the fish are not so small, as was the case in Marksville, Louisiana, on October 23, 1947, when as a biologist for the Department of Wildlife and Fisheries reminisced, "I was in the restaurant with my wife having breakfast, when the waitress informed us that fish were falling from the sky. We went immediately to collect some of the fish."

The fish were apparently native to the local waters and ranged from two to nine inches in length. They fell in a narrow strip 80 feet wide and some 1,000 feet long. Some were reported frozen, evidence that they had been lifted to a considerable height in the clouds.

Considerable fish falls have also been recorded in Louisville, Kentucky (1837), Boston, Massachusetts (1841), Providence, Rhode Island (1900), and Buffalo, New York (1900).

Falls of frogs and snails appear to be the next-most-common form of animal fall. A fall of frogs was reported in Trowbridge, England, on June 16, 1939, and, more recently, thousands of small frogs inundated Canet-Plage, France (near Perpignan on the Spanish Mediterranean border) on August 28, 1977. Small, living frogs were found encased in hailstones at Dubuque, Iowa, on June 16, 1882. Periwinkles once fell on Bristol, England (summer of 1821), clams on the Philadelphia suburb of Chester (June 6, 1869), jellyfish on Bath, England (August, 1894), living mussels on Paderhorn, Germany (August 9, 1892), and lizards or salamanders on Montreal, Quebec, in December of 1857. Probably the foulest of falls ever reported was that which engulfed Bucharest, Romania, on July 25, 1872:

> During the day [of July 25] the heat was stifling, and the sky cloudless. Towards nine o'clock a small cloud appeared on the horizon, and a quarter of an hour afterwards rain began to fall, when, to the horror of everybody, it was found to consist of black worms of the size of an ordinary fly. All the streets were strewn with these curious animals.

chapter 7
HURRICANES

urricanes are the most destructive storms on earth. Their waves have wiped out entire communities and forever altered coastal landscapes. Their winds humble even the sturdiest of man's structures. Their torrential rains transform gentle streams into raging rivers.

Tornadoes may be more violent than hurricanes, but hurricanes are more deadly. At least 20,000 Americans have been killed in hurricanes over the past 150 years and perhaps many more. The death tolls prior to 1900, and from the Galveston, Texas, and Lake Okeechobee, Florida, disasters of 1900 and 1928, are only estimates and probably way too conservative. In recent times, as forecasting techniques have improved death tolls have decreased, but damages have increased as a result of the massive development that has taken place along hurricane-prone coastlines and flood plains.

Saffir-Simpson Hurricane Scale

Category	Pressure	Wind Speed	Storm Surge	Damage
1	28.94" or higher 980 mb or higher	74–95 mph 119–153 km	4–5 ft 1.2–1.5 m	Trees and shrubs lose leaves and twigs.
2	28.50 - 28.93" 965 - 979 mb	96–110 mph 154–177 km	6–8 ft 1.8–2.4 m	Small trees blow down. Exposed mobile homes severely damaged. Chimneys blown from roofs.
3	27.91-28.49" 945-964 mb	111–130 mph 178–209 km	9–12 ft 2.7–3.6 m	Leaves stripped from trees. Large trees blown down. Mobile homes demolished. Small buildings damaged structurally.
4	27.17-27.90" 920-944 mb	131–155 mph 210–249 km	13–18 ft 3.9–5.4 m	Extensive damage to windows, roofs, and doors. Flooding to 6 miles inland. Severe damage to lower areas of buildings near exposed coast.
5	27.16" or lower 920 mb or lower	155 mph or more 250 km or more	18 ft or more 5.4 m or more	Catastrophic. All buildings severely damaged, small buildings destroyed. Major damage to lower areas of buildings less than 15 feet above sea level to .3 mile inland.

Hurricane Andrew's storm surge and winds carried the 210-ton vessel Seward Explorer some 500 feet inland from Biscayne Bay on August 24, 1992. (Carl Seibert, South Florida Sun-Sentinel)

Measuring Hurricane Intensity

The intensity of hurricanes is measured by the Saffir-Simpson Scale, which ranges from Category 1, the lowest hurricane rating with sustained winds of 74–95 mph, to Category 5, the highest rating, with sustained winds in excess of 155 mph. In theory, the maximum sustained wind a tropical storm might attain at sea level is about 200 mph. Above that, friction on the earth's surface begins to create a drag on the speed of the wind and prevents it from increasing further.

Pacific Typhoon Nancy was reported to have attained sustained winds of 213 mph (185 knots) at its peak strength on September 12, 1961, but this record was discredited after researchers concluded that the top wind speeds of typhoons in the 1950s and 1960s had been routinely exaggerated. Nevertheless, higher winds have been recorded in gusts at high elevations as demonstrated by the 234-mph reading attained at the 6,288-foot summit of Mt. Washington on April 12, 1934 (not measured during a tropical storm, however).

Only three Atlantic hurricanes have made landfall with winds estimated to have reached the rarefied extreme of 200 mph, at least in gusts. These include the Labor Day Hurricane of 1935 that passed over the Florida Keys (inspiring the classic Humphrey Bogart movie *Key Largo*); Hurricane Camille, which roared ashore at Pass Christian, Mississippi in 1969; and Hurricane Andrew in 1992, which struck the lower Florida peninsula. Some top-wind-speed gusts from Atlantic hurricanes actually measured by anemometer include the following (in most cases the instruments were destroyed before they measured the worst of their respective storms).

Highest Measured Wind Speed in Atlantic Tropical Storms

Speed	Date	Location	Storm name
186mph	9/21/1938	Blue Hill Observatory, Massachusetts	Great New England
178mph	9/6/1965	Great Abaco Island, Bahamas	Betsy
175mph	9/11/1961	Port Lavaca, Texas	Carla
175mph	9/27/1955	Chetumal, Mexico	Janet
174mph	8/24/1992	Coral Gables, Florida	Andrew
172mph	8/17/1969	Boothville, Louisiana	Camille
163mph	10/18/1944	Havana, Cuba	unnamed
160mph	9/13/1928	San Juan, Puerto Rico	unnamed
155mph	9/17/1947	Hillsboro Lighthouse, Florida	unnamed

*A wind gust of 190 mph was measured by the weather station on the island of Miyakojima in the Pacific Ryukyu Island chain on September 5, 1966. This is the highest wind speed ever measured in a tropical storm anywhere in the world.

The list above does not necessarily represent the most intense storms to have occurred in the Atlantic Basin. It is a collection of lucky (or unlucky for the observers) instrument readings. Very few anemometers are capable of accurately measuring the winds of a Category 5 hurricane. However, wind measurements made by the flight crews of reconnaissance aircraft, the so-called "hurricane hunters" of the U.S. Air Force, have often recorded winds in excess of 155 mph while tracking approaching Category 4 or 5 hurricanes in the Atlantic.

Major Hurricanes

A hurricane is considered "major" if it reaches Category 3 on the Saffir-Simpson scale. Winds in this category must have sustained speeds over 110 mph and thus be capable of producing "extensive" damage. In a typical year, about 10 to 11 tropical storms (winds over 60 mph) form in the North Atlantic Basin. Of these, five to six become hurricanes, and two usually strike the U.S. However, a major hurricane is likely to hit the U.S. shoreline only once every two years.

As of this writing, no major hurricane has hit the U.S. shoreline since Hurricane Brett struck an unpopulated stretch of the south Texas coast on August 22, 1999. Brett was a minimal Category 3 storm and did little damage.

The last truly destructive hurricane classified as "major" was Hurricane Fran, which came ashore just north of Wilmington, North Carolina, on September 5, 1996. It was also a minimal Category 3 storm with top winds of 115 mph, but Fran wreaked havoc on the heavily developed (some might say over developed) coast of North Carolina. Thirty-four people died and damages reached $3.2 billion.

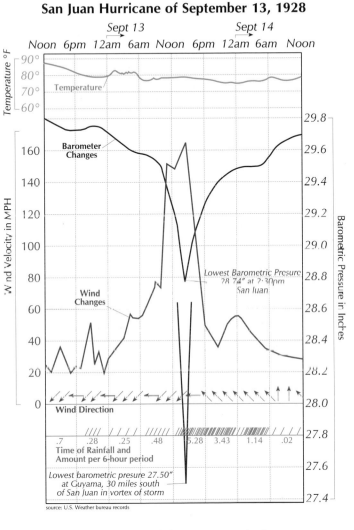

A storm chart analysis of the passage of a hurricane in San Juan, Puerto Rico, on September 13, 1928. The anemometer trace indicates wind gusts over 160 mph.

Even weak tropical storms are capable of causing tremendous destruction, as was the case with tropical storm Allison when it deluged the Houston area with up to 37" of rain (36.99" officially measured at the Port of Houston) in June 2001. Forty-one deaths and $5 billion in damage resulted. Curiously, this was the second time a tropical storm named Allison deluged the Houston area during June, the previous occasion being in 1989.

As of 2004, the United States was overdue for a major hurricane disaster. The last five-year period to go by without a major hurricane strike on the United States was 1921–1926 (*see chart opposite*).

COASTLINES OVERDUE FOR A MAJOR HURRICANE

1   Northeast

In the Northeast no tropical storm of consequence has occurred since Gloria in 1985 (and, to a lesser extent, Bob in 1991).

2   Texas Coast from Corpus Christi to Freeport

This stretch of coast is normally subjected to a 10–15% annual risk of being struck by a hurricane (4% annual risk of a major hurricane)—one of the highest risks anywhere on the Gulf Coast. Its last direct blow was Hurricane Celia in 1970.

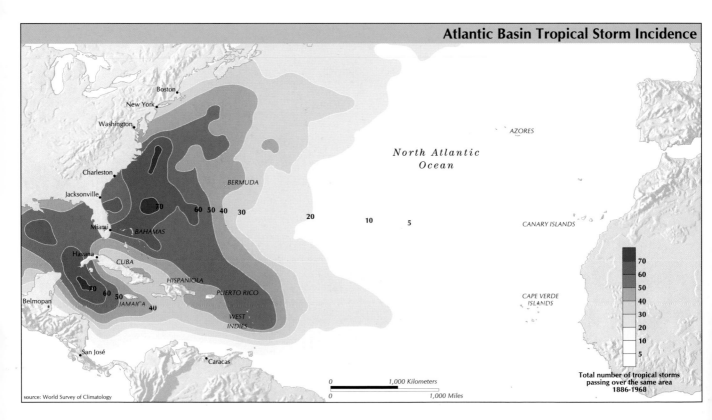

Atlantic Basin Tropical Storm Incidence

Total number of tropical storms passing over the same area 1886-1968

source: World Survey of Climatology

Major Hurricanes to Strike U.S. since 1900

Year	Category	Area Affected
1900-1909		
1900	4	North TX
1906	weak 3	MS, AL
1906	3	SC, NC
1909	weak 3	FL Keys
1909	4	LA
1909	weak 3	North TX
1910-1919		
1910	weak 3	Southwest FL
1915	4	North TX
1915	4	LA
1916	3	MS, AL
1916	3	South TX
1917	weak 3	Northwest FL
1918	weak 3	LA
1919	4	FL Keys, South TX
1920-1929		
1921	3	FL
1926	4	FL, MS, AL
1926	weak 3	LA
1928	4	South FL
1929	3	FL
1930-1939		
1932	4	North TX
1933	3	Southeast FL
1933	weak 3	NC
1933	3	South TX
1934	weak 3	LA
1935	5	FL Keys
1936	weak 3	Northwest FL
1938	3	Northeast U.S.
1940-1949		
1941	weak 3	North TX
1942	3	Central TX
1944	3	Northeast U.S.
1944	weak 3	FL
1945	3	Southeast FL
1947	4	FL, LA, MS
1948	weak 3	FL
1949	weak 3	Southeast FL

Weather Bureau began to name tropical storms in 1950

Year	Name	Category	Area Affected
1950-1959			
1950	King	weak 3	Southeast FL
1950	Easy	weak 3	Northwest FL
1954	Carol	weak 3	Northeast U.S.
1954	Hazel	4	SC, NC, Northeast U.S.
1954	Edna	weak 3	New England
1955	Diane	3	NC
1955	Connie	weak 3	NC, VA
1955	Ione	weak 3	NC
1957	Audrey	4	North TX, LA
1959	Gracie	3	SC
1960-1969			
1960	Donna	4	FL, Eastern U.S.
1961	Carla	4	North and Central TX
1964	Hilda	3	LA
1965	Betsy	3	Southeast FL, LA
1967	Beulah	3	South TX
1969	Camille	5	MS, AL
1970-1979			
1970	Celia	3	South TX
1974	Carmen	weak 3	LA
1975	Eloise	weak 3	Northwest FL
1979	Frederic	3	AL, MS
1980-1989			
1980	Allen	3	South TX
1983	Alicia	weak 3	North TX
1985	Gloria	3	Eastern US
1985	Elena	weak 3	MA, AL, FL
1989	Hugo	4	SC
1990-1999			
1992	Andrew	5	FL, LA
1993	Emily	weak 3	NC (did not amke landfall)
1995	Opal	3	FL, AL
1996	Fran	weak 3	NC
1999	Bret	weak 3	South TX

No major hurricanes made landfall between 2000-2003

Hurricane Strike Probability on Gulf and Atlantic Coastlines

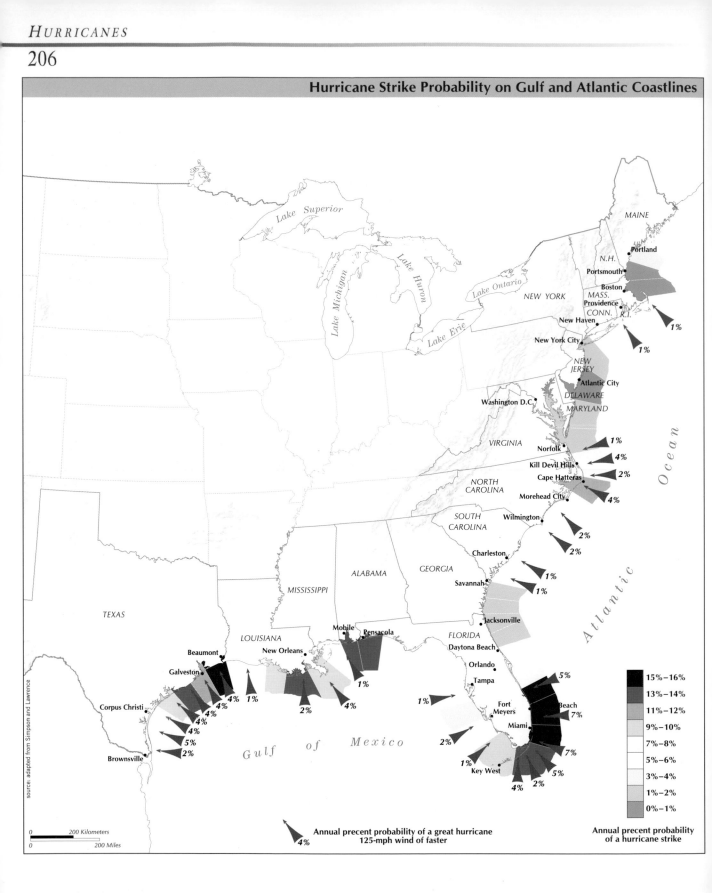

source: adapted from Simpson and Lawrence

Lake Superior
Lake Michigan
Lake Huron
Lake Ontario
Lake Erie

MAINE
N.H.
Portland
Portsmouth
Boston
MASS.
Providence
CONN.
R.I.
New Haven
New York City
NEW YORK

NEW JERSEY
Atlantic City
DELAWARE
Washington D.C.
MARYLAND
VIRGINIA
Norfolk
Kill Devil Hills
Cape Hatteras
Morehead City
NORTH CAROLINA
Wilmington
SOUTH CAROLINA
Charleston
Savannah
GEORGIA
ALABAMA
MISSISSIPPI
Jacksonville
FLORIDA
Daytona Beach
Orlando
Tampa
Fort Meyers
Beach
Miami
Key West

TEXAS
Beaumont
LOUISIANA
New Orleans
Mobile
Pensacola
Galveston
Corpus Christi
Brownsville

Gulf of Mexico

Ocean
Atlantic

1%
1%

1%
4%
2%
4%
2%
2%
1%
1%
5%
7%
7%
5%
4%
2%
1%
2%
1%
4%
2%
4%
1%
4%
4%
4%
5%
2%

0   200 Kilometers
0   200 Miles

4%
Annual precent probability of a great hurricane
125-mph wind of faster

**Annual precent probability
of a hurricane strike**

	15%–16%
	13%–14%
	11%–12%
	9%–10%
	7%–8%
	5%–6%
	3%–4%
	1%–2%
	0%–1%

3 Mississippi Delta of Louisiana

This area, which includes the city of New Orleans, hasn't been hit head on by a major hurricane since Betsy in 1965. New Orleans is exceptionally vulnerable to tropical storms because of its low elevation (sections of the city are below sea level). The city's traditional architecture reflects the fact that the city has often been inundated—most houses are elevated above a first floor that can sustain some flooding. Although dikes protect the city from the waters of nearby Lake Pontchartrain, these would almost certainly be overwhelmed by a Category 4 or 5 hurricane storm surge. Should this occur, much of downtown New Orleans, including the French Quarter could be swamped under 20 feet of water.

4 Tampa–St. Petersburg Area of Florida

The Tampa Bay area is past due for a direct hit. Hurricanes that come ashore here either form in the Gulf of Mexico or track across the Atlantic, pass to the south of Florida, and then curve back up Florida's western shoreline. Not since 1921, has Tampa been struck directly by a major storm.

5 Atlantic Coastline from Palm Beach to Savannah

From Palm Beach, Florida, north to Savannah, Georgia, the coast is long overdue for a major storm. Although it is relatively rare for hurricanes to strike the northern coastline of Florida, the stretch of beach between Melbourne and Palm Beach is particularly vulnerable and historically hurricane-prone.

Although the above five stretches of coastline may be overdue for a major hurricane they are not neccessarily the coastlines of greatest historical risk from such (*see map for annual risk assessment by section of coastline on opposite page*). The coastlines most likely to be struck during any given year are the following.

COASTLINES AT HIGHEST ANNUAL RISK FOR HURRICANES

1 Key West to Palm Beach, Florida

The coast between Miami and Fort Lauderdale is at the epicenter of this risk area. This area's last major hurricane was Hurricane Andrew in 1992, recently upgraded as one of the three Category 5 hurricanes to strike the United States.

2 Galveston, Texas, to the Louisiana border

Hurricane Alicia in 1983, was the last major hurricane to come ashore along this stretch of coastline.

3   **Mississippi Delta of Louisiana to Walton Beach, Florida**

This includes the Gulf Coast of Alabama and, to a lesser extent, the coast of Mississippi. Hurricane Elena in 1985 was the last really big one to come ashore here.

4   **Outer Banks of North Carolina south of Nags Head**

The Outer Banks have suffered a series of major hurricanes in recent years, the last one being Fran in 1996. (Hurricane Isabel in 2003 was strong, but not strong enough to be categorized as "major"). Hopefully, this region has fulfilled its hurricane quota for the time being.

Other stretches of the U.S. Atlantic and Gulf coastlines have a lesser risk of hurricane strikes than the above, but nowhere from Brownsville, Texas, to Eastport, Maine, is entirely free from some risk.

Atlantic hurricane activity appears to be cyclical, with some decades more active than others. To confuse matters, a given year may be very active with many tropical storms developing in the Atlantic Basin but with few storms making landfall. This was the case in 1995, the second most active year on record, when 18 tropical storms developed over the Atlantic Basin, but only two hurricanes actually struck the United States.

The most active year on record was 1933, when 21 tropical storms formed in the Atlantic. The least active years were 1890 and 1914, when only one storm reportedly formed. The most active decades were the 1930s, 40s, and 50s. The least active were the 1970s and 1980s. The 1990s saw an increase in activity but few storms making landfall in the U.S., with the notable exception of Hurricane Andrew, the costliest storm of any kind in U.S. history. In fact, during the 1990s the United States had fewer tropical storms making landfall than in any other decade of the 20th century.

Number of Tropical Storms in Atlantic Basin since 1900 by Decade

Decade	Total number of Tropical Storms		Total number of Hurricanes		Total Major Hurricanes
	All Areas	Landed U.S.	All Areas	Landed U.S.	Landed U.S.
1900-10	76	39	35	15	6
1911-20	51	34	35	20	8
1920-29	54	24	40	15	5
1930-39	98	40	47	18	8
1940-49	93	41	50	23	8
1950-59	104	37	69	18	10
1960-69	93	28	61	15	6
1970-79	82	22	49	12	4
1980-89	88	28	51	16	5
1990-99	99	19	65	14	5
2000-03	57	14	27	3	0

The chart on page 208 illustrates tropical storm activity by decade in the Atlantic Basin since 1900. It should be noted that prior to the 1930s it is likely that some tropical storms that may have formed in the Atlantic went unrecorded.

Historic Hurricanes: The Worst of the Worst

Deadliest Hurricane in U.S. History: Galveston, Texas, 1900

America's worst natural disaster was the Galveston, Texas, hurricane of September 8, 1900. The thriving city of Galveston was considered the richest and most sophisticated town in the state of Texas at this time. With its burgeoning population of some 20,000, it had aspirations to rival New Orleans as the preeminent Gulf trading port. Unfortunately, it was built on a barrier island along the Gulf Coast, the worst place to be during a hurricane of any strength, let alone a major hurricane with its deadly storm surge.

The hurricane that struck Galveston on the night of September 8 had winds faster than 120 mph and a storm surge over 15 feet high. The entire city was submerged by this flood of water. At one point the water rose four feet in as many seconds. Isaac M. Cline, the local Weather Bureau official, reported

> At 7:30 p.m. I was standing at my front door, which was partly open, watching the water, which was flowing with great rapidity from east to west. The water at this time was about eight inches deep in my residence, and the sudden rise of four feet brought it above my waist before I could change my position.

Fifty people had taken refuge in Mr. Cline's sturdy residence when, at 8:30 p.m., it collapsed under a barrage of debris and sea water. Mr. Cline survived, but 32 of his refugees did not. In

Deadliest Tropical Storms in U.S. History

Deaths	Location	Date
8000+	Galveston, TX	9/8-9/1900
2000+	Lake Okeechobee, FL	9/16/1928
1800-2000	Coastal LA and MS	10/1-2/1893
1000-2500	SC, GA	8/27-28/1893
700+	GA, SC	8/27/1881
638	New England	9/20-21/1938
600+	FL (marine)	9/10-11/1919
500+	GA, SC	9/7/1804
450+	Corpus Christi, TX	9/14/1919
408	Keys of FL	9/2-4/1935
400	Ile Derniere, LA	8/10-11/1856
390	New England	9/14-15/1944
390	Hurricane Audrey, west LA	6/27/1957
350+	Grand Isle, LA	9/20-21/1909
275	New Orleans, LA	9/9-10/1915
275	Upper coast of TX	8/16-17/1915
256	Hurricane Camille, MS, AL	8/17/1969
243	FL	9/19-20/1926
200+	SC	9/27-28/1822
184	Hurricane Diane, NC to ME	9/9-13/1955
179	GA coast	10/2/1898
176	Indianacola, TX	9/16/1875
164	Southeast FL	10/18/1906
134	FL, AL & MS	9/27/1906
122	Trop. storm Agnes, PA, NY	6/20-21/1972
100	FL	9/28-29/1896
100+	Sabine, TX	10/12/1886
95	Hurricane Hazel, NJ, NY	10/15/1954
90+	SC, NC (marine)	10/9/1837
75	Hurricane Betsy	9/7-8/1965
70	Brownsville, TX	8/6/1844
68	FL	10/8/1896
60	Hurricane Carol, MA, ME	9/11/1954
57	Hurricane Hugo, SC, NC	9/20-21/1989
56	Hurricane Floyd, NC to NJ	9/16-18/1999
53	NC	9/11/1883
51	San Antonio, TX	9/23/1921
51	Southeast FL, LA, MS	9/17-20/1947
50	Cameron Parish, LA	10/11/1886
50	South FL	11/30-12/1/1925

Even the strongest masonry structures were damaged or destroyed in Galveston, Texas, by the hurricane of September 8, 1900. At least 8,000 people perished in what remains America's deadliest natural disaster. (NOAA)

fact, every residence in Galveston was destroyed, and at least 8,000 people died, perhaps as many as 12,000, since no one knows how many non-residents were in the busy city at the time of the disaster or how many were lost at sea.

Costliest Hurricane in U.S. History: Hurricane Andrew, 1992

One of only three Category 5 hurricanes ever to strike the United States came ashore just south of Miami at Homestead, Florida, on the night of August 23–24, 1992. Winds as high as 175 mph, and perhaps some gusts of even 200 mph, along with a storm surge 16.9 feet high, ripped through the area leaving a path of destruction from Florida City in the south to Coconut Grove to the north.

The densely populated residential subdivisions in this area sustained damages unprecedented in Florida history, amounting to over $26.5 billion in 1992 dollars. Aside from the destruction of some 25,000 homes, 90% of the 2,100 businesses listed in the Homestead and South Dade County chambers of commerce were wrecked. Twenty thousand national guard, army, and marine corps personnel were dispatched to aid the more than 250,000 people made homeless and (because of a looting spree in some neighborhoods) to restore order. It is estimated that 80,000 former residents of Dade County, Florida, moved away following the storm and never returned.

Homestead Air Force Base was heavily damaged, as was (ironically) the National Hurricane Center in Coral Gables. The anemometer at the center registered winds of 164 mph before its equipment was blown away.

After traversing the southern Florida peninsula across the Everglades, Hurricane Andrew emerged into the Gulf of Mexico and took aim at Louisiana. The storm came ashore near New Iberia at 2 a.m. on August 26, with winds of 120 mph. The damage stretched from Morgan City to Lafayette before the storm dissipated inland.

A composite satellite image illustrates the progress of Hurricane Andrew across the Florida peninsula and Gulf of Mexico. The photographs are shown in 24-hour intervals. (NOAA).

(following pages) Hurricane Andrew lashes Dinner Key, Florida, in August 1992. The storm was upgraded to a Category 5 hurricane by the National Weather Service in 2002, one of only three such storms ever to strike the U.S. (Warren Faidley/Weatherstock, Inc.)

Most Intense Hurricane in United States History:
Florida Keys Labor Day Hurricane, 1935

The intensity of a tropical storm is determined by the barometric pressure measured at its center. A storm's central pressure is directly correlated to the wind speeds it generates, which in truly intense hurricanes, are usually too high to measure.

The hurricane that passed over Long and Matecumbe Keys in Florida on September 2, 1935, registered a barometric pressure of 26.35", by far the lowest ever measured in the United States. (Hurricane Camille had a central pressure of 26.84" when it came ashore in Mississippi in 1969, the second-lowest pressure on record). Known as the Labor Day hurricane (hurricanes were not given names until 1950), the storm that struck the Florida Keys in 1935 was very small and tightly wound but had winds estimated over 200 mph.

Unfortunately, the Florida East Coast Railroad was being worked on at the time by some 700 veterans living in a relief camp on Long Key. A train was sent to evacuate them along with the local civilian population. The train was delayed, and by the time it arrived in the Keys it was caught in the center of

The train sent to rescue veterans working in the Florida Keys was swept off its tracks by the Labor Day Hurricane's 200-mph winds. (Miami Daily News)

DEDICATED WEATHER OBSERVER

J. E. Duane was the official observer for the U.S. Weather Bureau at Long Key, Florida, on September 2, 1935, when the most intense hurricane in U.S. history roared across the Florida Keys. That he survived the storm's 200-mph winds and 20–30-foot storm surge on the tiny islet is a miracle; that he steadfastly continued to monitor the weather instruments is beyond belief. More than 400 people died in the Keys and a rescue train was blown off its tracks. We pick up his observations when the eye of the hurricane was directly over his position:

During this lull [in the eye of the hurricane] the sky is clear to northward, stars shining brightly and a very light breeze continued; no flat calm. About the middle of the lull, which lasted a timed 55 minutes, the sea began to lift up, it seemed, and rise very fast; this from the ocean side of the camp. I put my flashlight out to sea and could see walls of water which seemed many feet high. I had to race fast to regain the entrance of the cottage, but water caught me waist deep, although writer was only about 60 feet from doorway of cottage. Water lifted cottage from foundations and it floated.

10:10 p.m. — Barometer now 27.02 inches; wind beginning to blow from SSW.

10:15 p.m. — The first blast from SSW, full force. House breaking up—wind seemed stronger than any time during storm. I glanced at barometer which read 26.98 inches, dropped it in the water and was blown outside into sea; got hung up in broken fronds of coconut tree and hung on for dear life. I was struck by some object and knocked unconscious.

2:25 a.m. (September 3) — I became conscious in tree and found I was lodged about 20 feet above ground.

—J. E. Duane, 1935

the storm. All its cars, except for the engine and tender, were swept off the tracks. Over 400 of the veterans and civilians died, some of them literally sand-blasted to death by the hurricane's 200-mph winds.

Intense hurricanes often suck sand up off beaches and wind-whip the particles together so fast that they're charged electrically, transforming ordinary sand into what appears to be millions of sparking fireflies. A stunning display of this phenomenon was apparently observed during the Labor Day Hurricane.

TROPICAL STORM SURGES

The storm surge that accompanies hurricanes often overwhelms and devastates coastal areas. Storm surges are walls of sea water that rise up high before hurricane-force winds. Shallow water near the coastline causes the water to build up many feet higher than normal and flood inland as an unstoppable wave.

Extremely low barometric pressure common to strong tropical storms also causes the sea level to swell upwards. In major hurricanes the storm surge may exceed 10 feet, extending hundreds of yards inland and submerging shoreline structures. The most extreme example of this occurred during the passage of Hurricane Camille, when a storm surge of 24.2 feet came ashore at Pass Christian, Mississippi, on August 17, 1969. Among the 22 people attending a "hurricane party" in the Richelieu apartment building, which stood near the beach of Pass Christian, the lone survivor was swept away and clung to floating debris for ten miles before he was finally rescued.

Flooding from the surge traveled up to eight miles inland in the Waveland Bay–St. Louis area of Mississippi and 20–30 miles up river estuaries.

(opposite) Downtown Providence went underwater as a result of a storm surge coming in off Narragansett Bay during Hurricane Carol on August 31, 1954. (Providence Journal)

(below) An enormous storm surge roaring ashore on the Gulf Coast was captured in this unusual photograph.(NOAA)

The storm surge that drowned over 8,000 people at Galveston, Texas, in 1900 was 14.5 feet high. In 1961, a surge associated with Hurricane Carla reached 16.6 feet at Port Lavaca, Texas.

The configuration of the coastline plays an important part in determining the intensity of storm surges, as was seen in Rhode Island's Narragansett Bay during the Great New England Hurricane of 1938. Ocean water that funneled into the bay surged up to 17 feet deep near Providence, and many people on nearby, low-lying Prudence Island in the bay itself, lost their lives.

North Carolina's highest storm surge was that associated with Hurricane Hazel in October 1954, when a 40-mile stretch of coastline from Sunset Beach to Long Beach was submerged under 16 to 18 feet of water.

Nowhere does the topography of the coastline conspire to create storm surges as disastrous as those that occur in South Asia's Bay of Bengal. Cyclones charging northward cause the sea water to pile up to phenomenal heights, engulfing the delta islands of the Brahmaputra and Ganges Rivers in Bangladesh and India. Storm surges of 40 feet occurred at the mouth of the Hooghly River on October 7, 1737, drowning 300,000 people.

More recently, a storm surge of 40 feet was recorded on Hatia Island, Bangladesh, during the disastrous cyclone of November 1970, resulting in the deaths of 300,000 to 500,000 people.

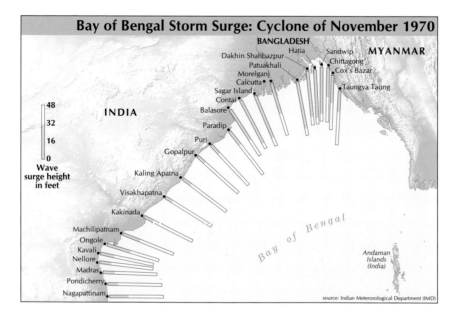

Bay of Bengal Storm Surge: Cyclone of November 1970

source: Indian Meteorological Department (IMD)

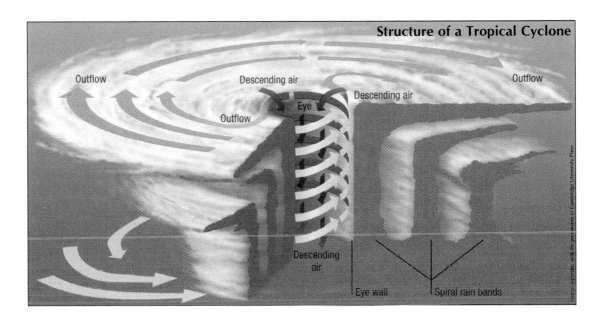

Perhaps the greatest storm surge ever roared up Bathurst Bay, Queensland, on the night of March 5, 1899, when a terrific cyclone produced a wall of water 42 feet high according to witnesses. Another contender for a "greatest" title is the reported storm surge of 46 feet in the Marshall Islands during a typhoon on June 30, 1905.

Waves in the open sea reach even higher than this during tropical storms, the highest on record being a monstrous 112 feet measured by the USS *Ramapo* during a typhoon in the western North Pacific on February 7, 1933. The North Rankin A gas platform off the coast of Western Australia was almost overwhelmed by 70-foot waves during Tropical Cyclone Orson on April 22, 1989.

TROPICAL STORM RAINFALLS

Torrential rains often ride with the violent winds of tropical storms. When the remnants of Hurricane Camille passed over Virginia on August 20, 1969, a deluge of 27" of rain fell over Nelson County, most of it in just eight hours. The floods that followed killed 113 people. More recently, Hurricane Danny meandered over Mobile Bay for two days, unleashing phenomenal amounts of rainfall. Dauphin Island received 32.52" in 24 hours, 26" of which fell in just seven hours. Doppler radar indicated that as much as 45" of rain may have fallen over the bay itself.

Selected Rainfalls Associated with Tropical Storms

Amount	Length	Date	Location
112.80″	5 days	11/5-9/1909	Silver Hill Plantation, Jamaica
108.31″	3 days	10/17-19/1967	Xinliao, Taiwan
99.09″	3 days	1/4-6/1979	Bellenden Ker, Queensland, Australia
79.72″	3 days	10/4-6/1962	Tacajo, Cuba
79.12″	2 days, 15 hours	7/14-16/1911	Baguio, Philippines
49.13″	24 hours	9/10-11/1963	Paishih, Taiwan
41.70″	21 hours	9/8/1986	Kadena Air Force Base, Okinawa

Tropical storm Claudette is credited with producing the United States' heaviest 24-hour rainfall of 43″ in Alvin, Texas, on July 25–26, 1979. Alvin is in the greater Houston area, which, as mentioned earlier, has had its share of tropical storm flooding.

Hurricanes carry their tropical moisture well beyond the tropics or sub-tropics as they head up the Atlantic Coast. When New England received the double punch of hurricanes Connie and Diane in the single week of August 12–19, 1955 rain totals of 8–26″ deluged all of southern New England and eastern Pennsylvania. The waters of Broadhead Creek near Stroudsburg, Pennsylvania swept away a summer camp and drowned all 75 of its visitors.

La Guardia Airport measured 12.20″ in 36 hours on August 12–13, probably New York City's greatest single rainstorm. A tropical storm that tracked across New Jersey on September 1, 1940, deluged Ewan with 24″ in just eight hours. Trenton was flooded with almost 4″ in a single hour.

Not all hurricanes produce torrential amounts of rain. A ferocious storm that passed just south of Miami on October 6, 1941, brought wind speeds as high as 123 mph at Dinner Key but dropped only .35″ of precipitation. Terrible damage was done to vegetation by sea spray, even well inland, because the usual rainfall was not there to wash the salt off the leaves. Nevertheless, this was a very unusual incident, and the norm is for torrential rainfall to occur even with weak tropical storms.

The most intense rainfalls ever measured in the world are the result of tropical storms passing over the mountainous terrain of Reunion Island in the Indian Ocean. *(see sidebar on p.120 in Rain & Floods)*. When Hurricane Mitch stalled off the coast of Honduras in October of 1998, an estimated 75″ of rain deluged a large area of the Honduran highlands, resulting in devastating floods throughout the region. As many as 18,000 people may have perished in the floods and mud slides, in what was the worst hurricane-related disaster to occur during the 20th century in the Americas. Some other phenomenal rainfalls caused by tropical storms around the world include those listed in the table above.

Keep in mind that the average annual rainfall for New York City is just 44″, and 20–25″ is normal for London, Paris, or San Francisco. In these cities, no single storm would normally produce more than two to three inches of rainfall.

TROPICAL STORMS AROUND THE WORLD

Meteorologically there is no difference between a hurricane, typhoon, and cyclone—they are all severe tropical storms that share the same fundamental characteristics aside from the fact that they rotate clockwise in the southern hemisphere and counterclockwise in the north.

Hurricanes are known as typhoons if they form in the western or southern Pacific Ocean but retain the classification of "hurricane" should they form in the eastern Pacific off the coast of Mexico. Occasionally, a hurricane will form in the eastern Pacific and migrate into the western Pacific, where upon crossing the 180° longitude, they become typhoons. In the Indian Ocean or Coral Sea near Australia, hurricanes are called "cyclones".

Classification of Tropical Storms Worldwide

Region	Wind Speed, mph								
		20	30	40	55	75	100	125	150+
Western North Pacific and South China Sea	Low Pressure Area	Tropical Depression		Tropical Storm	Severe Tropical Storm	Typhoon			Super Typhoon
Atlantic and Northeast Pacific Caribbean and Gulf of Mexico	Tropical Disturbance	Tropical Depression		Tropical Storm		Hurricane (Cat 1, Cat 2, Cat 3, Cat 4, Cat 5)			
North Indian Ocean, Bay of Bengal and Arabian Sea	Low Pressure Area	Depression	Deep Depression	Cyclonic Storm	Severe Cyclonic Storm	Very Severe Cyclonic Storm		Super Cyclonic Storm	
Southwest Indian Ocean	Tropical Disturbance			Moderate Tropical Storm	Severe Tropical Storm	Tropical Cyclone	Intense Tropical Cyclone	Very Intense Tropical Cyclone	
South Pacific Ocean & Southeast Indian Ocean	Tropical Disturbance	Tropical Depression		Tropical Cyclone		Tropical Cyclone / Severe Tropical Cyclone			

source: World Meteorological Organization

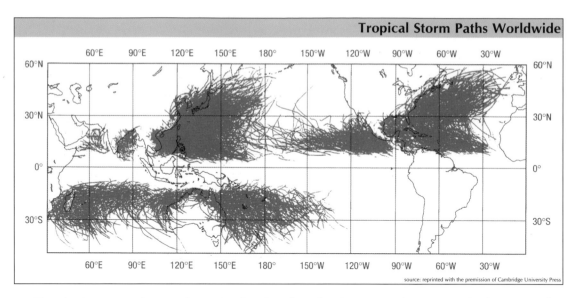

Tropical Storm Paths Worldwide

source: reprinted with the premission of Cambridge University Press

Tropical storms do not form in the South Atlantic or southeast Pacific because the ocean is not warm enough to allow for their generation. A notable and unique exception to this rule occurred in March 2004 when a hurricane formed off the coast of Brazil and came ashore with 90 mph winds.

Tropical storms are most frequent and at their most violent over the western Pacific, a vast region of warm water. The season of peak activity in the western Pacific is similar to that of the North Atlantic, except that about twice as many storms on average form

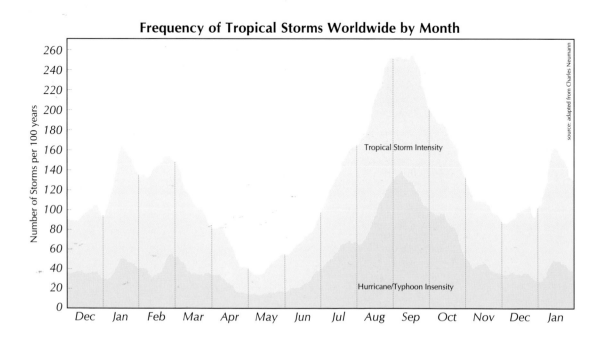

Frequency of Tropical Storms Worldwide by Month

Number of Storms per 100 years

Tropical Storm Intensity

Hurricane/Typhoon Insensity

source: adapted from Charles Neumann

in the Pacific. Thus, even during the Pacific's off-peak months of December through May, more storms are generated there than in the Atlantic.

All of the world's most intense tropical storms have been western Pacific typhoons. When a typhoon's winds exceed 150 mph it becomes classified as a "supertyphoon." In fact, there have been at least 14 supertyphoons that were stronger than the strongest Atlantic Category 5 hurricane (which was hurricane Gilbert in September 1988).

The western North Pacific is the most active tropical storm region in the world, with an average of 32 such storms forming every year. Since careful records have been kept (beginning in 1959), the maximum annual total has been 44 (compared to a maximum of 21 in the North Atlantic basin).

Probably the most tropical-storm-prone inhabited places in the world are the Caroline and Marianas island chains in the western North Pacific and the northeastern coastline of Luzon Island in the Philippines. The inhabitants of these places, who for obvious reasons are few, may expect about 10 typhoons every year.

A typhoon rips across Polynesia's Arutua Atoll in 1983. (Philippe Mazellier)

GREATEST STORM ON EARTH

The most powerful storm ever recorded on the earth's surface (since meteorologists have been able to keep track of such things), was Supertyphoon Tip, which formed in the western Pacific on October 5, 1979. Slow to develop and exceedingly erratic in its early movement, Tip eventually grew into a monster with a circulation and cloud formation 1,350 miles in diameter.

If such a storm were centered in the Gulf of Mexico, it would extend from Miami, Florida, to Amarillo, Texas. Tip's gale-force winds extended out from its eye in a radius of 683 miles, about five times greater than a typical Atlantic hurricane. At its peak on October 12, the air pressure in the eye fell to 25.69" (870 millibars), the lowest ever measured at sea level on the planet, and the equivalent of what normal air pressure would be at an altitude of about 2,500 feet. Winds circulating around Tip's eye were blowing at a sustained rate of 190 mph with gusts probably well over 200 mph. The eye wall extended up to 55,000 feet, where infrared temperatures were measured at an incredible −135°F. Fortunately, Tip never made landfall, although it took a shot at Guam, swerving at the last moment to the west and thus sparing the small and vulnerable island.

By October 18, Tip had accelerated towards the northwest and was rapidly losing power. By the time it brushed Japan on October 19 and 20, it was much tamer with the wind blowing in gusts at only 88 mph along the runways at Tokyo's airport.

Taiwan is also positioned in a particularly vulnerable location. About five typhoons strike or brush by the island on an annual basis.

Japan and Korea are frequently visited by typhoons, with often devastating results. Japan's deadliest typhoon disaster on record occurred on September 17, 1828, when a huge storm surge submerged the city of Nagasaki, drowning an estimated 15,000 people. Typhoons, however, have sometimes had a positive effect on Japan. In 1281, the famous Hakata Bay typhoon sank 90% of Kublai Khan's 2,200-ship invasion fleet, drowning between 45,000 and 65,000 of his troops and saving the Japanese homeland from foreign occupation. This event caused the word *kamikaze* to enter the national lexicon, its meaning being "divine wind."

If the western Pacific is home to the strongest and most frequent tropical storms, the Bay of Bengal is home to the deadliest. This is because of the population density of the coastal areas of India and Bangladesh, the relative flimsiness of the structures in the region, and insufficient warning systems. Well over one million people have lost their lives along the shores of the Bay of Bengal as a consequence of cyclones in recent centuries.

Fortunately, given the terrible death tolls, cyclones in the Bay of Bengal are not all that common, with an average of only about four forming in any given year.

After the northwest Pacific, the northeast Pacific leads the world in tropical storm formation. These storms usually develop off the southwest coast of Mexico and head harmlessly out to sea. Occasionally, one of these hurricanes churns onto Mexico's coastline causing havoc. The section of coast between the Gulf of Tehuantepec and Mazatlan is most commonly affected, although in recent years powerful hurricanes have swept across the Baja Peninsula.

Bay of Bengal Deadliest Cyclone Disasters

Deaths	Location	Date
300-500,000	Bangladesh	11/12-13/1970
300,000	Hooghly River, India	10/7/1737
200,000	South of Calcutta, India	10/31/1876
139,000	Bangladesh	4/30/1991
50,000	Calcutta, India	10/5/1864
40,000	Contai, India	10/16/1942

Ever so rarely, one of these hurricanes makes it all the way to Hawaii, as the residents of Kauai recall when Hurricane Iniki passed directly over the island on September 11, 1992, and wind gusts reached 142 mph. Rarer still are tropical storms that struggle far enough north to affect California or Arizona.

No tropical storm has ever survived the cold waters of the Pacific off California's coast, and the worst effects of the remnants of tropical storms are usually flooding rains in Southern California and Arizona. Probably the strongest tropical storm to affect California was that of September 25, 1939, when winds of 50 mph and rainfalls of up to 10.6" deluged the southern part of the state, resulting in the deaths of 45 people.

Australia is exposed to tropical storms originating from two different regions. The Coral Sea coastline of Queensland is occasionally visited by powerful cyclones coming in from the southwestern Pacific Ocean and traversing the Coral Sea before making landfall north of Brisbane. The second, and more active region for cyclone formation, is off the northwestern coast of Australia where cyclones form in the Indian Ocean and sometimes slam into the coast from Perth northward.

Australia's worst cyclone disaster was the result of a tropical storm that originated in the Coral Sea, passed over the top of Queensland, and came roaring ashore in the Northern Territory at Darwin on Christmas Day, 1974. Cyclone Tracy, as it was named, destroyed 80% of Darwin's structures, killing 66 people in addition to another 160 lost at sea. Winds were clocked at 137 mph, with estimated gusts as high as 175 mph.

Australia's most intense cyclone occurred relatively recently on April 22, 1989, when

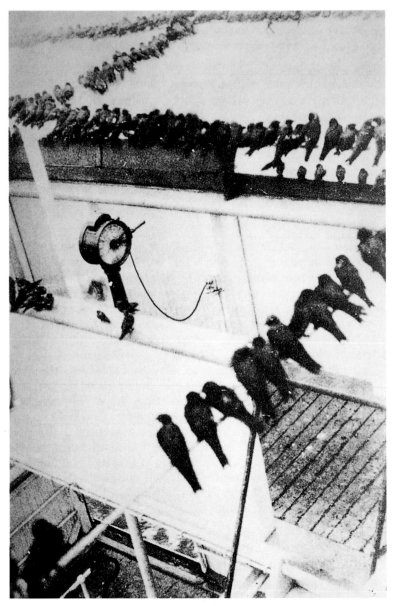

Birds sometimes become trapped in the calm air of the eye of a hurricane. This was the case in August 1926, when these sought refuge on the deck of the S.S. West Quechee *as it sailed through the eye of a hurricane in the Gulf of Mexico.*

Cyclone Orson grazed the coast of Western Australia near Port Headland and Dampier. A barometric pressure reading of 26.72″ was measured on the offshore North Rankin A gas platform and wind speeds probably reached close to 180 mph (the station's anemometer system was blown away).

Another region active with tropical storms is the southwestern Indian Ocean. On average, about a dozen cyclones form during the storm season of December through April. The most affected populated regions are the islands of Mauritius, Reunion, and Madagascar. The African nation of Mozambique relies on these storms for much of its precipitation (unfortunately, severe flooding often results). Rainfalls of 20–30″ have been recorded when a cyclone comes ashore here; its winds, however, usually weaken before reaching Mozambique.

Tropical Storm Superlatives

Largest and Most Intense
Supertyphoon Tip (see *Greatest Storm on Earth*, page 224).

Deadliest
Probably the deadliest, single tropical storm in recorded history was the cyclone of November 1970 that swept over the Brahmaputra River Delta of Bangladesh on November 12, 1970. A storm surge of at least 40 feet in height overran the fertile, low-lying, and densely populated islands lying at the head of the Bay of Bengal, and an estimated 300–500,000 lives were lost (also see pages 218 and 225).

Fastest Intensification
Supertyphoon Forrest, which formed in the Western Pacific in September 1983, intensified from 28.82" of pressure to 25.87", a drop of 3.00", in just 24 hours. Its winds increased from 75 mph to 175 mph at the same time.

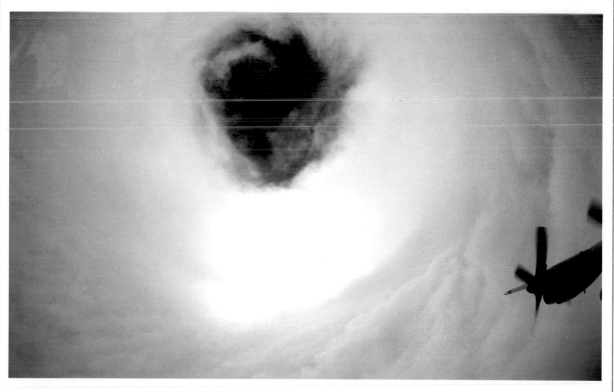

The eye of Supertyphoon Forrest as photographed from the window of a hurricane hunter aircraft over the Pacific in 1983. (Scott A. Dommin)

Largest Eye

Supertyphoon Carmen had developed an eye 230 miles in diameter when it passed over Okinawa on August 20, 1960. The storm was relatively weak, however, by the time it hit Okinawa, with a central barometric pressure of just 28.47" and sustained winds of 90 mph.

Smallest Eye

Tropical Cyclone Tracy, which devastated Darwin, Australia, on Christmas Day, 1974, was a very compact, intense storm, and its eye had contracted to just four miles in diameter the day before it made landfall on Christmas Eve. The storm had only a 31-mile radius of gale-force winds (34 mph+) at the time, while winds were gusting as high as 175 mph at its center.

Greatest Storm Surge

Australia's famous Bathurst Bay cyclone of March 5, 1899, had a storm surge of 42 feet, according to eyewitnesses. Another contender is the storm surge of 46 feet that swept over the Marshall Islands during a typhoon on June 30, 1905.

Several Bay of Bengal cyclones have produced storm surges of at least 40 feet along the Ganges and Brahmaputra River deltas of India and Bangladesh.

Most Powerful Eastern Pacific Storm

The most powerful hurricane on record to form in the eastern Pacific was Hurricane Linda, which formed off the coast of Mexico and reached a maximum intensity on September 12, 1997, with a central barometric pressure of 26.58". Its winds became sustained at 185 mph with estimated 220-mph gusts. The storm never made landfall. El Niño was credited with fueling the energy of this storm, and it is interesting to note that during the El Niño 1997–1998 period, five of the 14 strongest Pacific typhoons and hurricanes on record developed.

TROPICAL STORM RECORDS OF THE ATLANTIC BASIN

Deadliest

The deadliest Atlantic hurricane in history was the Great Hurricane of 1780. Between October 10 and 17, 1780, it swept through the Caribbean devastating the islands of Barbados, St. Vincent, St. Lucia, Martinique, Guadaloupe, and Dominica. The British fleet was almost completely wiped out, and some 22,000 sailors and townspeople perished.

Hurricane Mitch killed an unknown number of Hondurans and Nicaraguans in October 1998. The estimates of 11,000–18,000 would make it the deadliest tropical storm in recent history.

An enhanced satellite image of Hurricane Gilbert shows it at its most intense while churning slowly to the northwest just south of Cuba and towards the Yucatan Peninsula on September 14, 1988. (NOAA)

Most Intense

The most intense Atlantic hurricane of record was Hurricane Gilbert, which attained its lowest barometric pressure reading of 26.23" on September 12–14, 1988, while it was south of Cuba in the western Caribbean. Sustained winds were estimated at 180 mph with gusts to 200 mp.h. Fortunately, the storm weakened considerably prior to making landfall on Mexico's Yucatan Peninsula.

Costliest

Hurricane Andrew has often been cited as the costliest hurricane disaster in U.S. history. Striking Florida and Louisiana in August of 1992, its total damages were estimated at $26.5 billion. However, if one were to impose a "wealth normalization" factor, which takes into account not only inflation but what property values would be worth in 2000 as opposed to 1992, Hurricane Andrew's total cost would be $35 billion, and the costliest hurricane disaster ever would have been the Florida and Alabama hurricane of September 1926, with a total of $87.1 billion in adjusted, "normalized" costs.

Latest and Earliest in Season

Hurricane Alice persisted from December 31, 1954, to January 5, 1955, thus becoming both the earliest and latest hurricane ever to develop in the Atlantic Basin.

Most in a Season

The busiest Atlantic tropical storm season was that of 1933, when 21 tropical storms formed. The most hurricanes in a single season was 12 in 1969, and the most to hit the United States in a single season was seven in 1886.

Fewest in a Season

Apparently, only one tropical storm formed in the Atlantic during the seasons of 1890 and 1914. However, it is possible that other storms may have formed beyond the range of the shipping lanes and thus went unreported.

Longest Lived

The longest-lived tropical storm in Atlantic Basin history was the third storm of the 1899 season, known as the San Ciriaco hurricane after a town it devastated on Puerto Rico. It was a tropical depression from August 3 until September 4, a total of 33 days. Of these, it was a tropical storm for 28 days, tying Hurricane Ginger of 1971 for top honors.

Greatest Number Active at the Same Time

On August 22, 1893, four hurricanes raged simultaneously in the Atlantic Ocean. One was approaching Nova Scotia, another was centered between Bermuda and the Bahamas, a third was just northeast of the Lesser Antilles, and yet a fourth was churning west off Africa's Cape Verde Islands. Two of these storms were destined to strike the United States, including a direct hit on New York City by the Bahamas storm. The Cape Verde storm eventually came ashore in Georgia and South Carolina on August 27, killing as many as 2,000 residents of low-lying islands along the coastline.

Most Names

There has been one tropical storm that had three lives and three names. Tropical Storm Hattie developed off the Caribbean coast of Nicaragua on October 28, 1961, and drifted north and then west before crossing Central America at Guatemala. It re-emerged into the Pacific Ocean on November 1 and was re-christened Simone. Two days later it curled back towards the coastline of Central America and crossed over to the Atlantic via Mexico, re-emerging into the Gulf of Mexico as Inga.

The eye of Hurricane Andrew passed directly over Homestead Air Force Base, where winds of 160 mph were registered before the anemometer blew away. (John Curry, courtesy of the South Florida Sun-Sentinel*)*

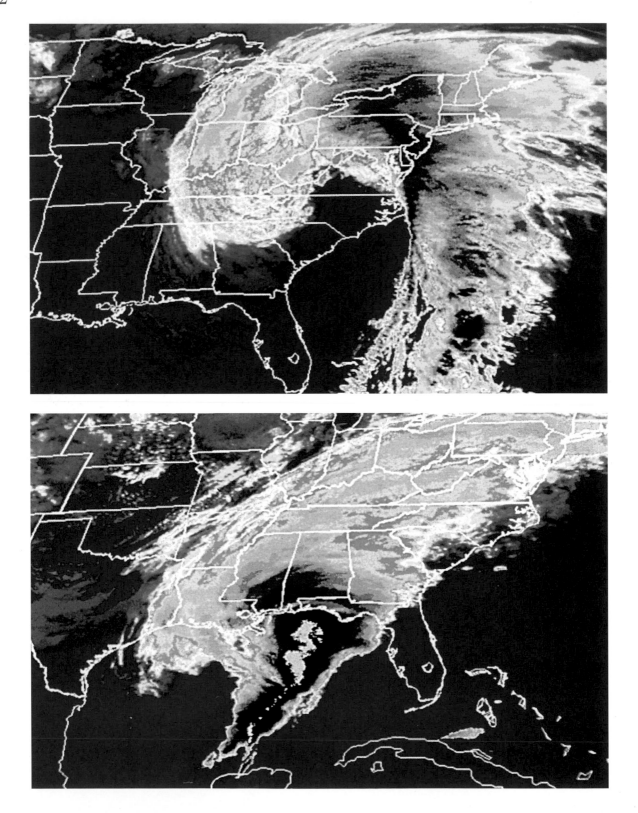

chapter 8
WINDSTORMS & FOG

EXTRA-TROPICAL STORMS

Most Americans never experience a major hurricane or tornado. The type of storm they are most likely to encounter is the extra-tropical storm, or your regular, old-fashioned, low-pressure system that provides the bulk of the nation's precipitation and day-to-day, run-of-the-mill, bad weather. Every now and then, however, these garden-variety, low-pressure systems become monster storms in their own right. Not as spectacular, perhaps, as tropical storms, they are often no less damaging or dangerous, and typically they affect a much larger swath of territory. The classic example of this kind of storm was the so-called Super Storm of March 1993, which virtually closed down the entire eastern third of the United States from Florida to Maine.

Extra tropical storms are the only storms of significance to affect the West Coast. Rotating in off the Pacific Ocean during the wet season of October through April, they provide the only "extreme weather" the big urban areas from San Diego to Seattle ever see.

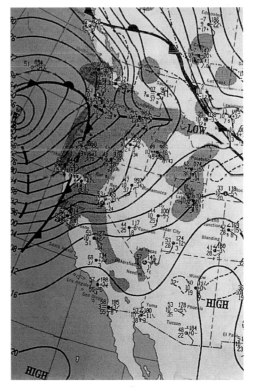

A National Weather Service map shows the intense Pacific Coast storm of December 12, 1995, about to come ashore in Oregon. San Francisco experienced its highest winds ever measured during the passage of the cold front depicted on the map.

(opposite) Enhanced infrared satellite images show the progression of the Super Storm of March 1993, one of the most powerful extra-tropical storms ever to develop in the U.S. The bottom image was taken at about midnight on March 12 when the center of the low pressure system was still over the Gulf of Mexico. A powerful cold front with tornadoes and 100 mph winds can be seen approaching the west coast of Florida. The top image is the storm 12 hours later as it's center passed by the coast of Virginia. Blizzard conditions were raging from North Carolina to New York at this time. (NOAA)

Pressure

Inches of HG	Millibars of HG
32.10	1087.0
32.00	
31.90	1080.0
31.80	
31.70	
31.60	1070.0
31.50	
31.40	
31.30	1060.0
31.20	
31.10	
31.00	1050.0
30.90	
30.80	
30.70	1040.0
30.60	
30.50	
30.40	1030.0
30.30	
30.20	
30.10	1020.0
30.00	
29.90	
29.80	1010.0
29.70	
29.60	
29.50	1000.0
29.40	
29.30	
29.20	990.0
29.10	
29.00	
28.90	980.0
28.80	
28.70	
28.60	970.0
28.50	
28.40	
28.30	960.0
28.20	
28.10	
28.00	950.0
27.90	
27.80	940.0
27.70	
27.60	
27.50	930.0
27.40	
27.30	
27.20	920.0
27.10	
27.00	
26.90	910.0
26.80	
26.70	
26.60	900.0
26.50	
26.40	
26.30	890.0
26.20	
26.10	
26.00	880.0
25.90	
25.80	
25.70	870.0
25.60	

The severity of extra-tropical, low-pressure systems, like that of tropical storms, is measured according to the barometric pressure of the storm center. The earth's average sea-level barometric pressure is about 30". Strong extra-tropical storms usually have a central minimum pressure of around 29"–29.50". When the pressure falls below 29" you have a major storm on your hands. The lower the central pressure, the more intense the storm.

World Barometric Pressure Records

Lowest Pressure Records

Record	Pressure	Date	Location	Event
Tropical Storm Records				
World Record	25.69"	10/12/1979	W. Pacific	Typhoon Tip
Atlantic Record	26.22"	12/13/1988	Gulf of Mexico	Hurricane Gilbert
Indian Ocean Record	26.30"	1833	Bay of Bengal	S.S. Duke of York
U.S.A. Record	26.35"	9/5/1935	Florida Keys	Hurricane
Eastern Pacific Record	26.58"	9/12/1997	E. Pacific	Hurricane Linda
Australian Record	26.72"	4/22/1989	W. Australia	Cyclone Orson

Record	Pressure	Date	Location
Extra–Tropical Storm Records			
Atlantic Lowest	26.96"	1/11/1993	Off Shetland Isles (est at storm center)
(27.05" actual measurement made aboard the S.S. Braer)			
Iceland Lowest	27.28"	1/3/1933	Reykjavik, Iceland
Great Britain Lowest	27.33"	1/26/1884	Ochertyre (near Crieff, Perthshire)
Alaska Lowest	27.35"	10/25/1977	St. Paul Island, Alaska
Ireland Lowest:	27.38"	12/8/1886	Belfast
Southern Hemi. Lowest	27.59"	8/11/1994	Halley Bay, Antarctica
Finland Lowest	27.75"	3/1/1990	
Canada Lowest	27.76"	1/20/1977	St. Anthony, Newfoundland
Holland Lowest	28.17"	11/27/1983	
U.S. Lowest	28.20"	1/3/1913	Canton, New York
	28.20"	3/6/1932	Nantucket, Massachusetts
Ohio Valley Lowest	28.28"	1/26/1978	Cleveland, Ohio
West Coast Lowest			
At Sea	28.20"	1/9/1880	Coastal OR, Umpqua R. mouth
On Land	28.40"	12/1/1987	Quillayute, Washington
Midwest Lowest	28.43"	11/10/1998	Albert Lea, Minnesota

Highest Pressure Records

Record	Pressure	Date	Location
World Record	32.06"	12/19/2001	Tonsontsengel, Mongolia
	32.01"	12/31/1968	Agata, USSR (2nd highest in world)
North American Highest	31.88"	2/2/1989	Dawson, Yukon
U.S. Highest	31.85"	12/31/1989	Northway, Alaska
Contiguous U.S. Highest	31.42"	12/24/1983	Miles City, Montana
United Kingdom Highest	31.15"	1/31/1902	Aberdeen, Scotland

PACIFIC NORTHWEST STORMS

The three contenders for "biggest storm ever" to strike the Pacific Northwest from San Francisco northward are: the Storm King of January 9, 1880; the Big Blow of Columbus Day on October 12, 1962; and the Great West Coast Windstorm of December 12, 1995. All three blew in from the southwest over the Pacific Ocean, with their centers making landfall along the coast of Oregon or Washington.

Storm King of 1880

Little data is available for the so-called "Storm King" of January 1880, but it appears the storm center came ashore just south of Astoria, Oregon, on January 9, when a barometric pressure of 28.45" was registered in the town. Portland bottomed out at 28.56", and for both locations these remain the lowest barometric-pressure readings on record. A ship near the mouth of the Umpqua River, where the town of Reedsport, Oregon, is today, reported a pressure of 28.20", the equivalent strength of a major Category 3 hurricane. Winds gusted over 70 mph in Portland, causing extensive damage and several deaths. Along the coast, wind gusts probably exceeded 100 mph. Enormous damage was done to the forests of both Oregon and Washington. Just outside of Portland, 500 to 600 trees were blown down over the railroad tracks between Beaverton and Hillsboro, a distance of just 10 miles. In the Seattle-Olympia region, record snowfalls were reported.

Following the storm, snow lay four to six feet deep in Seattle, collapsing many structures. In Tacoma, 54" lay on the ground, although it isn't clear just how much of this snow fell during the Storm King event and how much might already have been on the ground. In any case, no such depths have been approached since.

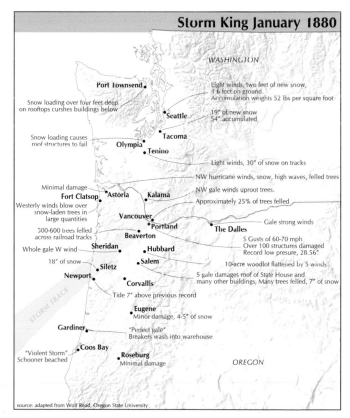

Storm King January 1880

WASHINGTON

Port Townsend
Snow loading over four feet deep on rooftops curshes buildings below
Light winds, two feet of new snow, 1 6 feet on ground. Accumulation weights 52 lbs per square foot
Seattle — 19" of new snow, 54" accumulated
Snow loading causes roof structures to fail
Olympia
Tacoma
Tenino
Light winds, 30" of snow on tracks
NW hurricane winds, snow, high waves, felled trees
NW gale winds uproot trees.
Minimal damage
Fort Clatsop, Astoria, Kalama
Approximately 25% of trees felled
Westerly winds blow over snow-laden trees in large quantities
Vancouver
Gale strong winds
500-600 trees felled across railroad tracks
Portland
The Dalles
Beaverton
S Gusts of 60-70 mph
Over 100 structures damaged
Record low presure, 28.56"
Whole gale W wind
Sheridan
Hubbard
18" of snow
Salem
10-acre woodlot flattened by S winds
Siletz
Newport
S gale damages roof of State House and many other buildings. Many trees felled, 7" of snow
Corvallis
Tide 7" above previous record
Eugene
Minor damage, 4-5" of snow
Gardiner
"Perfect gale" Breakers wash into warehouse
Coos Bay
"Violent Storm" Schooner beached
Roseburg
Minimal damage
OREGON
STORM TRACK

source: adapted from Wolf Read, Oregon State University

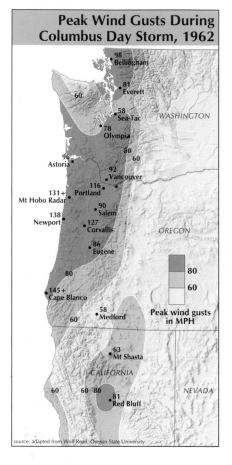

Peak Wind Gusts During Columbus Day Storm, 1962

98 • Bellingham
81 • Everett
60
58 • Sea-Tac
WASHINGTON
78 • Olympia
80
96 • Astoria   60
92 • Vancouver
116 • Portland
131 + Mt Hobo Radar
90 • Salem
138 • Newport
127 • Corvallis
OREGON
86 • Eugene
80
145 + Cape Blanco
58 • Medford   Peak wind gusts in MPH
60
63 • Mt Shasta
CALIFORNIA
60   60   80
81 • Red Bluff   NEVADA

80
60

source: adapted from Wolf Read, Oregon State University

Big Blow of 1962

The most destructive storm in Washington and Oregon's recent history roared ashore on Columbus Day, October 12, 1962. Although the lowest offshore barometric pressure associated with the Columbus Day storm was, at 28.34", higher than either the Storm King or the 1995 windstorm, the winds were much stronger. The reason for this was the amazing rapidity with which the storm intensified as it approached the coastline—a cyclogenic bomb, as meteorologists call it. Winds gusted to 116 mph on the Morrison Bridge in downtown Portland; 138 mph at Newport, Oregon, on the coast; and, according to a National Weather Service assessment of a damaged anemometer at Cape Blanco, a scarcely believable 179 mph at that location. In Washington, the wind reached 78 mph in Seattle and 160 mph along the coast. An astonishing 11 to 15 billion board feet of timber was lost in what remains the greatest blowdown in Pacific Northwest history. Forty-six deaths were reported.

Great Windstorm of 1995

One of the strongest low-pressure systems ever to strike the West Coast came ashore near Astoria, Oregon, during December 11 and 12 of 1995, and, as is usual with Pacific storms, the highest winds occurred to the south of the storm center. Barometric pressure bottomed out at 28.53" at Astoria, very close to the Storm King record of 1880. In fact, at Seattle the reading of 28.65" was the lowest on record there. Wind gusts of 110 to 120 mph were reported from Cannon Beach, Oregon, south to Cape Blanco.

The worst damage, however, was in the San Francisco Bay Area, where the city experienced its highest winds on record, with gusts to 74 mph at the airport, 85 mph downtown, and 103 mph on Angel Island in the Bay itself. Hundreds of trees were blown down in Golden Gate Park, and the Conservatory of Flowers, standing since 1878, was almost completely destroyed. (It reopened in October of 2003.) Although the storm moved quickly through the area, torrential amounts of rain fell. The Alameda Air Station near Oakland recorded 2.66" of rain in one

45-minute period. San Francisco collected a 5.33" storm total, one of its heaviest rainfalls ever. In the Russian River region north of San Francisco, more than 15" fell in a 48-hour period.

Southern California is rarely visited by strong cyclonic storms. The El Niño season of 1982–1983 brought San Diego's strongest storm system ashore when the barometer fell to a record low of 29.38", a relatively modest figure and equivalent to what is experienced in at least a dozen typical Atlantic Coast storms every year. In Southern California, flooding usually accompanies these storms during El Niño years.

During the last major El Niño (1997–98), 20.51" of rain fell in Los Angeles in the month of February at the University of California station.

Oregon and Washington's worst wind storm on record, the Big Blow of October 1962, topples the Campbell Hall Tower on the Western Oregon State College campus in Monmouth. (Wes Luchau)

The Armistice Day Storm of November 1940 was the worst blizzard in the history of Minnesota. (Minnesota Historical Society)

STORMS OF THE MIDWEST

All three of the worst cyclonic storms to affect the Midwest were, not coincidentally, among the worst blizzards ever to strike the region; consequently the author appologizes for some repetition here (*see* SNOW & ICE, *p. 93*).

Armistice Day Blizzard November 1940

Minnesota's worst blizzard, and one of its most intense cyclonic storms on record, hit on Armistice Day (now known as Veterans Day), November 11, 1940. The lowest barometric pressures measured with this storm reached 28.57" on Michigan's Upper Peninsula and 29.66" in Duluth, Minnesota. The storm was famous for the sudden drop of temperature, from 60° to zero in 12 hours, taking many people by surprise. Minneapolis received 16.8" of snowfall with amounts over 26" falling elsewhere. Drifts reached 25 feet across much of the state. Winds over 80 mph raked Lake Superior, sinking several ore ships with the loss of 59 crew.

(opposite) Although the Armistice Day Storm produced Minnesota's worst blizzard, the most intense cyclone to cross the state was that of November 1998. The cyclone reached its most developed stage at 6 a.m. (CST) on November 10, as depicted on this National Weather Service map.

January 1975 Storm

The intensity of the Armistice Day storm was surpassed on January 10–11, 1975, when another storm followed an almost identical path. This time the barometer bottomed out at 28.52″ at Basswood Lake, Minnesota, on the Canadian border, a new record for the Midwest. Duluth dropped to 28.55″ and Minneapolis 28.62″. Once again, a blinding blizzard accompanied the storm system. Wind gusts of 80 to 90 mph swept across portions of Iowa and Minnesota, while temperatures fell below zero, creating wind chills of –60° to –80°, one of the coldest storms in the region's history. Snowfalls of up to 23″ in Minnesota were blown into drifts 20 feet deep.

Storm of November 10, 1998

The lowest barometric readings ever registered in the Midwest occurred during a powerful cyclone that burst through Iowa and Minnesota on almost exactly the same date in November as the Great Armistice Day Storm of November 11, 1940. In this case, however, cold air was not present, and so snowfalls and true blizzard conditions did not develop. The lowest pressure reading measured 28.43″ in Albert Lea, Minnesota, with Minneapolis at 28.55″ and Duluth at 28.47″, all their lowest pressure readings ever.

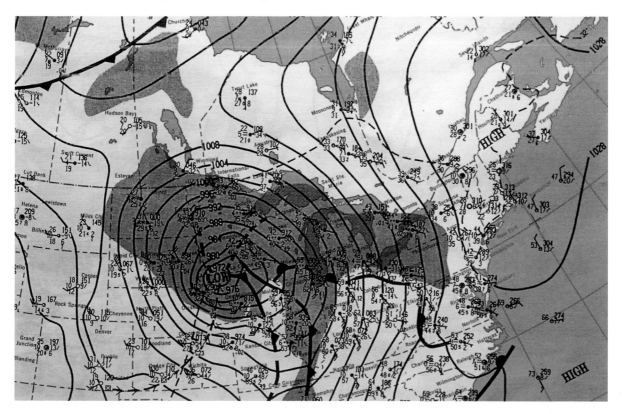

APPALACHIANS AND LOWER GREAT LAKES STORMS

Probably the two most intense cyclonic storms to ravage the Appalachians and lower Great Lakes were the Freshwater Fury of November 1913 and the Great Blizzard of January 1978.

Freshwater Fury of November 1913

As songwriter Gordon Lightfoot sang in his song *"The Edmund Fitzgerald,"* the gales of November are the dread of all Great Lakes sailors. The single deadliest storm ever for the sailors of the Great Lakes occurred on November 9–10, 1913. (The date of November 10th seems to come up again and again. The *S.S. Edmund Fitzgerald* sank during a storm on Lake Superior on November 10, 1975). In 1913, a low-pressure center formed in northern Georgia on a cold front that stretched along the Appalachian Mountains. The low deepened dramatically and moved north to a position over Lake Erie by the evening of November 9. The barometric pressure dropped to 28.61" at Erie, Pennsylvania, and hurricane-force winds blasted Lakes Huron, Erie, and Ontario. Winds of 79 mph whipped through Cleveland along with a record 22.2" of snow. Buffalo, New York, registered wind gusts of 80 mph. Ten large, 300-foot-long ore ships on Lake Huron and Lake Erie were lost, with another seven totally wrecked after being driven ashore by the gales. No one knows the exact number of sailors lost, but it has been estimated to be as many as 300.

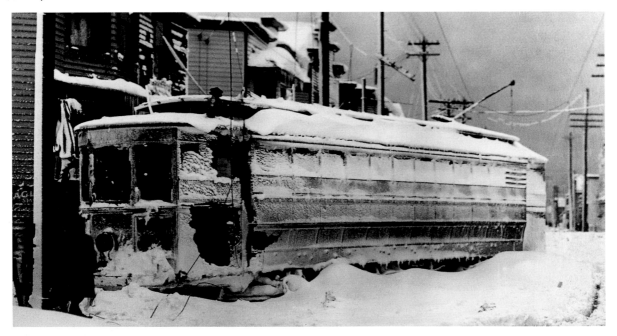

Cleveland was brought to a standstill by the "White Hurricane" of November 10, 1913.
(Western Reserve Historical Society)

The blizzard of January 25–26, 1978, created snowdrifts deep enough to bury northern Ohio homes up to their eaves. (Stephen Chang, courtesy of Kent State University Press and Thunder in the Heartland)

Great Blizzard of January 1978

An extraordinary cyclone developed on January 25–26, 1978, the likes of which have never been equaled anywhere in the United States west of the Atlantic seaboard. As was the case in the 1913 storm, a low-pressure system over Georgia developed explosively, the pressure falling 1.18″ in just 24 hours, as the storm moved north to a position over Lake Erie on January 26. Cleveland reported a barometric pressure of 28.28″and wind gusts to 82 mph, the highest on record. Over Lake Erie, the ore carrier *J. Burton Ayers* measured sustained winds of 86 mph and gusts to 111 mph. The temperature in Cleveland dropped from 44° at 4 a.m. with rain, to 7° by 10 a.m. with blizzard conditions. Snowdrifts approached 25 feet in both Ohio and Michigan.

The storm reached its maximum intensity over southern Ontario in Canada, where the barometric pressure bottomed out at 28.21″ in Sarnia. Toronto registered 28.40″, its lowest reading in 160 years of records, and this by a margin of .17″, a considerable amount by meteorological standards.

Just one week following this historic storm, between February 5 and 7, New England was hit by its severest blizzard in 90 years.

GREAT EXTRA-TROPICAL CYCLONES
OF THE ATLANTIC SEABOARD

Damaging cyclones occur almost every year on the Atlantic Coast. Among the most intense and unforgettable was the Super Storm of March 1993, with its extraordinary size and record-low barometric-pressure readings. The only other storm to rival it in terms of intensity (lowest barometric pressure) would be the storm of March 7, 1932. As for wind velocity, few storms can match the Great Appalachian Storm of November 25–26, 1950. For coastal flooding and destruction, the Great Storm of March 5–8, 1962, is unsurpassed.

Great Cyclone of March 1932

This storm did not bring the snowfall or disruption of the 1993 Super Storm, but it did develop a central low pressure even deeper, with barometric readings as low as 28.20" registered at Block Island, Rhode Island. This equals the lowest pressure reading ever measured in the lower 48 states for a non-tropical storm (*see table on p. 234*). Other record low-pressure readings included 28.78" at Charleston, South Carolina; 28.37" at Cape Henry, Virginia; and 28.37" in Atlantic City, New Jersey.

Great Appalachian Storm of November 1950

Although this storm tracked along the spine of the Appalachian mountain chain, dropping phenomenal amounts of snow in West Virginia (a 63" total storm accumulation in Coburn Creek), it was the high winds to the east of the storm center that were most notable. A vast, southerly air flow ahead of the storm system on November 25–26, produced hurricane-force winds from Virginia to New England. Sustained wind speeds of 80 mph were measured in Newark and Boston, with the former recording a gust of 108 mph. Concord, New Hampshire, had gusts up to 110 mph, while winds on Mt. Washington registered 160 mph. The wind remained at gale force for more than 12 hours in most areas, an extraordinary length of time to maintain such intensity. Even in Manhattan's Central Park, usually protected from strong winds by the surrounding skyscrapers, sustained winds of 70 mph were measured. Needless to say, damage was widespread from North Carolina to New Hampshire and west into Ohio.

THE PERFECT STORM

On October 28, 1991, an extra-tropical storm developed along a cold front that had moved off the northeast coast of the United States, setting the stage for one of the strangest weather events in U.S. storm history. As the front stalled off Nova Scotia, it began to suck tropical moisture into its circulation from a dying hurricane, named Grace, that was located east of Bermuda. The extra-tropical low along the front off Nova Scotia quickly deepened into a major storm and began to retrograde to the southwest and towards the coast of New England.

By October 30, sustained winds of 70 mph and sea waves of 30 to 40 feet were being reported in the open sea south of Nova Scotia. A strong high-pressure area in eastern Canada established a steep pressure gradient between itself and the low-pressure area, setting the stage for phenomenally high winds and waves.

As the low pressure slid slowly to the southwest, damaging winds and surf began to pound the eastern seaboard. Peak wind gusts of 78 mph were reported at the town of Chatham on Cape Cod. Gusts of 74 mph were recorded on Thatcher Island near the town of Gloucester—from whence sailed the ill-fated fishing boat the *Andrea Gail* (described in Sebastian Junger's book, *The Perfect Storm*). A third wind record of 68 mph was set at Marblehead, Massachusetts.

More destructive than the winds, however, was the surf of 10 to 30 feet, which hammered the seaboard from North Carolina to Maine and caused extensive coastal flooding. Tides rivaled those of the Great Atlantic Hurricane of 1944 and the famous March 1962 storm. Total damage topped $207 million, and four people died, but it was the loss of the swordfishing boat *Andrea Gail* with her crew of six, somewhere near the Grand Banks, that ultimately made this storm famous.

For meteorologists the unusual facet of this storm was not its strength but the fact that it eventually developed back into a true hurricane after the low's center drifted farther south and over a section of the Gulf Stream that had 80° sea-surface temperatures. This hurricane was never named, an unusual circumstance, perhaps unique since the weather bureau began naming tropical storms in 1950. Instead it became known as "The Perfect Storm," a label provided by the National Weather Service.

Great Atlantic Storm of March 1962

From March 5 to 8, 1962, a persistent and powerful low-pressure system off the Mid-Atlantic Coast coincided with the advent of abnormally high tides to produce the most widespread coastal erosion and damage to shore facilities in the history of the Mid-Atlantic states. No other storm, including hurricanes, wreaked such havoc to the shorelines of Delaware and New Jersey. Waves of 40 feet just offshore, and a surf of 20 to 30 feet, came atop a record storm surge of eight to13 feet along both coastlines. The U.S. Navy destroyer, USS *Monssen* was washed onto the beach at Holgate, New Jersey.

The high tides and surf, along with 70-mph winds, damaged or destroyed every pier in coastal Delaware and New Jersey, as well as over 5,000 buildings. In shoreline communities, at least 40 residents drowned. The coastline was transformed forever, with five new inlets forming along New Jersey's barrier islands. The cost of the damage from this storm was around $150 million, or close to $1 billion in 2003 dollars, the most destructive storm in both New Jersey's and Delaware's history.

A New Jersey Shore home floats out to sea during the Great Atlantic Storm of March 1962. This storm was the most destructive ever (including tropical storms) to affect the coast of New Jersey and Delaware. Should a similar event occur today, property damages would perhaps equal the $20 billion that resulted from Hurricane Andrew's landfall on Florida during August of 1992. (Urban Archives, Temple University)

Attempts are made to free the destroyer USS Monssen, beached near Holgate, New Jersey, during the Great Atlantic Storm. (Karl Von Schuler)

Super Storm of March 1993

Overall, this must be considered the most intense and extensive extra-tropical storm ever to rake the Atlantic seaboard. Forecasters were able to see this storm coming many days in advance. Computer models were forecasting a "perfect storm," meaning one with the greatest possible ferocity, and such a storm began developing in the Gulf of Mexico on March 12. The storm continued to intensify and move up the Atlantic Coast on March 13, spreading record snowfall from Alabama to New York. Barometric-pressure records were broken from Asheville, North Carolina (28.89"), to White Plains, New York (28.38"). Single-snowstorm state records were broken in Tennessee (56" on Mt. LeConte), North Carolina (50" on Mt. Mitchell), and Maryland (47" in Grantsville). From Birmingham, Alabama (13"), to Syracuse, New York (43"), single-snowstorm city records were shattered.

Fifteen tornadoes raked Florida when the cold front associated with the storm raced across the state, killing 44. Another 274 died from storm-related accidents, including 48 lost at sea. Virtually every airport on the East Coast was closed.

GREAT CYCLONES OF EUROPE

The most destructive storms throughout European history have been monstrous extra-tropical cyclones that have blown in from the Atlantic during the winter months. These storms are similar to the ones that rake the U.S. Pacific Northwest the same time of year. However, the northerly latitude of northwestern Europe is closer to that of the Canadian Pacific than the U.S. Pacific, and hence the low-pressure systems striking Europe tend to be more intense than those experienced in the U.S. Pacific states. The lowest pressure reading of 28.40" recorded along the Washington coast can hardly compare to Britain's lowest reading of 27.33".

Sea flooding, especially in the aptly named Low Countries of Holland and Belgium, has long been the the most devastating aspect of these storms. In 1228 some sources estimate 100,000 people drowned (a figure difficult to believe) in Holland's Friesland, when a storm surge flooded a large portion of the country. As recently as January 1953, a sea flood once again broke through the dikes of Holland drowning 1,800. Another 307 lives were lost in England to coastal flooding. The winds during this great storm reached 125 mph in the Orkney Islands. Winds up to 175 mph in the North Sea sank several cargo ships, and tragically, the ferry *Princess Victoria* went down with 132 passengers and crew in the Irish Sea.

Following the 1953 storm, Holland undertook a massive engineering project to protect its vast miles of reclaimed land behind a complex series of dikes and dams.

Great Britain's most infamous cyclone occurred on November 26–27, 1703, and has since been known as the Great Storm. It was documented by then journalist Daniel Defoe (author of *Robinson Crusoe*), who traveled the countryside in a vain attempt to count the number of trees blown down. He forsook this fruitless task upon reaching the number 17,000. Hundreds if not thousands of homes, windmills, barns, and churches were destroyed across southern England. But the worst of the disaster was at sea where hundreds of ships, many from the Royal Navy, were lost along with some 8,000 sailors.

A very similar storm cut a swath across southern Britain on the night of October 15–16, 1987, causing widespread destruction. In London's parks, thousands of century-old trees were lost. Winds were clocked at 115 mph at Shoreham-by-Sea, and the barometric pressure fell to 28.23" in Bristol.

An even more devastating storm swept across France and Germany on December 26–28, 1999, with winds up to 135 mph. Orly Airport near Paris recorded a maximum gust of 108 mph. The historic gardens of Versailles were almost completely destroyed and lost 10,000 trees, some of which had stood since the French Revolution. All in all, 140 people were killed in storm-related accidents and avalanches in France, Germany, Belgium, and Austria. This cyclone has since been dubbed Europe's "Storm of the Century."

The windstorm of October 16, 1987, that swept across southern Great Britain, destroyed many historic gardens such as those surrounding the landmark Emmetts House and Garden in Kent.
(Mike Howarth, National Trust)

EXTREME WINDS

Ironically, the highest (measured) wind speeds in the world have not been associated with hurricanes or tropical storms at all. The earth's highest surface wind on record was a gust of 231 mph recorded at the weather station on the summit of Mt. Washington, New Hampshire (elev. 6,288 feet), on April 12, 1934. A five-minute, sustained wind from the southeast of 188 mph was also observed during this storm.

The storm associated with this event was a not a particularly impressive one, with valley wind speeds only in the 30–50 mph range. It is the topography of the White Mountains and Mt. Washington's exposure that conspire to make such winds possible. The storm on April 12, had a tight pressure gradient oriented along a southeast-northwest axis. The pressure gradient, and the configuration of the surrounding mountains and valleys, produced something of a wind tunnel over the summit of Mt. Washington.

Mt. Washington is a phenomenally windy place. Its average annual wind speed is a gale-like 35.3 mph, with the month of January averaging 46.3 mph. An average of 104 days of the year experience hurricane winds over 75 mph.

Mt. Washington's summit holds the world-record wind speed of 231 mph established on April 12, 1934, during the passage of a relatively minor extra-tropical low-pressure system.
(Mt. Washington Observatory)

HIGHEST MEASURED WIND SPEED IN THE WORLD?

In its October 1972 issue, *Weatherwise* magazine reported the measurement of a near-world-record wind speed at Thule Air Force base in Greenland. The storm produced a wind gust of 207 mph, second in the world only to Mt. Washington's famous gust of 231 mph in 1934. After tracking up the Atlantic Coast of the U.S. on March 4–6, 1972, it intensified as it skirted Canada's Labrador Province and passed into the Davis Strait and Baffin Bay off Greenland's west coast. The 207-mph reading, in fact, was probably greater in force than the Mt. Washington reading because it was measured at the low-level elevation of 990 feet versus Mt. Washington's 6,200 feet, and thus occurred in air of greater density. Mt. Washington's 231 mph would be the equivalent of about 180 mph at sea level.

Thule Air Force Base is a large facility manned by some 5,000 American and Danish military and civilian personnel. Located on Greenland's northwest coast, it is at a latitude of 77°N, some 750 miles above the arctic circle. Scattered around the base are several off-base facilities such as radar sites and survival shelters (known as phase shacks). It was at Phase Shack Number 7, about six miles east of the base, that the wind gust of 207 mph was measured.

Knud Rasmussen, a heating plant operator and weather observer at P Mountain (10 miles south of the base) since July 1966, and at Thule since 1965, gave this account:

> This storm was the worst I have seen. During the height of the storm, the sides and, for the first time, the roof were constantly pelted with rocks and chunks of ice. All of us became very worried when three windows scattered throughout the complex were smashed by rocks and ice. I estimate the wind reached 140 knots [162 mph] up here.

John Kurasiewicz, one of the J-Site dispatchers who observed the record winds at Phase Shack Number 7, said, "The walls of J-Site fire station where we work were flapping and constantly being peppered by rocks and ice."

Cannon Mountain, about 20 miles southwest of Mt. Washington, registered several wind gusts over 199.5 mph on its 4.186-foot summit on the morning of April 2, 1973. It is likely some wind gusts exceeded 200 mph during this event since the anemometer was not capable of measuring higher than 199.5 mph, and yet it hit this figure four times, occasionally getting stuck at that figure for several long seconds.

The only other officially measured surface wind speed over 200 mph occurred at Thule Air Force Base on the northwest coast of arctic Greenland. A storm on March 8, 1972, produced a wind gust of 207 mph at a phase shack (an off-base survival shelter), and the winds were sustained at over 146 mph for a full three hours during the storm. To make matters worse, the temperature averaged –15°F during the period of peak wind speed.

Dwelling Damage vs Wind Speed

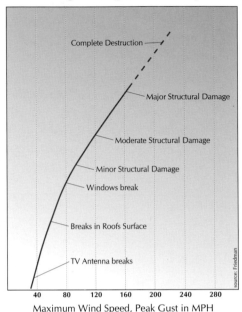

Maximum Wind Speed, Peak Gust in MPH

Prepared by the Institute of Behavioral Science at the University of Colorado in Boulder in 1971, this graph represents an attempt to establish a ratio of wind speed to residential home damage for the insurance industry.

Great Britain's highest measured wind speed was a gust to 172 mph recorded by a remote anemometer on the summit of Scotland's Cairngorm on March 20, 1986. At low-altitude sites, the fastest wind speed recorded was a reading of 141 mph at Kinnaird's Head Lighthouse near Fraserborough along Scotland's east coast on February 13, 1989. This same storm brought wind gusts estimated at over 150 mph on the Isle of Skye, where the damage was severe and all the island's inhabitants lost their power supply.

Gales along the coastline of Antarctica have reputedly produced winds as high as 200 mph, although there have been no measurements of such. A French expedition based at Port Martin in Antarctica during the early 1950s recorded a mean 24-hour wind speed of 108 mph on March 21–22, 1951, and a monthly average of 65 mph. These are the highest such values ever measured anywhere in the world.

Beaufort Wind Scale

Force	Speed in MPH	Name	General Description
0	.1 or less	Calm	Air feels still. Smoke rises vertically.
1	1–3	Light air	Wind vanes and flags do not move. Rising smoke drifts.
2	4–7	Light breeze	Drifting smoke indicates the wind direction.
3	8–12	Gentle breeze	Leaves rustle, small twigs move, and lightweight flags stir gently.
4	13–18	Moderate breeze	Loose leaves and pieces of paper blow about.
5	19–24	Fresh breeze	Small trees that are in full leaf sway in wind.
6	24–31	Strong breeze	It becomes difficult to use an open umbrella.
7	32–38	Moderate gale	The wind exerts strong pressure on people walking into it.
8	39–46	Fresh gale	Small twigs torn from trees.
9	47–54	Strong gale	Chimneys are blow down. Slates and tiles are torn from roofs.
10	55–63	Whole gale	Trees are broken or uprooted.
11	64–75	Storm	Trees are uprooted and blown some distance.
12	76 and greater	Hurricane	Buildings are badly damaged and many trees are uprooted.

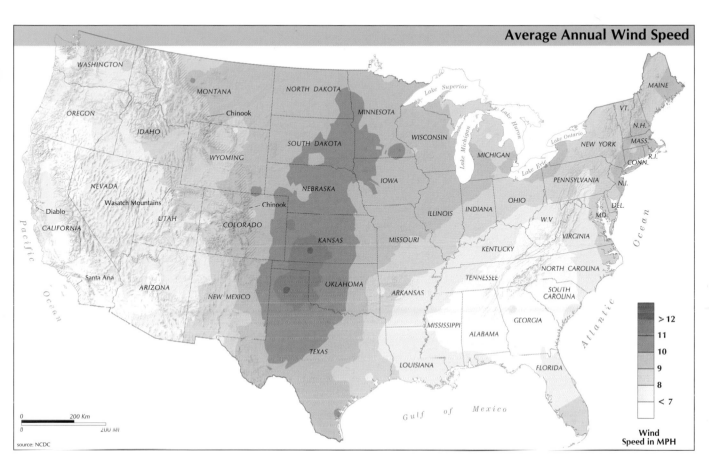

Average Annual Wind Speed

source: NCDC

Wind Speed in MPH

> 12
11
10
9
8
< 7

DANGEROUS WINDS OF THE UNITED STATES

Regional winds of several varieties occur in the United States and often create headaches for homeowners and their insurers. These include the chinook winds along the front range of the Rocky Mountains; the Wasatch Mountain winds affecting communities along the front range of those mountains in Utah; and the (sometimes) fire-producing Santa Ana and Diablo winds of Southern and Central California.

Chinooks

The most intense manifestation of chinook winds occurs in and around the city of Boulder, Colorado. Canyons to the west of town funnel the winds directly over the downtown area and severe damage occurs when the conditions conspire to maximize the effect. A gust of 143 mph was registered in one of these events during January 1971. The winds are most common during January, when high pressure settles over western Colorado or Utah while at the same time a low-pressure system tracks across the Northern Plains. A chinook on January 17, 1982, caused more than $10 million in damage in the Boulder area.

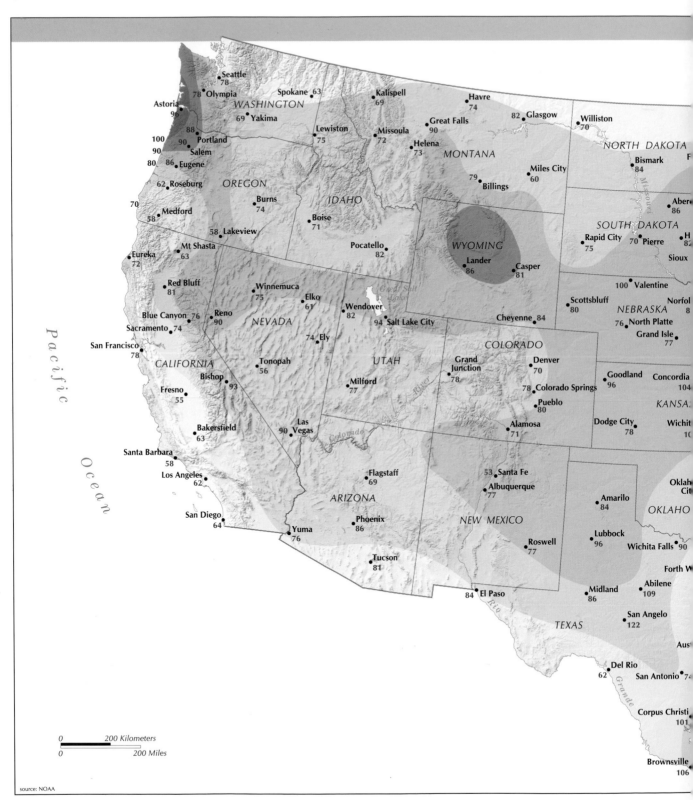

Seattle 78
78 Olympia
Astoria 96
Spokane 63
WASHINGTON
69 Yakima
88
100
90 Portland
90 Salem
80 86 Eugene
62 Roseburg
OREGON
70
Medford 58
Lewiston 75
Burns 74
58 Lakeview
Mt Shasta 63
IDAHO
Boise 71
Pocatello 82
Kalispell 69
Missoula 72
Helena 73
Havre 74
Great Falls 90
Glasgow 82
MONTANA
Miles City 60
Billings 79
Williston 70
NORTH DAKOTA
Bismark 84
Aber 86
SOUTH DAKOTA
Rapid City 75
Pierre 70
H 82
Sioux
Eureka 72
Red Bluff 81
Winnemuca 75
Elko 61
Wendover 82
Great Salt Lake
94 Salt Lake City
WYOMING
Lander 86
Casper 81
Cheyenne 84
Valentine 100
Scottsbluff 80
NEBRASKA
Norfol 8
Blue Canyon 76
Reno 90
Sacramento 74
San Francisco 78
NEVADA
74 Ely
CALIFORNIA
Tonopah 56
Bishop 93
Fresno 55
UTAH
Milford 77
Grand Junction 78
COLORADO
Denver 70
78 Colorado Springs
Pueblo 80
Alamosa 71
North Platte 76
Grand Isle 77
Goodland 96
Concordia 104
KANSAS
Dodge City 78
Wichit 10
Bakersfield 63
Santa Barbara 58
Los Angeles 62
San Diego 64
Las Vegas 90
Colorado River
Flagstaff 69
ARIZONA
Phoenix 86
Yuma 76
Tucson 81
Santa Fe 53
Albuquerque 77
NEW MEXICO
Roswell 77
El Paso 84
Amarilo 84
OKLAHO
Oklah Cit
Lubbock 96
Wichita Falls 90
Forth W
Midland 86
Abilene 109
San Angelo 122
Aus
TEXAS
Del Rio 62
San Antonio 74
Rio Grande
Corpus Christi 101
Brownsville 106

Pacific Ocean

0   200 Kilometers
0   200 Miles

source: NOAA

Highest Wind Speed to be Expected in any 50-Year Period

	110
	100
	90
	80
	70
	60

Highest wind speed to be expected
in any 50-year period in mph

•86 - Record maximum wind speed

International Falls 62
Duluth 78
MINNESOTA
St Cloud 78
Minneapolis 110
Rochester 85
Sioux City
Waterloo 83
Dubuque 74
IOWA
Des Moines 83
Omaha 109
WISCONSIN
Marquette 81
Escanaba 56
Green Bay 109
La Crosse 93
Milwaukee 81
Sault St Marie 72
Alpena 60
Houghton Lake 60
MICHIGAN
Muskegon 67
Grand Rapids 80
Flint 81
Detroit 95
Chicago 115
South Bend 92
Fort Wayne 78
INDIANA
Moline 81
Peoria 75
ILLINOIS
Springfield 98
Indianapolis 111
Terre Haute 58
Dayton 78
Kansas City
Topeka 75
St Louis 120
MISSOURI
Evansville 113
Louisville 78
Lexington 72
KENTUCKY
Cario 84
Tulsa
Fort Smith 85
Little Rock 87
ARKANSAS
Memphis 75
Nashville 73
TENNESSEE
Knoxville 86
Bristol 66
Shreveport 83
Jackson 69
Meridian 68
MISSISSIPPI
Birmingham 89
ALABAMA
Montgomery 72
Macon 87
LOUISIANA
Baton Rouge 70
Lake Charles 130
New Orleans 69
Houston 84
Galveston 120
Mobile 107
Pensacola 114
Apalachicola 85
Tallahassee 83
GEORGIA
80
90 Savannah
Jacksonville 82
Daytona Beach 77
Orlando 70
Tampa 99
FLORIDA
Fort Meyers 92
W Palm Beach 86
Miami 132
100
110

Lake Superior
Lake Michigan
Lake Huron
Lake Ontario
Buffalo 96
Rochester 89
Syracuse 77
NEW YORK
Erie 71
Cleveland 82
Toledo 87
Youngstown 66
Akron 75
Columbus 84
OHIO
Parkersburg 81
W. V.
Charleston 66
Pittsburg 83
Williamsport 78
Scranton 74
PENNSYLVANIA
Harrisburg 80
Allentown 94
Newark 108
New York City 113
Philadelphia 71
Wilmington
N.J.
DEL
Baltimore 80
MD.
Washington D.C. 98
Richmond 79
VIRGINIA
Lynchburg 74
Roanoke 77
Norfolk 80
Raleigh 153
Cape Hatteras 110
NORTH CAROLINA
Asheville 72
Charlotte 87
Greenville 71
Columbia 78
SOUTH CAROLINA
Athens 83
Augusta 62
Charleston 121
Wilmington 88
Atlanta 77

Caribou 56
MAINE 70
Eastport 80 72
Mt Washington 231
Burlington 72
VT.
N.H.
Concord 110
Portland 76
Boston 87
MASS.
Worcester 94
Providence 105
Hartford 100
CONN.
Block Island 130
R.I.

Ocean
Atlantic

Gulf of Mexico

River
Ohio River
Mississippi

Wasatch Winds

Communities along the front of the Wasatch Mountain Range in Utah occasionally suffer from chinook-like winds when high pressure over Wyoming sets up a steep pressure gradient with a low pressure centered in Arizona. The Salt Lake City and Ogden area have recorded wind gusts of 95 mph associated with these down-slope winds. Unlike the Rocky Mountain Front Range winds, the Wasatch winds normally occur during the spring, from March through May.

Santa Ana and Diablo Winds

The Santa Ana winds of Southern California and the Diablo winds of the San Francisco Bay Area bring hot, dry air to the normally cool coastlines of these two regions, and are associated with devastating wildfires. Temperatures have soared during these windstorms, and the 40- to 70-mph winds are so dry (humidity as low as 3% has been observed), they desiccate all vegetation in their path.

It was during Diablo winds in October 1991 that the disastrous Oakland Hills fire occurred, killing 25 people, burning down 3,500 homes, and causing $1.5 billion in damage. At the time, this was the third costliest natural disaster in United States history (hurricanes Andrew and Hugo were first and second).

EXTREME SANTA ANA

Santa Barbara, California, like all of coastal Southern California, suffers from the occasionally violent Santa Ana winds. Perhaps the most phenomenal event ever reported here occurred on June 17, 1850, as recalled by two doctors. Coincidentally, Santa Barbara's modern record temperature of 115° also occurred on June 17, in 1917, and under the influence of a Santa Ana wind. The doctors wrote:

> About 1 o'clock p.m. [on June 17, 1850] a blast of hot air from the northeast swept suddenly over the town, and struck the inhabitants with terror. It was quickly followed by others. At two o'clock the thermometer exposed to air rose to 113° and continued at or near that point for nearly three hours, while the burning wind raised dense clouds of implacable dust. No human being could withstand the heat. All betook themselves to their dwellings....Calves, rabbits, birds, etc., were killed, trees blighted, fruit was blasted and fell to the ground, burned only on one side; the gardens were ruined. At five o'clock the thermometer fell to 122° [sic], and at seven it stood at 77°.

> —Related by Walter Lindley, M.D. and Joseph Widney, M.D.

October 20, 1991: A resident of Oakland's Rockridge district watches in disbelief as his neighbors' homes are consumed in Diablo Wind–fed flames during the most destructive urban wildfire in U.S. history. (Jim Pire)

Southern California's worst fire disaster caused by Santa Ana winds happened in October of 2003, when 1,100 square miles were charred and yet another 3,500 homes were lost along with 20 lives. These winds may occur any time of the year in Southern California; in the San Francisco region they usually arrive in September and October. They are at their most destructive for both regions in October, after six or more rainless months have passed and vegetation is tinder dry. In spite of the hazards they pose, the Santa Ana and Diablo winds are usually a welcome phenomenon for the residents of coastal California. The warm, dry air pushes the smog and fog far out to sea, providing crystal-clear days and rare warm evenings. After the rainy season begins in November, the land becomes wet and green and the threat of wildfires recedes.

FOG

Fog gets the last word in this book. The foggiest places in the United States (excluding mountaintops and passes) are along the coasts of the Pacific and Atlantic Oceans, with a secondary area being the low valleys of the central Appalachian Mountains.

Foggiest of the foggy is Cape Disappointment at the mouth of the Columbia River between Oregon and Washington. Heavy fog is reported on average 2,552 hours a year, the equivalent of 106 complete days of dense fog (dense meaning visibility is less than an eighth of a mile). Willapa, on the coast of Washington, reported an average of 3,863 hours of fog a year (161 days) over a four-year period, according to an official with the U.S. Environmental Data Service. One of those years had no less than 7,613 hours of fog (317 days!).

San Francisco, the foggy city, averages about 40 days a year of dense fog in the downtown area, but only 17 days at the airport just a few miles to the south but protected from the sea by

An astonishingly dense bank of fog follows ferries across Puget Sound near Seattle. (Dave Torchia, courtesy of Weatherwise *magazine)*

San Francisco's natural air conditioner kicks in as advection fog rolls under the Golden Gate Bridge. (Kerrick James)

the coastal hills. The Ocean Beach district of San Francisco, which lies entirely along the Pacific Coast, is normally socked in with fog two days out of three in August.

Most of San Francisco's foggy days occur between June and August when the heat of California's interior valleys sucks the cool air of the Pacific Ocean inland through the one-mile-wide gap of the Golden Gate and across a middle stretch of the 50-mile-long bay.

On the Atlantic Coast, Nantucket Island has an average of 85 days with dense fog every year. The coast of Maine is also very foggy, with Moose Peak Lighthouse on Mistake Island averaging 65 days of heavy fog (1,580 hours). In fact, the entire immediate coastline of Maine from Mt. Desert Island to Eastport is exceptionally foggy, with 55 or more days of fog.

The foggiest inland area in the United States (exclusive of mountaintops, which, being uninhabited we won't bother with here—Mt. Washington, for instance, has fog an average 308 days a year) are the Appalachian valleys of West Virginia, especially in the Monongahela National Forest in the eastern part of the state. Cool overnight temperatures create inversions and deep condensation in the valley bottoms, and fog forms in these valleys an average 50 to 60 days of the year.

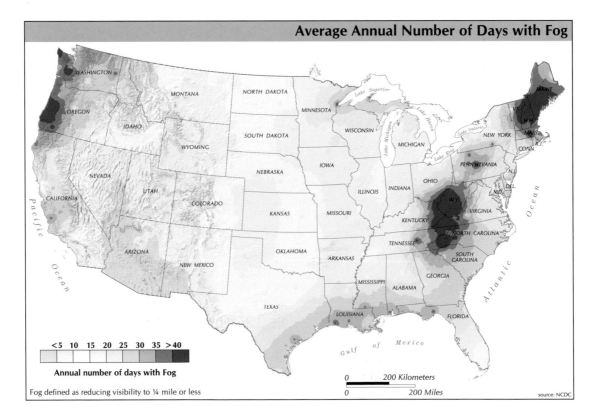

Average Annual Number of Days with Fog

<5 10 15 20 25 30 35 >40

Annual number of days with Fog

Fog defined as reducing visibility to ¼ mile or less

0   200 Kilometers
0   200 Miles

source: NCDC

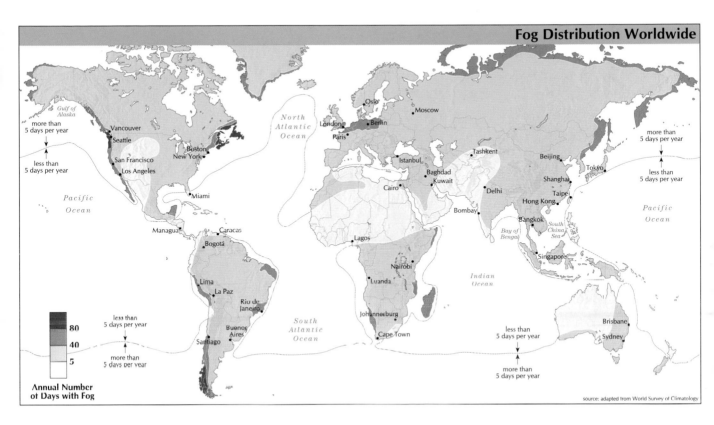

Fog Distribution Worldwide

Annual Number
of Days with Fog

80
40
5

less than
5 days per year

more than
5 days per year

source: adapted from World Survey of Climatology

The least foggy town in the U.S. is Key West, Florida, where only one foggy day each year might be expected to spoil a sunny day by the sea.

The foggiest place in the world is probably Newfoundland, Canada, and the Grand Banks in the North Atlantic. Cape Race, Newfoundland, reports an average of 158 foggy days a year.

Other extremely foggy locations include the southern coast of Chile, the coastline of Namibia in Africa, and the Severnaya Zemlya Islands in Russia's Kara Sea. All of these locations experience at least 80 days of dense fog annually.

London, like San Francisco, is forever associated in the popular imagination with fog. Although London does average about 30 days a year of dense fog, it is the smogs of the past that gave the city its reputation. Until the early 1960s, Londoners heated their homes with soft-coal stoves. Winter temperature inversions sometimes trapped these effluents at ground level for days on end with catastrophic results. In one week, that of December 3–10, 1952, more than 4,000 people died in London from inhaling the fumes of coal smoke and sulfur dioxide, which mixed with the fog to form a lethal soup of air. Great Britain passed its Clean Air Act in 1956, largely as a consequence of this disaster.

El Niño and La Niña

Meteorologists have become increasingly aware of the influence that anomalies in Pacific sea-surface temperatures have on the world's climate. When the sea-surface temperature in the eastern Pacific around the equator becomes much warmer than normal it is called El Niño, and conversely when the temperatures are cooler than normal it is called La Niña.

The term El Niño dates back to Peru in the 19th century, when the correlation between warm ocean waters and floods was first noticed. Since the event normally peaks around Christmastime, the Spanish term for Christ Child came to be applied to the phenomenon. The term La Niña was more recently coined to denote its converse relationship to El Niño with cooler than normal ocean temperatures in the same region of the Pacific normally affected by El Niños.

Strong El Niño events occur about every 15 years and are the catalyst for extreme weather around the world. Normally these events are manifested in flooding rains in Peru, Chile, Argentina, California, and East Africa. Droughts and heat waves are likely to occur in Australia, Southeast Asia, and Southern Asia.

The last major El Niño event was during the winter season of 1997–1998. This event was not quite as powerful as the 1982–1983 El Niño but received much more attention because it was the first major event to be both well forecast and well researched.

To say that El Niños are disastrous is inaccurate, since the El Niño itself does not create catastrophe (except for Peruvian fishermen) but rather influences individual storms, causing them to become stronger than normal; or it causes long spells of dry weather that induce drought. One might argue that El Niños do as much good as harm. California is certainly better off in years of ample rainfall than in years of scant rainfall.

La Niñas often, but not always, follow El Niños. La Niñas usually result in cooler and wetter weather in south and southeast Asia, drier and warmer weather in the southeastern U.S., cooler weather in Japan and Korea, and cooler- and drier-than-normal weather in Peru.

The next big El Niño event should occur sometime between 2010 and 2013.

One of the most intense El Niño events on record reached its peak in November 1997, as illustrated by the white swath (warm water) across the Pacific equatorial zone in the top frame of this composite satellite image. The situation had reversed itself by February 1999, as a well-defined La Niña (upwelling of cold water) began to take shape. This is depicted in the lower frame by the purple swath spread across the same region of the Pacific. (NASA)

El Niño / La Niña

TOPEX/POSEIDON and Jason-1

Nov '97

Feb '99

http://topex-www.jpl.nasa.gov

-18 -14 -10 -6 -2 2 6 10 14 cm

TOPEX/POSEIDON maps of sea surface height relative to normal

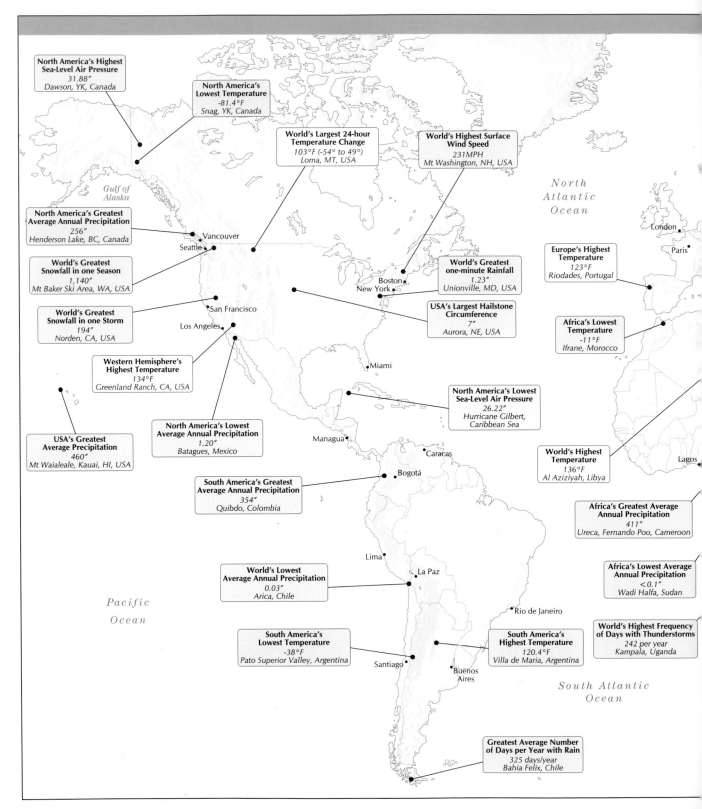

North America's Highest Sea-Level Air Pressure
31.88"
Dawson, YK, Canada

North America's Lowest Temperature
-81.4°F
Snag, YK, Canada

World's Largest 24-hour Temperature Change
103°F (-54° to 49°)
Loma, MT, USA

World's Highest Surface Wind Speed
231MPH
Mt Washington, NH, USA

North Atlantic Ocean

Gulf of Alaska

North America's Greatest Average Annual Precipitation
256"
Henderson Lake, BC, Canada

Vancouver
Seattle

London

Paris

World's Greatest Snowfall in one Season
1,140"
Mt Baker Ski Area, WA, USA

Boston
New York

World's Greatest one-minute Rainfall
1.23"
Unionville, MD, USA

Europe's Highest Temperature
123°F
Riodades, Portugal

World's Greatest Snowfall in one Storm
194"
Norden, CA, USA

San Francisco

Los Angeles

USA's Largest Hailstone Circumference
7"
Aurora, NE, USA

Africa's Lowest Temperature
-11°F
Ifrane, Morocco

Western Hemisphere's Highest Temperature
134°F
Greenland Ranch, CA, USA

Miami

North America's Lowest Average Annual Precipitation
1.20"
Batagues, Mexico

North America's Lowest Sea-Level Air Pressure
26.22"
Hurricane Gilbert, Caribbean Sea

USA's Greatest Average Precipitation
460"
Mt Waialeale, Kauai, HI, USA

Managua

Caracas

World's Highest Temperature
136°F
Al Aziziyah, Libya

Bogotá

Lagos

South America's Greatest Average Annual Precipitation
354"
Quibdo, Colombia

Africa's Greatest Average Annual Precipitation
411"
Ureca, Fernando Poo, Cameroon

Lima

La Paz

Africa's Lowest Average Annual Precipitation
<0.1"
Wadi Halfa, Sudan

World's Lowest Average Annual Precipitation
0.03"
Arica, Chile

Pacific Ocean

Rio de Janeiro

World's Highest Frequency of Days with Thunderstorms
242 per year
Kampala, Uganda

South America's Lowest Temperature
-38°F
Pato Superior Valley, Argentina

Santiago

South America's Highest Temperature
120.4°F
Villa de Maria, Argentina

Buenos Aires

South Atlantic Ocean

Greatest Average Number of Days per Year with Rain
325 days/year
Bahia Felix, Chile

World Weather Extremes

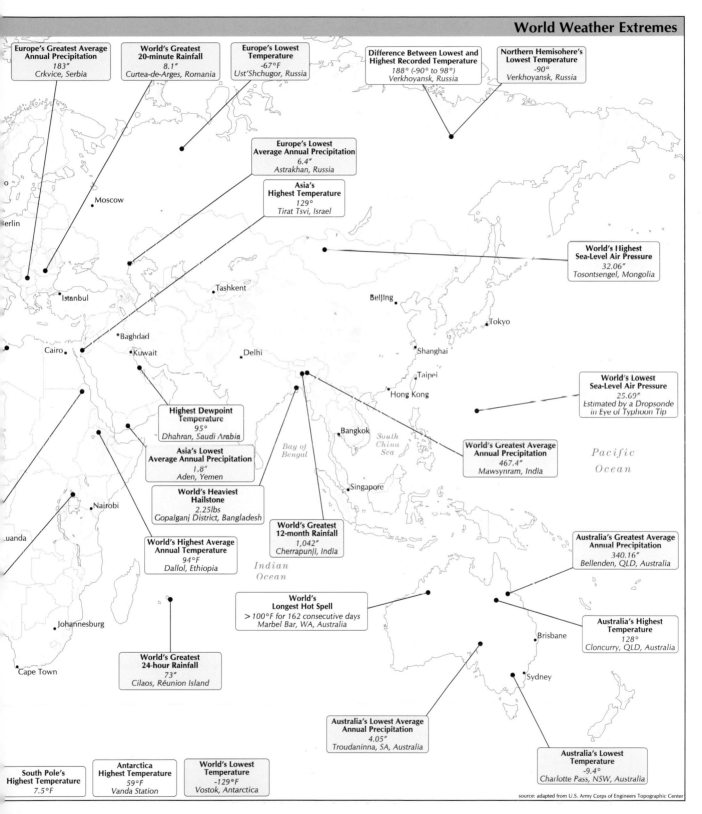

Europe's Greatest Average Annual Precipitation
183"
Crkvice, Serbia

World's Greatest 20-minute Rainfall
8.1"
Curtea-de-Arges, Romania

Europe's Lowest Temperature
-67°F
Ust'Shchugor, Russia

Difference Between Lowest and Highest Recorded Temperature
188° (-90° to 98°)
Verkhoyansk, Russia

Northern Hemisohere's Lowest Temperature
-90°
Verkhoyansk, Russia

Europe's Lowest Average Annual Precipitation
6.4"
Astrakhan, Russia

Asia's Highest Temperature
129°
Tirat Tsvi, Israel

World's Highest Sea-Level Air Pressure
32.06"
Tosontsengel, Mongolia

Moscow

Berlin

Istanbul

Tashkent

Beijing

Tokyo

Shanghai

Taipei

Hong Kong

Cairo

Baghdad

Kuwait

Delhi

World's Lowest Sea-Level Air Pressure
25.69"
Estimated by a Dropsonde in Eye of Typhoon Tip

Highest Dewpoint Temperature
95°
Dhahran, Saudi Arabia

Asia's Lowest Average Annual Precipitation
1.8"
Aden, Yemen

World's Heaviest Hailstone
2.25lbs
Gopalganj District, Bangladesh

World's Highest Average Annual Temperature
94°F
Dallol, Ethiopia

World's Greatest 12-month Rainfall
1,042"
Cherrapunji, India

Bangkok

South China Sea

Bay of Bengal

Singapore

World's Greatest Average Annual Precipitation
467.4"
Mawsynram, India

Pacific Ocean

Nairobi

uanda

Indian Ocean

Johannesburg

Cape Town

World's Greatest 24-hour Rainfall
73"
Cilaos, Réunion Island

World's Longest Hot Spell
>100°F for 162 consecutive days
Marbel Bar, WA, Australia

Australia's Greatest Average Annual Precipitation
340.16"
Bellenden, QLD, Australia

Australia's Highest Temperature
128°
Cloncurry, QLD, Australia

Brisbane

Sydney

Australia's Lowest Average Annual Precipitation
4.05"
Troudaninna, SA, Australia

South Pole's Highest Temperature
7.5°F

Antarctica Highest Temperature
59°F
Vanda Station

World's Lowest Temperature
-129°F
Vostok, Antarctica

Australia's Lowest Temperature
-9.4°
Charlotte Pass, NSW, Australia

source: adapted from U.S. Army Corps of Engineers Topographic Center

APPENDICES

EXPLANATION OF TABLES

The records on the following tables represent the temperature and precipitation extremes for over 300 U.S. towns and cities and are based, with a few exceptions, on the data collected by the U.S. Weather Bureau cum National Weather Service since the inception of the period of record for each station. That being said, there are a handful of records which pre-date the official period of record but are included because of the extreme nature of the weather event and the fact that the extremes were well recorded by a large number of observers from many locations. The events in question are the cold waves of January and February 1835 and the winters of 1856–57. Thermometers were commonplace in the first half of the 19th century and careful observations were made on a regular basis by men of science. The fact that the temperature extremes were widely recorded by many observers tends to corroborate the accuracy of the readings.

Because the periods of record for the towns in the tables vary widely from place to place, they are not useful for the purpose of comparisons, but rather for determining what the actual coldest, hottest, wettest, or snowiest record is for any given locale. The beginning of the period of record is noted next to the location's name for both the temperature and precipitation tables.

For the larger cities that have more than one official weather station, I have tried to use the most extreme measurement made by any one of the stations in that city. For instance, New York City's hottest reading of 107° was recorded at La Guardia Airport, its coldest reading (-15°) in Central Park, its fastest wind speed (113 mph) at the Battery, and its snowiest month (37.9") at the Lower Manhattan station (now closed). Chicago's hottest temperature of record, 109°, was recorded at what is now called Midway Airport in 1934. The rainfall during Los Angeles's wettest month of record (20.51"), February 1998, was recorded at the UCLA campus site, one of three stations (UCLA, Civic Center, LAX Airport) used to formulate the tabulations in that sprawling city.

By the time this book is published a few of these records will most likely already have fallen. They are current as of **June 1, 2004**. Updates and corrections will be published monthly (to the best of my ability) on our web site:

www.extremeweatherguide.com

I also encourage vigilant readers to point out errors, omissions, and new records.

Sources of Data

David M. Ludlum's *Weather Record Book* published by *Weatherwise* magazine in 1971, laid the foundation for the data. It has been updated and expanded using official NCDC (National Climate Data Center) statistics augmented by figures from the on-line six Regional Climate Centers (RCC's) Historical Climate Information files. The six centers cover the following regions:

Western	www.wrcc.dri.edu
High Plains	www.hprcc.unl.edu
Midwestern	www.mcc.sws.uiuc.edu
Southern	www.srcc.lsu.edu
Southeast	www.sercc.com
Northeast	www.nrcc.cornell.edu

Older data was culled from the following dusty volumes of the Department of Agriculture, Weather Bureau:

Climatological Data by Sections U.S. Dept. of Agriculture, Weather Bureau, Washington D.C., 1890-present

Climatology of the United States, 1906 U.S. Dept. of Agriculture, Weather Bureau Bulletin Q, 1906

Summaries of Climatological Data by Sections, 1912 Vols. 1-2 U.S. Dept. of Agriculture, Weather Bureau Bulletin W, 1912

Summaries of Climatological Data by Sections, 1926 Vols. 1-3 U.S. Dept. of Agriculture, Weather Bureau Bulletin W, 1926

Monthly temperature averages and monthly precipitation data predating establishment of the Weather Bureau statistics can be found in this hefty volume:

World Weather Records; Collected from Official Sources, Volume 79 Smithsonian Institution, Washington D.C., 1927

In addition to the above, data was researched from hundreds of other publications which are listed in the bibliography on pp. 293-298.

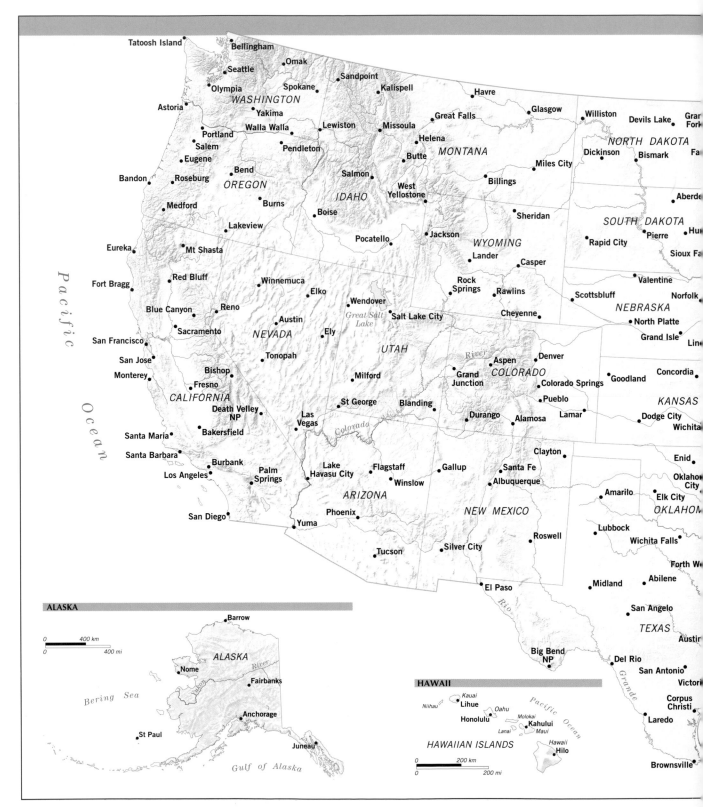

STATION LOCATIONS

STATE PRECIPITATION RECORDS

State	Maximum 24-hour Precipitation in Inches, Location and Date			Maximum Monthly Precipitation in Inches, Location and Date		
ALABAMA	32.52	Dauphin Island	7/19-20/1997	36.71	Dauphin Island	7/1997
ALASKA	15.20	Angoon	10/12/1982	70.99	MacLeod Harbor	11/1976
ARIZONA	11.40	Workman Creek	9/5/1970	16.95	Crown King	8/1951
ARKANSAS	*14.06	Big Fork	12/3/1982	23.86	El Dorado	12/1931
CALIFORNIA	26.12	Hogees Camp	1/22/1943	71.54	Helen Mine	1/1909
COLORADO	*11.08	Holly	6/17/1965	23.28	Ruby	2/1897
CONNECTICUT	12.77	Burlington	8/19/1955	27.70	Torrington	8/1955
DELAWARE	12.36	Vernon	9/15-16/1999	17.69	Bridgeville	8/1967
WASHINGTON D.C	7.31	24th and M Street	8/11/1929	17.45	24th and M Street	9/1934
FLORIDA	38.70	Yankeetown	9/5/1950	42.33	Ft. Lauderdale	10/1965
GEORGIA	21.10	Americus	7/5-6/1994	31.61	Americus	7/1994
HAWAII	38.00	Kilauea Plantation	1/24-25/1956	107.00	Kukui, Maui	3/1942
IDAHO	7.17	Rattlesnake Creek	11/23/1909	28.23	Roland	12/1933
ILLINOIS	16.94	Aurora	7/17-18/1996	20.03	Monmouth	9/1911
INDIANA	10.50	Princeton	8/6/1905	21.39	Evans landing	1/1937
IOWA	*16.70	Decatur County	8/5-6/1959	22.18	Red Oak	6/1967
KANSAS	17.00	Pittsburg	9/24-25/1993	24.56	Ft. Scott	6/1845
KENTUCKY	*10.48	Louisville	3/1/1997	22.97	Earlington	1/1937
LOUISIANA	*22.00	Hackberry	8/29/1962	37.99	Lafayette	8/1940
MAINE	19.19	Saco	10-21-22/1996	20.0+	Saco	10/1996
MARYLAND	14.75	Jewell	7/26/1897	20.35	Leonardtown	7/1945
MASSACHUSETTS	18.15	Westfield	8/18-19/1955	26.85	Westfield	8/1955
MICHIGAN	*9.78	Bloomingdale	9/1/1914	19.44	Muskegon	2/1912
MINNESOTA	*10.84	Fort Ripley	7/21-22/1972	16.52	Alexandria	8/1900
MISSISSIPPII	15.68	Columbus	7/9/1968	30.75	Merill	7/1916
MISSOURI	18.18	Edgerton	7/20/1965	25.54	Joplin	5/1943
MONTANA	*11.50	Circle	6/19-20/1921	16.79	Circle	6/1921
NEBRASKA	*13.15	York	7/8/1950	25.29	Falls City	7/1993
NEVADA	8.06	Cloverdale Ranch	5/29/1896	33.03	Mount Rose	12/1964
NEW HAMPSHIRE	*11.07	Mount Washington	10/21-22/1996	25.56	Mount Washington	2/1969
NEW JERSEY	*14.81	Tuckerton	8/19/1939	25.98	Patterson	9/1882
NEW MEXICO	11.28	Lake Maloya	5/19/1955	16.21	Portales	5/1941
NEW YORK	13.70	Brewster	9/16/1999	25.27	West Shokan	10/1955
NORTH CAROLINA	22.22	Altapass	7/1516/1916	37.40	Gorge	7/1916
NORTH DAKOTA	*8.10	Litchville	6/29/1975	14.01	Mohall	6/1944
OHIO	10.75	Lokington Dam	8/7-8/1995	17.33	Carthagenia	6/1887
	10.75	Sandusky	7/12/1966			
OKLAHOMA	*15.68	Enid	10/10-11/1973	23.95	Miami	5/1943
OREGON	11.65	Port Orford	11/19/1996	52.78	Glenora	12/1917
PENNSYLVANIA	34.50	Smethport	7/17/1942	23.66	Mount Pocono	8/1955
RHODE ISLAND	12.13	Westerly	9/16-17/1932	15.00	Rocky Hill	8/1955
SOUTH CAROLINA	17.00	Antreville	8/27/1995	31.13	Kingstree	7/1916
SOUTH DAKOTA	15.00	Nemo, Box Elder Creek	6/9/1972	18.61	Deadwood	5/1946
TENNESSEE	11.00	McMinnville	3/28/1902	23.90	McKenzie	1/1937
TEXAS	43.00	Alvin	7/25-26/1979	45.00	Alvin	7/1979
UTAH	8.40	North Ogden	9/7-8/1991	25.45	Alta	12/1983
VERMONT	9.92	Mt. Mansfield	9/17/1999	16.99	Mays Hill	10/1955
VIRGINIA	27.35	Massies Mill	8/20/1969	24.98	Glasgow	6/1995
WASHINGTON	14.26	Mt. Mitchell	11/23-24/1986	57.04	Peterson's Ranch	12/1933
WEST VIRGINIA	19.00	Rockport	7/18/1889	26.66	Rockport	7/1889
WISCONSIN	11.72	Mellen	6/24/1946	18.33	Port Washington	6/1996
WYOMING	*6.06	Cheyenne	8/1/1985	12.78	Alva	5/1962

* Records exceeded at other locations, see page 117 (Army Corps of Engineers records)

STATE PRECIPITATION RECORDS

State	Maximum Annual Precipitation in Inches, Location and Date			Least Annual Precipitation in Inches, Location and Date		
ALABAMA	114.01	Evergreen	1975	22.00	Primrose Farm	1954
ALASKA	332.29	MacLeod Harbor	1976	1.61	Barrow	1935
ARIZONA	58.92	Hawley Lake	1978	.03	Yuma	2002
ARKANSAS	98.55	Newhope	1957	19.11	Index	1936
CALIFORNIA	153.54	Monumental	1909	.00	Death Valley	1929
COLORADO	92.84	Ruby	1897	1.69	Buena Vista	1939
CONNECTICUT	78.53	Burlington Dam	1955	23.60	Baltic	1965
DELAWARE	72.75	Lewes	1948	21.38	Dover	1965
WASHINGTON D.C.	61.33	24th and M Street	1889	18.79	Washington	1826
FLORIDA	127.00	Fort Barrancas	1877	19.99	Key West	1974
GEORGIA	122.16	Flat Top	1959	17.14	Swainsboro	1954
HAWAII	704.83	Kukui, Maui	1982	.19	Grandview, Hawaii	1953
IDAHO	81.05	Roland	1933	2.09	Grandview	1947
ILLINOIS	74.58	New Burnside	1950	16.59	Keithsborg	1956
INDIANA	97.38	Marengo	1890	18.67	Brookville	1934
IOWA	74.50	Muscatine	1851	12.11	Cherokee	1958
KANSAS	68.55	Blaine	1993	4.77	Johnson	1956
KENTUCKY	79.68	Russellville	1950	14.51	Jeremiah	1968
LOUISIANA	113.74	New Orleans	1991	23.10	Shreveport	1899
MAINE	75.64	Brunswick	1945	23.06	Machias	1930
MARYLAND	72.59	Salisbury	1948	17.76	Picardy	1930
MASSACHUSETTS	76.49	New Salem	1996	21.76	Chatham Lighthouse	1965
MICHIGAN	71.19	Detroit	1855	15.64	Croswell	1936
MINNESOTA	52.36	Fairmont	1993	7.81	Angus	1936
MISSISSIPPII	104.36	Waveland	1991	25.97	Yazoo City	1936
MISSOURI	92.77	Portageville	1957	16.14	La Belle	1956
MONTANA	55.51	Summit	1953	2.97	Belfry	1960
NEBRASKA	64.52	Omaha	1869	6.30	Hull	1931
NEVADA	59.03	Mount Rose	1969	Trace	Hot Springs	1898
NEW HAMPSHIRE	130.14	Mount Washington	1969	22.31	Bethlehem	1930
NEW JERSEY	85.99	Patterson	1882	19.85	Canton	1965
NEW MEXICO	62.45	White tail	1941	1.00	Hermanas	1910
NEW YORK	90.97	Slide Mountain	1996	17.04	Rochester	1834
NORTH CAROLINA	129.60	Rosman	1964	22.69	Mt. Airy	1930
NORTH DAKOTA	37.98	Milnor	1944	4.02	Parshall	1934
OHIO	70.82	Little Mountain	1870	15.72	Cincinnati, Abbe Station	1901
OKLAHOMA	84.47	Kiamichi Tower	1957	6.53	Regnier	1956
OREGON	204.12	Laurel Mountain	1996	3.33	Warmspring Resevoir	1939
PENNSYLVANIA	81.64	Mount Pocono	1952	15.71	Breezewood	1965
RHODE ISLAND	70.21	Kingston	1983	24.08	Block island	1965
SOUTH CAROLINA	110.79	Jocasse	1994	20.73	Rock Hill	1954
SOUTH DAKOTA	48.42	Deadwood	1948	2.89	Ludlow	1936
TENNESSEE	114.88	Haw Knob	1957	25.23	Halls	1941
TEXAS	109.38	Clarksville	1873	1.64	Presidio	1956
UTAH	108.54	Alta	1983	1.34	Hanksville	1956
VERMONT	100.96	Mt. Mansfield	1996	20.99	Burlington	1881
VIRGINIA	83.70	Philpott Dam	1996	12.52	Moores Creek Dam	1941
WASHINGTON	184.56	Wynoochee Oxbow	1931	2.61	Wahluke	1930
WEST VIRGINIA	94.01	Romney	1948	9.50	Upper Tract	1930
WISCONSIN	62.07	Embarrass	1884	12.00	Plum Island	1937
WYOMING	55.46	Grassy Lake Dam	1945	1.28	Lysite	1960

STATE SNOW RECORDS

State	Maximum 24-hour Snowfall in Inches, Location and Date			Maximum Single Storm Snowfall in Inches, Location and Date		
ALABAMA	19.2	Florence	12/31-1/1/1964	19.5	Florence	12/31-1/1/1964
ALASKA	62.0	Thompson Pass	12/29/1955	175.4	Thompson Pass	12/26-31/1955
ARIZONA	38.0	Heber Ranger Station	12/14/1967	82.0	Alpine	2/23-26/1987
ARKANSAS	25.0	Corning	1/22/1918	25.0	Corning	1/22/1918
CALIFORNIA	68.0	Squaw Valley (near)	1/1/1997	194.0	Norden	4/20-23/1880
COLORADO	75.8	Silver lake	4/14-15/1921	141.0	Ruby	3/23-30/1899
CONNECTICUT	28.0	New Haven	3/12/1888	50.0	Middletown	3/11-14/1888
DELAWARE	25.0	Dover	2/18-19/1979	27.0	Middletown	3/20-21/1958
WASHINGTON D.C.	25.0		1/27-28/1922	28.0		1/27-28/1922
FLORIDA	4.0	Milton Exp. Station	3/6/1954	4.0	Milton Exp. Station	3/6/1954
GEORGIA	24.0	Mountain City	3/12-13/1993	24.0	Mountain City	3/12-13/1993
HAWAII	NA		NA	12.0	Mauna Loa, Hawaii	2/9/1922
IDAHO	38.0	Sun Valley	2/11/1959	60.0	Roland West Portal	12/25-27/1937
ILLINOIS	36.0	Astoria	2/27-28/1900	37.8	Astoria	2/27-28/1900
INDIANA	20.0	Evansville	1/14/1918	37.0	La Porte	2/14-19/1958
IOWA	21.0	Sibley	2/18/1962	30.8	Rock Rapids	2/17-21/1962
KANSAS	26.0	Fort Scott	12/28-29/1954	37.0	Olathe	3/23-24/1912
KENTUCKY	25.0	Hazard	3/14/1993	28.3	Phelps	1/5-8/1996
LOUISIANA	24.0	Rayne	2/14-15/1895	24.0	Rayne	2/14-15/1895
MAINE	40.0	Orono	12/29-30/1962	57.0	Harris Station	2/25-27/1969
MARYLAND	31.0	Clear Spring	3/29/1942	49.0	Keysers Ridge	2/15-17/2003
MASSACHUSETTS	33.0	Worcester	3/31-4/1/1997	47.0	Peru	3/2-5/1947
MICHIGAN	33.5	Baldwin	11/30/1960	61.7	Sault Ste. Marie	12/8-12/1995
MINNESOTA	36.0	Wolf Ridge	1/6-7/1994	46.5	Wolf Ridge	1/6-8/1994
MISSISSIPPI	18.0	Mt. Pleasant	12/23/1963	18.0	Mt. Pleasant	12/23/1963
MISSOURI	27.6	Neosho	3/16-17/1970	35.0	Bethany	1/22-25/1898
MONTANA	48.0	Shonkin	5/28-29/1982	77.5	Summit	1/17-20/1972
NEBRASKA	24.0	Hickman	2/11/1965	41.0	Chadron	1/2-4/1949
NEVADA	38.0	Incline	2/7-8/1985	75.0	Mt. Rose Resort	1/18-22/1969
NEW HAMPSHIRE	56.0	Randolf	11/23-24/1943	97.8	Mt. Washington	2/24-28/1969
NEW JERSEY	35.0	White House	1/6-7/1996	35.0	White House	1/6-7/1996
NEW MEXICO	36.0	Dulce	11/22/1931	51.0	Red River	2/4-6/1989
NEW YORK	68.0	Adams	1/9/1976	102.0	Oswego	1/27-31/1966
NORTH CAROLINA	31.0	Nashville	3/2/1927	60.0	Newfound Gap	4/2-5/1987
				60.0	Mt. Pisgah	5/5-8/1992
NORTH DAKOTA	24.0	Berthold Agency	2/25/1930	40.4	Grand Forks	3/2-5/1966
OHIO	22.0	Cleveland	1/31/1878	68.9	Chardon	11/9-14/1996
OKLAHOMA	23.0	Buffalo	2/21/1971	36.0	Buffalo	2/21-22/1971
OREGON	39.0	Bonneville Dam	1/9/1980	119.0	Crater Lake	3/13-16/1975
PENNSYLVANIA	40.0	Seven Springs	3/13-14/1993	50.0	Morgantown	3/19-21/1958
RHODE ISLAND	34.0	Foster	2/8-9/1945	38.0	Woomsocket	2/6-7/1978
SOUTH CAROLINA	24.0	Rimini	2/9-10/1973	28.9	Caesar's Head	2/15-16/1929
SOUTH DAKOTA	38.9	Lead	10/26/1996	114.6	Lead	2/25-3/2/1998
TENNESSEE	22.0	Morristown	3/9/1960	56.0	Mt. LeConte	3/12-14/1993
TEXAS	24.0	Plainview	2/3-4/1956	61.0	Vega	2/1-8/1956
UTAH	55.5	Alta	1/5-6/1994	105.0	Alta	1/24-30/1965
VERMONT	37.0	Peru	3/14/1984	50.0	Readsboro	3/2-6/1947
VIRGINIA	33.5	Luray	3/2-3/1994	48.0	Big Meadows	1/6-7/1996
WASHINGTON	65.0	Crystal Mountain	1/23-24/1994	129.0	Laconia	2/24-26/1910
WEST VIRGINIA	35.0	Flat Top	1/27-28/1998	63.2	Coburn Creek	11/23-30/1950
WISCONSIN	26.0	Neillsville	12/27/1904	39.0	Hurley	11/1-3/1989
WYOMING	41.0	Glenrock	4/19/1973	58.0	Glenrock	4/18-20/1973

STATE SNOW RECORDS

State	Maximum Monthly Snowfall in Inches, Location and Date			Maximum Seasonal Snowfall in Inches, Location and Date		
ALABAMA	24.0	Valley Head	1/1940	27.9	Florence	1963-1964
ALASKA	346.1	Thompson Pass	2/1964	974.4	Thompson Pass	1952-1953
ARIZONA	126.0	Flagstaff	1/1895	400.9	Sunrise Mountain	1972-1973
ARKANSAS	48.0	Calico Rock	1/1918	61.0	Hardy	1917-1918
CALIFORNIA	390.0	Tamarack	1/1911	884.0	Tamarack	1906-1907
COLORADO	249.0	Ruby	3/1899	838.0	Wolf Creek Pass	1978-1979
CONNECTICUT	73.6	Norfolk	3/1956	177.4	Norfolk	1955-1956
DELAWARE	36.5	Dover	2/1979	59.4	Wilmington	1995-1996
WASHINGTON D.C.	35.2		2/1899	61.9		1995-1996
FLORIDA	4.0	Milton Exp. Station	3/1954	4.0	Milton Exp. Station	1953-1954
GEORGIA	26.5	Diamond	2/1895	39.0	Diamond	1894-1895
HAWAII	N/A			N/A		
IDAHO	143.8	Burke	1/1954	441.8	Roland West Portal	1949-1950
ILLINOIS	52.8	McHenry Lock and Dam	1/1979	103.3	Antioch	1978-1979
INDIANA	86.1	South Bend	1/1978	172.0	South Bend	1977-1978
IOWA	43.7	Lake Park	12/1968	90.4	Northwood	1908-1909
KANSAS	55.9	Olathe	3/1912	103.6	McDonald	1983-1984
KENTUCKY	46.5	Benham	3/1960	108.2	Benham	1959-1960
LOUISIANA	24.0	Rayne	2/1895	24.0	Rayne	1894-1895
MAINE	89.0	Harris Station	2/1969	238.5	Long Falls Dam	1968-1969
MARYLAND	81.5	Eagle Rock	2/2003	174.9	Deer Park	1901-1902
MASSACHUSETTS	78.0	Monroe	2/1893	162.0	Monroe	1892-1893
MICHIGAN	129.5	Copper Harbor	1/1982	391.9	Delaware	1978-1979
MINNESOTA	66.4	Collegeville	3/1965	147.5	Pigeon River Bridge	1936-1937
MISSISSIPPI	23.0	Cleveland	1/1966	25.2	Senatobia	1967-1968
MISSOURI	47.5	Poplar Bluff	1/1918	74.0	Oregon	1888-1889
MONTANA	131.1	Summit	1/1972	418.1	Cooke City	1977-1978
NEBRASKA	59.6	Chadron	1/1949	112.0	Harrison	1972-1973
NEVADA	139.0	Daggett Pass	1/1969	412.0	Daggett Pass	1968-1969
NEW HAMPSHIRE	172.8	Mt. Washington	2/1969	566.4	Mount Washington	1968-1969
NEW JERSEY	50.1	Freehold	12/1880	122.0	High Point State Park	1995-1996
NEW MEXICO	144.0	Anchor Mine	3/1912	483.0	Anchor Mine	1911-1912
NEW YORK	192.0	Bennetts Ridge	1/1978	466.9	Hooker	1976-1977
NORTH CAROLINA	60.0	Mt. Pisgah	5/1992	103.4	Boone	1959-1960
NORTH DAKOTA	46.4	Grand Forks	12/1996	117.0	Fargo	1996-1997
OHIO	94.8	Chardon	11/1996	161.5	Chardon	1959-1960
OKLAHOMA	46.0	Kenton	2/1903	87.3	Beaver	1911-1912
OREGON	313.0	Crater lake	1/1950	903.0	Crater lake	1949-1950
PENNSYLVANIA	86.0	Blue Knob	12/1890	225.0	Blue Knob	1890-1891
RHODE ISLAND	62.0	Foster	3/1956	129.1	North Foster	1995-1996
SOUTH CAROLINA	33.9	Caesar's Head	2/1969	60.3	Caesar's Head	1968-1969
SOUTH DAKOTA	94.0	Dumont	3/1950	324.0	Lead	1996-1997
TENNESSEE	56.0	Mt. LeConte	3/1993	75.5	Mountain City	1959-1960
TEXAS	61.0	Vega	2/1956	65.0	Romero	1923-1924
UTAH	244.5	Alta	12/1983	846.8	Alta	1982-1983
VERMONT	90.3	Jay Peak	12/2000	*318.6	Mt. Mansfield	1970-1971
VIRGINIA	54.0	Warrentown	2/1899	124.2	Wise	1995-1996
WASHINGTON	363.0	Paradise Ranger Station	1/1925	1140.0	Mt. Baker	1998-1999
WEST VIRGINIA	104.0	Terra Alta	1/1977	301.4	Krumbrabow State Forest	1959-1960
WISCONSIN	103.5	Hurley	1/1997	277.0	Hurley	1996-1997
WYOMING	188.5	Bechler River	1/1933	491.6	Bechler River	1932-1933

*571.0 claimed as seasonal total at Jay Peak 2000-2001

CITY TEMPERATURE RECORDS

State, City and Beginning of Record	Highest Temperature and Date		Lowest Temperature and Date		Coldest Month and Date		Coldest Winter and Date		Warmest Month and Date		Warmest Summer and Date	
ALABAMA	112	9/5/1925	-27	1/30/1966								
Birmingham 1894-	107	7/29/1930	-10	2/13/1899	30.6	1/1940			84.8	8/1995		
Huntsville 1930-	111	7/29/1930	-16	1/30/1966	26.0	1/1940			83.9	7/1930		
Mobile 1871-	105	8/29/2000	-1	2/13/1899	39.6	1/1940			86.0	8/1951		
Montgomery 1871-	107	7/7/1881	-5	2/13/1899	36.0	1/1977			85.9	8/1954		
ALASKA	100	6/27/1915	-80	1/23/1971								
Anchorage 1916-	92	6/26/1931	-38	2/3/1947	-9.2	12/1917			62.6	7/1977		
Barrow 1921-	79	7/13/1993	-56	2/3/1924	-33.0	2/1983			46.8	8/1989		
Fairbanks 1904-	99	7/28/1919	-66	1/14/1934	-36.6	1/1906			68.4	7/1975		
Juneau 1899-	90	7/29/1975	-22	1/12/1972	6.8	1/1989			63.2	7/1907		
Nome 1907-	86	7/29/1977	-54	1/27/1989	-17.3	2/1990			56.4	8/1977		
St. Paul Island 1916-	66	8/25/1987	-26	1/1919	5.4	1/1919			51.1	7/1977		
ARIZONA	128	6/29/1994	-41	1/7/1971								
Flagstaff 1898-	97	7/5/1973	-30	1/22/1937	12.6	1/1937			70.0	7/2002		
Lake Havasu City* 1894-	128	6/29/1994	9	12/1911	38.9	1/1937			102.2	7/1996		
Phoenix 1878-	122	6/28/1990	16	1/7/1913	43.2	1/1937	48.5	1948-1949	97.6	7/2003	94.8	2002
Tucson 1878-	117	6/26/1990	6	1/7/1913	41.2	1/1937			90.4	7/1994		
Winslow 1898-	109	7/13/1971	-19	12/14/1898	12.9	1/1968			82.5	7/1971		
Yuma 1875-	124	7/28/1995	22	1/24/1937	44.9	1/1937	50.5	1948-1949	97.1	8/1995	94.5	2002
*early record Parker												
ARKANSAS	120	8/10/1936	-29	2/13/1905								
Fort Smith 1882-	113	8/10/1936	-15	2/12/1899	25.6	1/1979			89.2	7/1934		
Little Rock 1879-	112	7/20/1986	-13	2/12/1899	27.5	1/1979	37.3	1917-1918	89.4	8/2000	85.2	1954
Texarkana 1883-	117	8/10/1936	-9	2/13/1905								
CALIFORNIA	134	7/10/1913	-45	1/20/1937								
Bakersfield 1889-	118	7/28/1908	13	12/24/1905	40.4	1/1949			88.3	7/1970		
Bishop 1895-	112	9/11/1995	-15	1/30/1916	23.8	1/1949			80.1	7/2003		
Blue Canyon 1943-	104	8/10/1981	-14	12/9/1972	26.3	1/1949			72.8	7/1959		
Burbank 1931-	113	9/12/1971	21	1/9/1937	45.1	1/1949			81.4	9/1984		
Death Valley 1911-	134	7/10/1913	15	1/7/1913	43.3	12/1990			107.2	7/1917		
Eureka 1885-	87	10/26/1993	20	1/14/1888	40.3	1/1949	43.3	1948-1949	62.3	9/1979	58.3	1947
Fort Bragg 1948-	94	10/5/1985	18	12/21/1990	40.1	12/1990			61.5	7/1992		
Fresno 1887-	115	7/8/1905	17	1/6/1913	39.6	1/1949			87.0	7/1984		
Los Angeles 1877-	112	6/26/1990	23	1/9/1937	46.8	1/1949	50.9	1948-1949	81.3	9/1984	76.7	1967
Monterey 1894-	104	10/5/1987	20	12/22/1990	46.4	1/1950			67.7	9/1984		
Mt. Shasta City* 1888-	108	7/8/1905	-13	12/22/1990	20.8	1/1937			76.8	8/1898		
Palm Springs 1908-	123	8/1/1993	18	1/7/1913	43.9	1/1937			96.2	8/1998		
Red Bluff 1877-	121	8/7/1981	17	1/9/1937	35.5	1/1937			87.3	7/1988		
Sacramento 1877-	115	6/15/1961	17	12/11/1932	38.4	1/1949	43.1	1948-1949	81.6	7/2003	76.4	1961 & 1988
San Diego 1849-	111	9/15/1963	25	1/7/1913	47.8	1/1949	51.4	1948-1949	78.9	9/1984	74.6	1983
San Francisco 1850-	103	6/14/2000	27	12/11/1932	43.6	1/1937	46.9	1948-1949	69.4	9/1984	63.9	1995
San Jose 1873-	109	6/14/2000	17	1/10/1920	40.4	1/1882			76.1	6/1874		
Santa Barbara 1864-	115	6/17/1917	20	12/22/1990	43.2	1/1949			74.1	9/1984		
Santa Maria 1885-	109	6/21/1929	20	12/7/1978	43.3	1/1949			71.9	9/1984		
*formerly Sisson												
COLORADO	118	7/11/1888	-61	2/1/1985								
Alamosa 1933-	96	7/5/1989	-50	1/28/1948	1.4	1/1992	7.0	1991-1992	68.2	7/2003	63.6	1980
Aspen 1948-	93	6/23/1954	-33	1/12/1963	15.2	1/1979			67.9	7/2003		
Co Springs 1871-	101	6/7/1874	-32	1/20/1883	16.5	2/1933			75.8	7/2003		
Denver 1871-	105	8/8/1878	-30	2/8/1936	16.2	1/1937	24.6	1898-1899	78.5	7/1966	74.3	1954
Durango 1888-	102	5/5/1989	-30	1/13/1963	14.9	1/1937			73.3	7/2003		
Grand Junction 1892-	105	7/13/2003	-23	1/13/1913	11.5	1/1973			84.1	7/2003	79.5	1994
Lamar 1889-	111	7/13/1934	-30	1/30/1949	15.8	1/1940			85.2	7/1934		
Pueblo 1888-	109	7/13/2003	-31	2/1/1951	16.1	1/1979			81.2	7/2003		
CONNECTICUT	106	7/15/1995	-37	2/16/1943								
Hartford 1885-	102	7/3/1966	-27	1/5/1835	13.9	1/1857			77.1	7/1994		
New Haven 1780-	101	7/22/1926	-23	1/5/1835	16.7	1/1857	21.9	1835-1836	77.4	7/1876	74.2	1876

CITY TEMPERATURE RECORDS

State, City and Beginning of Record		Highest Temperature and Date		Lowest Temperature and Date		Coldest Month and Date		Coldest Winter and Date		Warmest Month and Date		Warmest Summer and Date	
DELAWARE		110	7/21/1930	-17	1/17/1893								
Wilmington	1894-	107	8/7/1918	-15	2/9/1934	20.2	2/1934			81.2	7/1955	76.7	1993
DC/Washington		106	7/20/1930	-16	1/5/1835	21.3	1/1856	28.3	1855-1856	82.9	7/1999	80.0	1943
Dulles	1830-			-18	1/22/1984								
FLORIDA		109	6/29/1931	-2	2/13/1899								
Apalachicola	1904-	104	6/25/1914	9	1/21/1985	43.6	1/1940			85.6	7/1998		
Daytona Beach	1923-	102	7/15/1981	15	1/21/1985	48.6	1/1940			84.5	6/1998		
Everglades City	1918-	100	5/11/1991	23	2/5/1996	55.7	1/1981			84.8	7/1942		
Fort Meyers	1851-	103	6/1/1981	24	12/29/1894	55.5	2/1958			85.9	6/1981		
Gainesville	1894-	104	6/27/1952	6	2/13/1899	46.0	1/1940			84.0	6/1998		
Jacksonville	1841-	105	7/21/1942	7	1/21/1985	44.0	1/1977	48.9	1976-1977	85.2	7/1875	83.6	1954
Key West	1830-	100	8/1886	41	1/13/1981	61.2	1/1852	65.8	1957-1958	86.8	7/1967	85.5	1951
Lakeland	1916-	105	6/4/1985	20	12/13/1962	51.9	1/1977			85.1	8/1987		
Melbourne	1948-	102	7/14/1980	17	1/19/1977	52.2	1/1981			85.0	6/1998		
Miami	1839-	100	7/21/1942	26	12/30/1934	59.2	1/1940	61.4	1957-1958	85.4	6/1998	84.6	1987
Orlando	1891-	103	9/7/1921	19	1/21/1985	50.4	1/1940			88.0	8/1943		
Pensacola	1879-	106	7/14/1980	5	1/21/1985	40.8	1/1940	47.5	1976-1977	85.7	8/1951	83.4	1980
Tallahassee	1803-	104	6/20/1933	-2	2/13/1899	41.8	1/1940			85.3	6/1998		
Tampa	1890-	99	6/5/1985	18	12/13/1962	50.4	1/1981	55.2	1957-1958	85.6	6/1998	84.2	1998
W. Palm Beach	1931-	101	7/21/1942	27	1/20/1977	58.2	2/1958			85.1	7/1942		
GEORGIA		112	8/20/1983	-17	1/27/1940								
Albany	1892	107	7/15/1980	-2	2/13/1899	38.4	1/1940			85.4	7/1932		
Athens	1891-	108	7/12/1930	-10	2/8/1835	31.2	1/1977			85.3	7/1993		
Atlanta	1878-	105	7/17/1980	-9	2/13/1899	29.3	1/1977	37.0	1976-1977	85.4	7/1993	82.7	1980
Augusta	1870-	108	8/21/1983	-4	2/8/1835	35.5	1/1977			86.3	7/1993		
Columbus	1892-	106	9/5/1925	-3	2/13/1899	36.2	1/1940			85.8	7/1993		
Macon	1894-	108	7/17/1980	-6	1/21/1985	35.2	1/1940			86.6	7/1986		
Rome	1856-	109	7/19/1913	-9	1/21/1985	28.9	1/1977			84.3	7/1954		
Savannah	1870-	105	7/20/1986	0	2/8/1835	39.9	1/1977			86.7	7/1993		
HAWAII		100	4/27/1931	12	5/17/1979								
Honolulu	1905-	95	9/1994	49	2/9/1981	67.2	1/1905			84.3	8/1994		
Hilo	1905-	94	5/20/1966	51	5/1910	68.2	3/1942			78.6	8/1994		
Kahului	1905-	97	8/31/1994	48	1/20/1969	67.9	2/1965			81.5	8/1982		
Lihue	1905-	91	9/1936	46	1/1946	65.8	1/1969			81.2	8/1980		
IDAHO		118	7/28/1934	-60	1/18/1943								
Boise	1877-	111	8/4/1961	-28	1/1888	10.3	1/1949	22.1	1948-1949	80.7	7/1960	75.7	1961
Lewiston	1889-	117	7/27/1939	-23	12/13/1919	14.3	1/1949			80.0	7/1985		
Pocatello	1899-	105	7/21/1931	-33	2/1/1985	4.8	1/1949			76.0	7/1936		
Salmon	1906-	106	7/21/1936	-37	1/23/1930	-2.4	1/1949			73.4	7/1985		
Sandpoint	1910-	104	7/20/1923	-37	12/30/1968	7.6	1/1937			71.0	7/1998		
ILLINOIS		117	7/14/1954	-36	1/5/1999								
Cairo	1871-	106	8/9/1930	-16	1/12/1918	21.0	1/1918			85.8	7/1980		
Chicago	1830-	109	7/23/1934	-27	1/20/1985	10.1	1/1977	17.3	1855-1856	81.3	7/1955	76.4	1955
Moline	1872-	111	7/14/1936	-28	2/3/1986	6.3	1/1979	16.1	1977-1978	81.0	8/1947	76.4	1955
Peoria	1855-	113	7/15/1936	-27	1/5/1884	8.6	1/1977			84.3	7/1936		
Quincy	1911-	114	7/15/1936	-22	12/22/1989	11.2	1/1977			88.4	7/1934		
Rockford	1889-	112	7/15/1936	-27	1/10/1982	5.9	1/1977			80.6	8/1947		
Springfield	1879-	112	7/14/1954	-24	2/13/1905	10.3	1/1977			86.2	7/1936		
Urbana	1902-	109	7/14/1954	-25	1/19/1994	10.7	1/1977			83.1	7/1936		
INDIANA		116	7/14/1936	-36	1/19/1994								
Evansville	1896-	111	7/28/1930	-23	2/2/1951	14.8	1/1977			85.1	7/1936		
Fort Wayne	1898-	106	7/14/1936	-24	1/12/1918	9.2	1/1977	19.9	1976-1977	79.6	7/1921	76.4	1983
Indianapolis	1871-	107	7/25/1934	-27	1/19/1994	10.3	1/1977	21.0	1976-1977	82.8	7/1936	78.6	1874
South Bend	1894-	109	7/24/1934	-22	1/20/1943	12.3	1/1977	20.6	1976-1977	80.1	7/1921	76.3	1983

CITY TEMPERATURE RECORDS

State, City and Beginning of Record		Highest Temperature and Date		Lowest Temperature and Date		Coldest Month and Date		Coldest Winter and Date		Warmest Month and Date		Warmest Summer and Date	
INDIANA Con't													
Terre Haute	1891-	110	7/14/1936	-24	1/17/1977	9.9	1/1977			85.4	7/1936		
IOWA		118	7/20/1934	-47	2/3/1996								
Burlington	1898-	111	7/14/1936	-27	2/13/1905	9.2	1/1977			86.6	7/1936		
Des Moines	1865-	110	7/25/1936	-30	1/5/1884	6.2	1/1912	14.4	1935-1936	85.8	7/1936	79.3	1934
Dubuque	1873-	110	7/14/1936	-32	1/7/1887	3.4	1/1977			83.3	7/1936		
Mason City	1897-	107	7/14/1936	-35	2/9/1899	-1.6	1/1875			80.2	7/1936		
Sioux City	1889-	111	7/11/1939	-35	1/12/1912	1.9	2/1936			87.2	7/1936		
Waterloo	1895-	112	7/14/1936	-34	3/1/1962	-0.1	1/1977			84.0	7/1936		
KANSAS		121	7/24/1936	-40	2/13/1905								
Concordia	1885-	116	8/12/1936	-26	12/22/1989	11.1	1/1886			88.4	7/1934		
Dodge City	1874-	110	6/29/1998	-26	2/12/1899	13.0	1/1875	22.0	1884-1885	87.3	7/1934	83.4	1934
Goodland	1906-	111	7/25/1940	-27	12/22/1989	14.6	12/1983			83.0	7/1934		
Topeka	1886-	114	7/24/1936	-26	12/22/1989	11.8	1/1979			88.4	7/1936		
Wichita	1888-	114	8/12/1936	-22	2/11/1899	16.2	1/1940			90.5	7/1980		
KENTUCKY		114	7/28/1930	-37	1/19/1994								
Bowling Green	1887-	113	7/28/1930	-21	1/23/1963	20.0	1/1977			84.7	7/1901		
Lexington	1872-	108	7/10/1936	-21	1/24/1963	17.8	1/1977			81.6	8/1936		
Louisville	1871-	107	7/14/1936	-22	1/18/1994	18.6	1/1977	27.8	1917-1918	84.2	7/1901	81.0	1936
LOUISIANA		114	8/10/1936	-16	2/13/1899								
Baton Rouge	1887-	110	8/18/1909	2	2/13/1899	40.1	1/1940			85.7	7/1960		
Lake Charles	1887-	106	6/27/1930	3	2/13/1899	39.8	1/1940			86.0	8/1951		
New Orleans	1871-	103	8/30/2000	7	2/13/1899	43.0	1/1940	47.8	1977-1978	87.1	8/1951	84.7	1951
Shreveport	1871-	110	8/18/1909	-5	2/12/1899	34.9	1/1978			88.5	7/1998		
MAINE		105	7/10/1911	-48	1/19/1925								
Bangor	1929-	104	8/19/1935	-40	1/5/1835								
Caribou	1939-	96	5/22/1977	-41	2/1/1955	-0.7	1/1994			69.7	7/1970		
Eastport	1873-	98	8/5/1975	-23	12/30/1933	11.5	1/1920	16.5	1904-1905	65.5	7/1977	62.4	1949
Portland	1871-	103	7/4/1911	-39	2/16/1943	12.2	1/1971			72.4	7/1994		
MARYLAND		109	7/10/1936	-40	1/13/1912								
Baltimore	1871-	107	7/10/1936	-10	1/4/1835	22.9	1/1977			81.5	7/1995		
Frederick	1898-	109	7/10/1936	-21	1/13/1912								
Salisbury	1906-	106	7/21/1929	-9	1/21/1918								
MASSACHUSETTS		107	8/2/1975	-40	1/22/1984								
Boston	1818-	104	7/4/1911	-18	2/9/1934	16.8	1/1857	23.9	1917-1918	78.0	7/1983	74.1	1983
Nantucket	1886-	100	8/2/1975	-6	2/5/1918	22.6	2/1934			71.8	7/1937		
Pittsfield	1894-	101	7/22/1926	-32	1/5/1835	10.2	2/1934						
Worcester	1881-	102	7/4/1911	-24	2/16/1943	14.9	1/1970			74.3	7/1952		
MICHIGAN		112	7/13/1936	-51	2/9/1934								
Alpena	1872-	106	7/13/1936	-37	2/17/1979	3.5	2/1875			73.6	7/1955		
Detroit	1840-	105	7/24/1934	-24	12/22/1872	12.2	2/1875	18.7	1903-1904	80.6	7/1850	77.5	1850
Escanaba	1871-	100	7/13/1936	-32	2/17/1979	1.1	1/1912			71.8	7/1955		
Flint	1889-	108	7/13/1936	-28	2/17/1916	9.4	1/1918			78.0	7/1921		
Grand Rapids	1887-	108	7/13/1936	-24	2/14/1899	12.7	1/1977			79.1	8/1947		
Houghton	1899-	104	7/13/1936	-31	1/28/1915	3.5	2/1905			70.9	7/1988		
Houghton Lake	1907-	107	7/13/1936	-48	2/1/1918	8.7	1/1977			73.3	7/1955		
Lansing	1887-	102	8/6/1918	-29	1/4/1981	11.1	1/1977			75.4	8/1955		
Marquette	1871-	108	7/15/1901	-34	2/17/1979	1.3	1/1912			72.2	7/1921		
Muskegon	1895-	99	8/3/1964	-30	2/11/1899	11.3	1/1912			76.2	7/1901		
Sault Ste. Marie	1888-	98	8/6/1947	-37	2/8/1934	0.2	1/1912	8.4	1903-1904	74.1	7/1921	67.2	1955
MINNESOTA		114	7/6/1936	-60	2/2/1996								
Bemidji	1897-	107	7/11/1936	-50	1/30/1950	-8.7	1/1966			75.5	7/1936		

CITY TEMPERATURE RECORDS

State,City and Beginning of Record	Highest Temperature and Date		Lowest Temperature and Date		Coldest Month and Date		Coldest Winter and Date		Warmest Month and Date		Warmest Summer and Date	
MINNESOTA Con't												
Duluth 1870-	106	7/31/1936	-42	1/2/1885	-7.2	1/1912			72.2	7/1980		
Int'l Falls 1896-	103	7/22/1923	-49	1/4/1896	-10.5	1/1982			70.2	7/1975		
Minn/St. Paul 1859-	108	7/14/1936	-41	1/13/1888	-2.7	1/1912	4.5	1874-1875	81.4	7/1936	75.2	1933
Rochester 1886-	108	7/14/1936	-42	1/7/1887	-1.8	1/1977			77.6	7/1936		
St. Cloud 1894-	107	7/13/1936	-43	1/9/1977	-5.6	1/1912			78.9	7/1936		
MISSISSIPPI	115	7/29/1930	-19	1/30/1966								
Jackson 1894-	107	7/29/1930	-5	1/27/1940	31.9	1/1940			86.0	8/1954		
Meridian 1889-	107	7/14/1980	-7	1/27/1940	33.6	1/1940			84.9	8/1951		
Vicksburg 1871-	104	9/6/1925	-1	2/12/1899	34.9	1/1940			84.6	7/1878		
MISSOURI	118	7/14/1954	-40	2/13/1905								
Columbia 1899-	113	7/12/1954	-26	2/12/1899	13.6	1/1977	24.3	1977-1978	87.0	7/1980	82.7	1936
Kansas City 1888-	113	8/14/1936	-23	12/23/1989	12.5	1/1979	21.4	1978-1979	90.2	7/1980	84.9	1934
St. Joseph 1907-	110	8/14/1936	-25	1/12/1974	10.7	1/1979			87.3	7/1934		
St. Louis 1836-	115	7/14/1954	23	1/1/1864	15.1	1/1977	23.8	1977-1978	87.4	7/1936	83.4	1936
Springfield 1887-	113	7/14/1954	-29	2/12/1899	16.0	1/1940			85.4	7/1934		
MONTANA	117	7/5/1937	-70	1/20/1954								
Billings 1894-	112	7/31/1901	-49	2/11/1899	-0.2	1/1916			81.5	7/1936		
Butte 1894-	100	7/22/1931	-52	2/9/1933	-5.5	1/1937			69.2	7/1919		
Glasgow 1893-	113	7/31/1900	-59	2/15/1936	-15.8	2/1936			81.2	7/1936		
Great Falls 1891-	107	7/10/1973	-49	2/15/1936	-5.2	2/1936			76.4	7/1936		
Havre 1879-	111	8/5/1961	-57	1/27/1916	-13.3	1/1916			79.9	7/1936		
Helena 1880-	105	7/12/2002	-42	1/25/1957	2.4	1/1937	9.3	1948-1949	76.3	7/2003	70.2	1961
Kalispell 1889-	105	8/4/1961	-38	1/31/1950	-0.2	1/1979			71.2	8/1971		
Miles City 1891-	112	7/1886	-49	2/11/1899	-6.4	1/1916			84.2	7/1936		
Missoula 1870-	105	8/4/1961	-42	1/1888	4.5	1/1949			74.8	7/1985		
NEBRASKA	118	7/24/1936	-47	2/12/1899								
Grand Isle 1891-	117	7/12/1936	34	2/12/1899	6.5	2/1936			87.2	7/1934		
Lincoln 1870-	115	7/25/1936	-33	1/12/1974	7.2	1/1979			87.9	7/1936		
Norfolk 1891-	116	7/17/1936	-39	1/12/1912	2.6	2/1936			86.4	7/1936		
North Platte 1874-	112	7/11/1954	-35	2/12/1899	6.0	1/1979	12.6	1978-1979	83.0	7/1934	78.2	1936
Omaha 1870-	114	7/25/1936	-32	1/5/1884	6.5	2/1936	14.7	1935-1936	86.7	7/1936	81.0	1934
Scottsbluff 1891-	110	7/11/1939	-45	2/12/1899	9.2	1/1949			80.9	7/1936		
Valentine 1885-	114	7/2/1990	-39	12/22/1989	1.8	2/1936			84.0	7/1936		
NEVADA	125	6/29/1994	-50	1/8/1937								
Austin 1889-	105	7/4/1922	-36	12/4/1940	14.0	1/1949			76.1	7/1931		
Elko 1888-	108	7/24/1889	-43	1/21/1937	4.7	1/1949			75.9	7/1985		
Ely 1888-	101	7/12/2002	-36	1/1888	5.8	1/1949			72.8	7/1908		
Las Vegas 1900-	118	7/26/1931	8	1/13/1963	31.2	1/1937			94.8	7/2003	92.2	1994
Reno 1888-	108	7/11/2002	-19	1/8/1890	14.0	1/1949			79.2	7/2003		
Tonopah 1906-	104	7/18/1960	-15	1/24/1962	11.7	1/1937			78.6	7/1931		
Winnemucca 1877-	109	7/11/2002	-37	12/22/1990	9.0	1/1949			77.2	7/1931		
NEW HAMPSHIRE	106	7/4/1911	-46	1/28/1925								
Concord 1871-	102	7/5/1966	-37	2/16/1943	10.6	1/1977			74.1	7/1955		
Mt. Washington 1871-	72	6/26/2003	-50	1/22/1885	-5.4	12/1989			53.6	7/1955		
NEW JERSEY	110	7/10/1936	-34	1/5/1904								
Atlantic City 1873-	106	6/28/1969	-11	2/12/1979	19.7	1/1977			78.7	7/1983		
Newark 1893-	105	8/9/2001	-14	2/9/1934	19.9	2/1934			82.6	7/1993		
Trenton 1865- Years	106	7/9/1936	-14	2/9/1934	18.5	2/1934			80.5	7/1955		
NEW MEXICO	122	6/27/1994	-50	2/1/1951								
Albuquerque 1893-	107	6/26/1994	-17	1/7/1971	25.2	12/1911	29.9	1912-1913	83.8	7/2003	80.4	1994
Clayton 1897-	105	7/31/1934	-21	1/4/1959	23.3	1/1979			79.2	7/1934		
Gallup 1921-	101	7/21/1979	-34	12/23/1990	20.3	12/1992			74.2	7/2003		

CITY TEMPERATURE RECORDS

State, City and Beginning of Record		Highest Temperature and Date		Lowest Temperature and Date		Coldest Month and Date		Coldest Winter and Date		Warmest Month and Date		Warmest Summer and Date	
NEW MEXICO Con't													
Roswell	1894-	114	6/27/1994	-29	2/13/1905	28.4	1/1949			85.4	7/1980		
Santa Fe	1849-	100	6/26/1994	-29	2/1/1951	21.0	1/1861			77.1	7/2003		
Silver City	1911-	106	6/26/1994	-13	1/11/1962	28.7	1/1949			77.4	7/1951		
NEW YORK		108	7/22/1926	-52	2/18/1979								
Albany	1820-	104	7/4/1911	-32	1/5/1835	9.7	1/1970	17.3	1967-1968	79.7	7/1868	76.4	1872
Binghampton	1891-	103	7/9/1936	-28	1/17/1893	11.4	2/1934			74.0	8/1937	70.3	1995
Buffalo	1870-	99	8/27/1948	-21	2/9/1934	11.6	2/1934	19.3	1917-1918	76.2	7/1921	72.7	1949
New York City	1821-	107	7/3/1966	-15	2/9/1934	19.6	1/1857	25.7	1917-1918	81.4	7/1999	77.3	1966
Rochester	1870-	102	7/9/1936	-22	2/9/1934	12.6	2/1934			75.8	7/1952		
Syracuse	1903-	102	7/9/1936	-26	2/18/1979	12.0	2/1934			76.7	7/1955		
NORTH CAROLINA		110	8/21/1983	-34	1/21/1985								
Asheville	1857-	101	7/28/1952	-16	1/21/1985	24.8	1/1977			78.0	7/1993		
Cape Hatteras	1874-	97	6/27/1952	6	1/21/1985	35.1	1/1893			81.9	7/1993		
Charlotte	1878-	104	9/6/1954	-5	1/21/1985	30.1	1/1977			85.5	7/1993		
Greensboro	1894-	105	8/31/1932	-8	1/21/1985	26.7	1/1977			82.0	7/1993		
Raleigh	1887-	105	8/18/1988	-9	1/21/1985	26.6	1/1977	35.9	1976-1977	82.5	7/1993	79.7	1900
Wilmington	1871-	104	6/27/1952	0	12/25/1989	35.7	1/1977			84.1	7/1993		
NORTH DAKOTA		121	7/6/1936	-60	2/15/1936								
Bismarck	1874-	114	7/6/1936	-45	2/16/1936	-11.4	2/1936	-0.6	1886-1887	83.4	7/1936	75.1	1936
Devils Lake	1870-	112	7/6/1936	-46	2/15/1936	-15.0	2/1936			79.0	7/1936		
Dickinson	1892-	114	7/6/1936	-47	2/16/1936	-12.2	2/1936			81.3	7/1936		
Fargo	1881-	114	7/6/1936	-48	1/8/1887	-9.8	2/1936			80.2	7/1936		
Grand Forks	1891-	109	7/12/1936	-44	1/30/2004	-13.1	2/1936			79.1	7/1936		
Williston	1879-	110	7/5/1936	-50	12/23/1983	-13.6	2/1936			80.8	7/1936		
OHIO		113	7/21/1934	-39	2/10/1899								
Akron	1887-	104	8/6/1918	-25	1/19/1994	11.4	1/1977			78.9	7/1999		
Cincinnati	1870-	109	7/21/1934	-25	1/19/1977	12.0	1/1977	21.8	1977-1978	82.5	7/1901	78.8	1934
Cleveland	1870-	104	6/25/1988	-20	1/19/1994	11.0	1/1977			79.1	7/1955		
Columbus	1878-	106	7/14/1936	-24	1/9/1856	11.4	1/1977			80.2	7/1999		
Dayton	1882-	108	7/22/1901	-28	2/13/1899	11.6	1/1977			81.5	7/1934		
Toledo	1870-	105	7/14/1936	-20	1/21/1984	9.6	1/1977			78.5	8/1995		
Youngstown	1885-	103	7/9/1936	-23	2/10/1899	10.3	1/1977			77.4	7/1934		
OKLAHOMA		120	6/27/1994	-27	1/18/1930								
Oklahoma City	1890-	113	8/11/1936	-17	2/12/1899	23.2	1/1930	31.2	1978-1979	88.7	8/1936	85.9	1980
Tulsa	1905-	115	8/10/1936	-16	1/22/1930	21.7	1/1918			91.7	7/1980		
OREGON		119	8/10/1898	-54	2/10/1933								
Astoria	1885-	101	7/1/1942	6	12/8/1990	31.1	1/1949			63.8	8/1997		
Bandon	1878-	100	9/21/1990	8	12/21/1990	37.2	1/1949			62.3	8/1997		
Bend	1902-	105	7/12/2002	-26	1/31/1950	15.2	1/1930			69.1	7/1998		
Burns	1892-	107	7/12/2002	-32	2/10/1933	10.0	1/1949			73.2	7/1930		
Eugene	1891-	108	8/9/1981	-12	12/8/1972	28.2	1/1949			72.0	8/1967		
Lakeview	1884-	108	8/10/1898	-24	1/16/1888	12.4	1/1949			71.6	7/1931		
Medford	1911-	115	7/20/1946	-10	12/13/1919	26.9	1/1937			77.9	8/1967		
Pendleton	1889-	119	8/10/1898	-28	1/27/1957	11.5	1/1930			79.7	8/1961		
Portland	1871-	107	8/9/1981	-3	2/2/1950	26.7	1/1950	34.9	1948-1949	75.0	8/1967	71.0	1967
Roseburg	1877-	109	7/20/1946	-6	1/16/1888	32.2	1/1949			74.6	7/1996		
Salem	1892-	108	8/9/1981	-12	12/8/1972	27.4	1/1930			72.0	7/1941		
PENNSYLVANIA		111	7/10/1936	-42	1/5/1904								
Allentown	1923-	105	7/3/1966	-15	1/21/1994	16.4	2/1934			79.0	7/1955		
Erie	1873-	100	6/25/1988	-18	1/19/1994	12.5	1/1977			77.6	7/1921		
Harrisburg	1888-	107	7/3/1966	-22	1/21/1994	19.1	1/1918			81.9	7/1999		
Philadelphia	1870-	106	8/7/1918	-11	2/9/1934	20.0	1/1977	28.0	1976-1977	84.1	7/1999	78.9	1995
Pittsburgh	1870-	103	7/16/1988	-24	1/19/1994	11.4	1/1977	20.7	1976-1977	80.3	7/1887	75.8	1900

CITY TEMPERATURE RECORDS

State, City and Beginning of Record	Highest Temperature and Date		Lowest Temperature and Date		Coldest Month and Date		Coldest Winter and Date		Warmest Month and Date		Warmest Summer and Date	
PENNSYLVANIA Con't												
WB-Scranton 1895-	103	7/9/1936	-21	1/21/1994	15.0	1/1977			77.4	7/1955		
Williamsport 1895-	106	7/9/1936	-22	1/21/1994	14.9	1/1977			79.6	7/1901		
RHODE ISLAND	104	8/2/1975	-25	2/5/1996								
Block Island 1880-	95	8/27/1948	-10	2/9/1934	20.3	2/1934			73.8	7/1952		
Providence 1880-	104	8/2/1975	-26	1/4/1835	17.4	2/1934			78.2	7/1952		
SOUTH CAROLINA	111	6/28/1954	-22	1/21/1985								
Charleston 1738-	105	8/1/1999	1	2/8/1835	38.7	1/1977	44.6	1976-1977	86.1	7/1986	82.8	1871
Columbia 1887-	107	8/21/1983	-2	2/14/1899	35.2	1/1940			86.2	7/1993		
Florence 1891-	108	7/27/1954	-1	2/14/1899	37.4	1/1981			86.6	7/1993		
Greenville 1887-	106	7/1887	-10	2/8/1835	30.7	1/1977			83.2	7/1993		
SOUTH DAKOTA	120	7/5/1936	-58	2/17/1936								
Aberdeen 1890-	115	7/9/1936	-47	2/9/1994	-7.3	2/1936	7.4	1935-1936	84.8	7/1936	77.1	1936
Huron 1881-	112	1/10/1966	-43	1/8/1987	-5.4	2/1936			84.4	7/1936		
Pierre 1891-	115	7/23/1940	-40	2/2/1905	-2.1	2/1936	11.1	1935-1936	87.6	7/1936	81.0	1936
Rapid City 1888-	110	7/8/1989	-40	2/2/1905	1.4	2/1936			82.4	7/1936		
Sioux Falls 1890-	110	6/21/1900	-42	2/9/1899	0.5	2/1936	8.0	1978-1979	84.6	7/1936	77.8	1936
TENNESSEE	113	8/9/1930	-32	12/30/1917								
Bristol 1928-	102	7/28/1952	-21	1/21/1985	22.1	1/1977			79.6	7/1993		
Chattanooga 1879-	106	7/28/1952	-10	1/21/1985	28.5	1/1977			85.2	7/1993		
Knoxville 1871-	104	7/12/1930	-24	1/21/1985	26.7	1/1940			83.0	7/1993		
Memphis 1871-	108	7/13/1980	-13	12/24/1963	27.2	1/1940	36.1	1917-1918	88.8	7/1980	84.2	1954
Nashville 1871-	107	7/28/1952	-17	1/21/1985	24.5	1/1977	33.4	1962-1963	84.7	6/1952	82.8	1952
TEXAS	120	8/12/1936	-23	2/8/1933								
Abilene 1885-	111	8/3/1943	-9	1/4/1947	34.0	2/1905			90.1	8/1952		
Amarillo 1892-	108	6/28/1998	-16	2/12/1899	23.9	2/1905	29.5	1898-1899	84.1	7/1934	81.7	1934
Austin 1856-	112	9/5/2000	-5	2/12/1899	36.6	1/1856			89.1	7/1860		
Brownsville 1880-	106	3/27/1984	12	2/13/1899	40.9	1/1881	44.6	1898-1899	87.5	7/1980	97.0	1998
Corpus Cristi 1887-	109	9/5/2000	11	2/12/1899	45.6	2/1905			87.1	7/1953		
Dallas 1882-	115	8/18/1909	-10	2/12/1899	34.4	1/1940			91.7	7/1998		
Del Rio 1906-	115	9/5/2000	10	12/23/1989	43.8	1/1930			91.7	7/1998		
El Paso 1878-	114	6/30/1994	-8	1/11/1962	35.6	1/1949	42.0	1912-1913	89.0	6/1994	87.8	1994
Fort Worth 1898-	113	6/27/1980	-8	2/12/1899	33.8	1/1978	39.3	1898-1899	92.0	7/1980	89.2	1980
Galveston 1871-	104	9/5/2000	8	2/12/1899	42.5	2/1895	50.5	1904-1905	87.4	7/1875	85.5	1875
Great Bend NP 1948-	105	6/30/1994	-3	1/30/1949								
Houston 1867-	109	9/4/2000	5	1/23/1940	40.8	1/1978			87.5	7/1980		
Laredo 1895-	115	6/11/1942	5	2/12/1899								
Lubbock 1911-	114	6/27/1994	-17	2/8/1933	31.2	1/1949			85.4	7/1966		
Midland 1906-	116	6/27/1994	-11	2/8/1933	35.5	1/1949			87.2	8/1964		
Port Arthur 1911-	108	8/30/2000	11	1/18/1930	41.6	1/1940			85.6	8/1962		
San Angelo 1906-	111	7/29/1960	-4	12/23/1989	36.2	1/1979			88.2	8/1952		
San Antonio 1877-	111	9/5/2000	0	1/31/1949	43.2	1/1930	47.8	1977-1978	88.1	7/1998	86.2	1994
Victoria 1898-	111	9/5/2000	9	1/18/1930	44.4	1/1940			87.9	8/1951		
Waco 1867-	112	8/11/1969	-5	1/31/1949	35.1	1/1940			90.8	7/1925		
Wichita Falls 1931-	117	6/28/1980	-12	1/4/1947	30.1	1/1978			91.9	7/1980		
UTAH	117	7/5/1985	-69	2/1/1985								
Blanding 1904-	110	6/22/1905	-23	2/8/1933	12.6	1/1937			81.6	7/2003		
Milford 1906-	107	7/18/1998	-35	12/23/1990	6.8	1/1949			79.5	7/2002		
Saint George 1892-	117	7/5/1985	-11	2/22/1937	22.0	1/1937			91.5	7/2003		
Salt Lake City 1874-	107	7/13/2002	-30	2/9/1933	11.6	1/1949	19.5	1932-1933	83.2	7/2003	78.6	1994
Wendover 1911-	112	7/13/1939	-19	2/10/1933	10.4	1/1937			87.7	7/1911		
VERMONT	105	7/4/1911	-50	12/30/1933								
Burlington 1871-	101	8/11/1944	-32	1/10/1859	3.6	1/1970	12.1	1917-1918	75.4	7/1921	72.2	1949

CITY TEMPERATURE RECORDS

State, City and Beginning of Record		Highest Temperature and Date		Lowest Temperature and Date		Coldest Month and Date		Coldest Winter and Date		Warmest Month and Date		Warmest Summer and Date	
VIRGINIA		110	7/15/1954	-30	1/22/1985								
Lynchburg	1871-	106	7/10/1936	-10	2/5/1996	23.0	1/1977			81.9	7/1934		
Norfolk	1870-	105	8/7/1918	-3	1/21/1985	29.1	1/1977			83.4	7/1995		
Richmond	1895-	107	8/6/1918	-17	1/21/1985	24.2	1/1940			82.5	7/1993		
Roanoke	1902-	105	8/21/1983	-12	12/30/1917	23.6	1/1977			80.3	7/1993		
WASHINGTON		118	8/5/1961	-48	12/30/1968								
Bellingham	1909-	97	6/1/1922	-4	1/30/1929	20.6	1/1950			68.0	7/1958		
Olympia	1877-	104	8/9/1981	-8	1/1/1979	25.9	1/1950			69.1	7/1958		
Omak	1909-	114	7/26/1928	-26	2/1/1950	5.1	1/1949			76.4	7/1994		
Seattle	1890-	100	6/9/1955	0	1/31/1950	24.6	1/1950	37.4	1928-1929	71.1	8/1967	68.4	1958
Spokane	1881-	108	8/4/1961	-30	1/16/1888	8.5	1/1949	18.8	1948-1949	75.9	7/1906	71.3	1922
Tatoosh Is	1884-1966	88	7/23/1924	7	1/31/1893	30.2	1/1950	41.2	1948-1949	59.0	7/1942	58.0	1936
Walla Walla	1872-	114	7/10/1975	-29	1/13/1875	12.9	1/1875			82.3	7/1985		
Yakima	1910-	111	7/26/1928	-25	2/1/1950	12.5	1/1950			78.8	7/1941		
WEST VIRGINIA		112	7/10/1936	-37	12/30/1917								
Charleston	1903-	108	7/3/1931	-24	1/23/1857	18.6	1/1977			81.4	7/1934		
Elkins	1899-	99	7/16/1988	-28	12/30/1917	15.0	1/1977			74.1	7/1934		
Huntington	1893-	108	7/28/1930	-24	2/10/1889	19.7	1/1977			82.0	7/1934		
Parkersburg	1888-	106	8/6/1918	-27	2/10/1899	17.4	1/1977			80.0	7/1999		
WISCONSIN		114	7/13/1936	-55	2/1/1996								
Bayfield	1891-	104	7/20/1901	-34	2/3/1996	2.6	1/1994			73.4	8/1900		
Eau Claire	1891-	111	7/13/1936	-45	1/30/1951	-0.7	1/1977			76.2	7/1988		
Green Bay	1886-	104	7/13/1936	-36	1/21/1888	-1.1	1/1912			77.4	7/1921	72.6	1995
La Crosse	1872-	108	7/13/1995	-43	1/18/1873	2.9	1/1977			79.5	7/1936	75.8	1995
Madison	1853-	107	7/14/1936	-37	1/30/1951	1.2	1/1912			79.5	7/1901	74.6	1995
Milwaukee	1870-	108	7/13/1995	-30	1/1/1864	8.0	1/1977			78.8	7/1999		
Wausau	1895-	107	7/13/1936	-40	1/30/1951	2.0	1/1977			74.1	7/1983		
WYOMING		115	8/8/1983	-63	2/9/1933								
Casper	1910-	105	7/12/1954	-41	12/21/1990	5.7	1/1949			76.1	7/1954		
Cheyenne	1870-	100	6/23/1954	-38	1/9/1875	12.0	2/1899			75.1	7/2003		
Jackson	1906-	101	7/17/1934	-52	12/20/1924	0.2	1/1937			65.5	7/1989		
Lander	1891-	102	8/5/1979	-40	2/8/1936	1.0	1/1937			75.9	7/2003		
Rawlins	1898-	102	7/1902	-36	2/6/1989	9.1	1/1979			71.8	7/2003		
Rock Springs	1904-	99	7/13/2002	-42	2/12/1905	5.1	1/1949			74.1	7/2003		
Sheridan	1893-	107	7/14/2002	-45	2/5/1899	1.8	1/1916			77.9	7/1936		
W. Yellowstone	1924-	99	7/13/2002	-60	1/12/1963	-4.5	1/1930			65.1	7/1936		
UNTIED STATES		134	7/10/1913	-80	1/23/1971	-48.4	12/1917			107.4	8/1897		
		Greenland Ranch, CA		Prospect Creek, AK		Fort Yukon, AK				Salton, CA			

*all figures in °F

CITY PRECIPITATION RECORDS

State, City and Beginning of Record		Maximum 24hr Precipitation and Date		Maximum Monthly Precipitation and Date		Minimum Monthly Precipitation and Date		Maximum Annual Precipitation and Date		Least Annual Precipitation and Date	
ALABAMA		32.52	7/19-20/1997	36.71	7/1997			114.01	1975	22.00	1954
Birmingham	1894-	9.22	12/2-3/1983	20.12	7/1916	.00	10/1924	81.82	1929	34.32	1904
Huntsville	1831-	9.07	12/22/1990	18.68	12/1990	.00	10/1963	81.02	1874	29.08	1839
Mobile	1840-	13.36	4/13/1955	26.67	6/1900	.00	10/1924	92.32	1881	33.49	1954
Montgomery	1873-	9.98	1/12-13/1882	21.32	11/1948	T	10/1904	77.89	1912	26.82	1954
ALASKA		15.20	10/12/1982	70.99	11/1976			332.29	1976	1.61	1935
Anchorage	1916-	4.12	8/25/1989	9.77	8/1948	T	many	27.55	1989	8.08	1969
Barrow	1921-	1.38	7/21-22/1987	3.19	7/1989	.00	many	9.77	1963	1.61	1935
Fairbanks	1904-	3.42	8/12/1967	6.88	8/1930	.00	2/1919	18.73	1907	5.50	1957
Juneau	1899-	5.54	9/25-26/1918	25.87	11/1936	.07	2/1989	119.48	1939	37.80	1951
Nome	1906-	2.99	8/1976	8.58	8/1998	.00	1/1917	29.49	1922	7.39	1962
St. Paul Island	1869-	2.00	8/23-24/1984	9.89	10/1878	.00	10/1924	47.08	1875	9.82	1977
ARIZONA		11.40	9/5/1970	16.95	8/1951			58.92	1978	.03	2002
Flagstaff	1888-	4.48	2/19/1993	12.60	1/1895	.00	many	36.59	1965	9.90	1942
Lake Havasu City*	1894-	3.43	9/5/1939	8.85	9/1939	.00	many	12.42	1941	.34	1956
Phoenix	1876-	4.98	7/1-2/1911	6.47	7/1911	.00	many	19.73	1905	2.82	2002
Tucson	1867-	3.93	7/1958	7.93	8/1955	.00	many	24.17	1905	5.16	1924
Winslow	1898-	2.43	8/1/1928	6.12	12/1888	.00	many	17.17	1923	4.45	1899
Yuma	1870-	5.25	8/9/1989	6.25	8/1909	.00	many	11.41	1905	.03	2002
*early record from Parker											
ARKANSAS		14.06	12/3/1982	23.86	12/1931			98.55	1957	19.11	1936
Fort Smith	1882-	8.58	6/9-10/1945	15.02	6/1945	T	10/1964	71.81	1945	19.80	1917
Little Rock	1879-	10.20	12/24-25/1987	18.04	1/1937	I	6/1952	75.55	1882	31.47	1924
Texarkana	1883	7.60	NA	16.65	4/1966	.00	many	76.73	1905	28.40	1893
CALIFORNIA		26.12	1/22/1943	71.54	1/1909			153.54	1909	.00	1929
Bakersfield	1889-	3.02	2/9-10/1978	5.35	2/1998	.00	many	14.66	1997-1998	2.26	1933-1934
Bishop	1895-	3.64	2/24/1969	9.67	1/1914	.00	many	18.02	1968-1969	1.69	1897-1898
Blue Canyon	1899-	9.33	12/22/1964	48.35	1/1909	.00	many	122.35	1994-1995	26.97	1976-1977
Burbank	1931	7.76	1/22/1943	15.92	1/1995	.00	many	39.25	1940-1941	5.12	2001-2002
Death Valley	1911-	1.47	4/15/1988	2.59	1/1995	.00	many	6.09	1997-1998	.00	1929 & 1953
Eureka	1887-	6.79	12/27/2002	27.25	2/1998	.00	many	74.10	1889-1890	17.56	1976-1977
Fort Bragg	1901-	3.84	1/9/1995	27.02	1/1909	.00	many	79.13	1997-1998	14.54	1976-1977
Fresno	1881-	2.86	11/16-17/1900	9.54	11/1885	.00	many	23.57	1982-1983	4.43	1933-1934
Los Angeles	1877-	7.36	12/31-1/1/1934	20.51	2/1998	.00	many	38.18	1883-1884	4.46	2001-2002
Monterey	1847-	3.85	12/23/1955	14.26	2/1998	.00	many	47.12	1997-1998	6.95	1897-1898
Mt. Shasta City*	1889-	6.97	1/15-16/1974	27.48	1/1995	.00	many	75.89	1997-1998	12.80	1976-1977
Palm Springs	1889-	4.57	1/23/1943	8.43	1/1943	.00	many	12.90	1931-1932	.00	1896-1897
Red Bluff	1877-	6.12	9/13-14/1918	21.49	1/1995	.00	many	53.22	1877-1878	11.27	1919-1920
Sacramento	1849-	7.24	4/20-21/1880	15.04	1/1862	.00	many	36.35	1852-1853	4.71	1850-1851
San Diego	1850-	3.62	12/23-24/1940	9.26	12/1940	.00	many	25.97	1883-1884	3.02	2001-2002
San Francisco	1849-	6.76	11/5-6/1994	24.36	1/1862	.00	many	49.27	1861-1862	6.96	1975-1976
San Jose	1874-	3.60	1/30/1968	10.55	12/1889	.00	many	30.30	1889-1890	5.62	1975-1976
Santa Barbara	1864-	7.45	1/9-10/1995	21.94	1/1995	.00	many	46.99	1997-1998	4.49	1876-1877
Santa Maria	1885-	7.52	1/9-10/1995	11.78	1/1995	.00	many	32.56	1997-1998	3.30	1989-1990
*formerly Sisson											
COLORADO		11.08	6/17/1965	23.28	2/1897			92.84	1897	1.69	1939
Alamosa	1948-	1.82	9/30/1959	5.40	8/1993	.00	many	11.55	1969	3.09	1989
Aspen	1914-	2.87	3/14/1960	5.80	9/1961	.00	6/2002	33.27	1984	12.22	1958
Colo Springs	1871-	4.21	8/4-5/1999	8.10	5/1935	.00	12/1970	27.58	1999	6.07	1939
Denver	1872-	6.53	5/22/1876	8.57	5/1876	.00	12/1881	23.84	1999	6.27	1954
Durango	1886-	3.68	9/9/2003	11.79	10/1972	.00	many	34.29	1911	8.90	1901
Grand Junction	1885-	3.03	9/23/1896	3.78	9/1896	.00	10/1952	15.69	1957	3.64	1900
Lamar	1889-	5.64	5/29/1964	9.59	5/1914	.00	many	24.51	1946	7.39	1898
Pueblo	1869-	3.77	10/7-8/1957	8.13	4/1900	.00	10/1933	23.09	1957	5.78	1934
CONNECTICUT		12.77	8/19/1955	27.70	8/1955			78.53	1955	23.60	1965
Hartford	1846-	12.12	8/18-19/1955	21.87	8/1955	.18	10/1924	64.55	1972	29.45	1965
New Haven	1804-	8.73	8/8-9/1974	17.08	7/1889	.12	6/1949	60.26	1888	27.68	1965

CITY PRECIPITATION RECORDS

State and City		Maximum 24hr Precipitation and Date		Maximum Monthly Precipitation and Date		Minimum Monthly Precipitation and Date		Maximum Annual Precipitation and Date		Least Annual Precipitation and Date	
DELAWARE		12.36	9/15-16/1999	17.69	8/1967			72.75	1948	21.38	1965
Wilmington	1894-	8.29	9/16/1999	14.91	8/1911	.06	10/1924	61.05	1945	24.90	1965
DC/Washington		7.31	8/11-12/1928	17.45	9/1934	T	10/1963	61.33	1889	18.79	1826
Dulles	1824-	11.88	6/21-22/1972								
FLORIDA		38.70	9/5/1950	42.33	10/1965			127.00	1877	19.99	1974
Apalachicola	1903-	11.71	9/13-14/1932	27.73	9/1924	T	5/1914	99.30	1959	30.69	1904
Daytona Beach	1922-	12.85	10/10/1924	24.82	10/1924	T	11/1967	74.71	1924	31.36	1956
Everglades City	1918-	12.50	6/14/1936	24.73	6/1936	.00	many	78.19	1947	32.54	1988
Fort Meyers	1851-	11.70	6/11-12/1901	26.91	6/1912	.00	1/1950	82.64	1853	32.83	1964
Gainesville	1856-	7.42	10/24/1938	19.91	9/1894	.00	10/1987	76.73	1885	32.97	1917
Jacksonville	1851-	11.40	9/24/1989	23.32	6/1932	T	11/1970	82.27	1947	31.20	1990
Key West	1832-	23.28	11/11/1980	27.67	11/1980	.00	4/1959	69.69	1870	19.99	1974
Lakeland	1914-	10.12	6/22-23/1945	16.03	5/1979	.00	4/1967	70.24	1959	34.93	1927
Melbourne	1948-	9.06	8/2/1995	19.68	9/1948	.00	11/1986	70.11	1995	31.97	1981
Miami	1855-	16.21	4/24-25/1979	27.86	10/1908	.00	12/1906	89.33	1959	28.66	1944
Orlando	1892-	9.67	9/15-16/1945	19.57	7/1960	T	12/1944	74.19	1905	30.38	2000
Pensacola	1879-	17.07	10/4-5/1934	21.43	8/1935	.00	10/1978	93.32	1953	28.52	1954
Tallahassee	1885-	10.13	6/12-13/2001	23.85	9/1924	.00	12/1899	104.18	1964	30.98	1954
Tampa	1840-	15.50	8/2/1915	24.52	7/1840	T	5/2001	89.86	1840	26.76	1956
W. Palm Beach	1939-	15.23	4/17-18/1942	24.86	9/1960	.04	4/1967	108.64	1947	37.31	1955
GEORGIA		21.10	7/5-6/1994	31.61	7/1994			122.16	1959	17.14	1954
Albany	1892-	7.60	4/23/1928	20.48	7/1916	.00	10/1889	71.46	1964	32.06	1954
Athens	1872-	9.93	6/8/1967	18.43	8/1908	T	10/1963	71.42	1964	28.61	1954
Atlanta	1865-	7.36	3/29/1886	17.71	7/1994	T	10/1963	71.45	1948	31.80	1954
Augusta	1870-	9.82	9/30-10/1/1929	14.82	10/1990	T	10/1963	73.82	1929	19.13	1845
Columbus	1892-	6.80	8/3-4/1977	16.14	7/1916	.00	10/1963	73.22	1964	26.39	1999
Macon	1899-	10.37	7/5/1994	20.52	8/1928	.00	10/1963	67.80	1929	26.05	1954
Rome	1855-	7.53	3/28-29/1951	17.98	3/1980	T	10/1963	77.65	1932	31.50	1867
Savannah	1840-	11.44	9/17-18/1928	22.88	9/1924	T	10/2000	73.36	1885	22.00	1931
HAWAII		38.00	1/24-25/1956	107.00	3/1942			704.83	1982	.19	1953
Honolulu	1874-	17.41	3/5-6/1958	24.90	2/1904	T	8/1974	55.91	1904	4.52	1998
Hilo	1880-	27.24	11/1-2/2000	66.96	3/1922	.13	1/1998	211.22	1990	68.09	1983
Kahului	1892-	7.01	1/1980	17.07	2/1904	.00	6/1957	40.63	1989	6.76	1998
Lihue	1904-	13.52	7/2/1929	27.98	1/1921	T	2/1983	74.40	1982	16.40	1983
IDAHO		7.17	11/23/1909	28.23	12/1933			81.05	1933	2.09	1947
Boise	1877-	2.72	3/6/1871	7.66	3/1871	.00	10/1988	25.80	1871	6.64	1966
Lewiston	1880-	2.34	NA	6.31	12/1880	.00	8/1939	21.71	1884	8.40	1935
Pocatello	1899-	4.31	3/23/1941	4.93	3/1941	.00	10/1988	22.43	1909	5.34	1966
Salmon	1906-	1.50	4/24/1971	4.32	6/1964	.00	10/1987	14.75	1977	6.73	1999
Sandpoint	1910-	2.95	5/27/1998	11.99	12/1933	.00	many	46.72	1996	18.03	1929
ILLINOIS		16.94	7/17-18/1996	20.03	9/1911			74.58	1950	16.59	1956
Cairo	1871-	7.56	8/11-12/1952	15.70	6/1928	T	10/1964	72.98	1957	22.75	1987
Chicago	1843-	9.35	8/13-14/1987	17.10	8/1987	T	9/1979	49.35	1983	21.19	1962
Moline		6.57	8/11/1949	15.23	8/1987	.01	10/1964	56.36	1973	17.33	1901
Peoria	1856-	5.52	5/18/1927	13.09	9/1961	.03	10/1964	55.35	1990	22.16	1988
Quincy	1905-	5.84	6/14/1950	15.71	9/1911	T	7/1916	66.60	1973	20.00	1973
Rockford	1873-	8.41	7/17-18/1952	14.16	6/1892	.01	10/1952	56.48	1973	22.87	1910
Springfield	1879-	6.12	12/2-3/1982	15.16	9/1926	T	9/1979	58.21	1882	21.95	1914
Urbana	1902-	5.32	8/12/1993	13.82	7/1992	T	11/1904	58.54	1993	18.28	1894
INDIANA		10.50	8/6/1905	21.39	1/1937			97.38	1890	18.67	1934
Evansville	1877-	7.92	4/28-29/1996	14.78	1/1937	.01	10/1964	70.61	1882	25.55	1930
Fort Wayne	1887-	4.93	8/1/1926	11.00	7/1986	.08	2/1907	54.58	1990	23.16	1897
Indianapolis	1871-	7.31	9/1-2/2003	14.55	7/1992	.03	10/1963	57.65	1876	27.72	1963
South Bend	1894-	4.88	8/2-3/1995	10.86	6/1993	.01	9/1979	55.61	1990	22.17	1917

CITY PRECIPITATION RECORDS

State and City		Maximum 24hr Precipitation and Date		Maximum Monthly Precipitation and Date		Minimum Monthly Precipitation and Date		Maximum Annual Precipitation and Date		Least Annual Precipitation and Date	
INDIANA Con't											
Terre Haute	1891-	6.74	9/1-2/2003	11.76	5/1981	.00	6/1933	56.97	1990	26.52	1963
IOWA		16.70	8/5-6/1959	22.18	6/1967			74.50	1851	12.11	1958
Burlington	1897-	6.28	6/29/1933	15.41	6/1882	.06	10/1964	51.55	1902	22.87	1956
Des Moines	1863-	6.18	8/27/1975	15.79	6/1881	.00	10/1952	56.81	1881	17.07	1956
Dubuque	1851-	8.96	8/21-22/2002	15.46	9/1965	.00	10/1996	63.39	1961	19.35	1894
Mason City	1893-	7.00	6/1/1900	15.73	8/1980	.00	10/1952	69.54	1943	15.10	1910
Sioux City	1890-	5.50	7/17/1972	11.78	5/1903	T	10/1958	41.10	1903	14.33	1976
Waterloo	1876-	9.31	7/16-17/1968	12.84	7/1999	T	11/1958	53.07	1993	17.35	1910
KANSAS		17.00	9/24-25/1993	24.56	6/1845			68.55	1993	4.77	1956
Concordia	1885-	6.46	5/7-8/1950	16.75	7/1993	.00	1/1986	44.79	1993	12.83	1956
Dodge City	1866-	6.03	6/7-8/1899	12.82	5/1881	.00	1/1986	34.29	1944	7.58	1868
Goodland	1906-	4.15	6/28/1989	10.10	7/1985	.00	1/1933	30.89	1915	9.19	1956
Topeka	1887-	8.08	9/11/1909	15.20	7/1967	T	11/1989	60.89	1973	19.07	1963
Wichita	1888-	7.99	9/6 7/1911	14.43	6/1923	.00	10/1952	50.48	1951	12.15	1966
KENTUCKY		10.48	3/1/1997	22.97	1/1937			79.68	1950	14.51	1968
Bowling Green	1887	6.15	12/7/1924	20.70	1/1937	.03	10/1924	76.56	1979	25.80	1999
Lexington	1887-	8.06	8/1-2/1932	17.52	3/1997	.11	10/1924	66.46	1935	24.89	1930
Louisville	1871-	10.48	3/1/1997	19.17	1/1937	.01	9/1995	63.76	1996	23.88	1930
LOUISIANA		22.00	8/29/1962	37.99	8/1940			113.74	1991	23.10	1899
Baton Rouge	1843-	12.08	4/14-15/1967	23.73	5/1907	.00	9/1924	88.32	1989	37.78	1924
Lake Charles	1888-	16.88	5/15-16/1980	25.33	6/1989	.00	many	85.16	2002	30.88	1954
New Orleans	1836-	14.01	7/24-25/1933	25.11	10/1937	.00	10/1963	113.74	1991	31.07	1899
Shreveport	1872-	12.44	7/24-25/1933	31.42	7/1937	.00	10/1963	81.99	1991	23.10	1899
MAINE		19.19	10/21-22/1996	20.00	10/1996			75.64	1945	23.06	1930
Bangor	1927-			11.61	11/1983						
Caribou	1939-	6.89	8/17/1981	12.09	8/1981	.12	1/1944	51.11	1954	27.92	1966
Eastport	1873-	5.48	5/17/1881	13.22	5/1881	.13	5/1911	64.53	1883	21.24	1924
Portland	1871-	13.32	10/21 22/1996	15.22	8/1991	.04	2/1987	66.33	1983	25.27	1941
MARYLAND		14.75	7/26/1897	20.35	7/1945			72.59	1948	17.76	1930
Baltimore	1817-	8.35	8/12-13/1955	18.35	8/1955	T	10/1963	62.66	2003	21.55	1930
Frederick	1854-			11.41	9/1859	.19	10/1892	51.98	1915	29.18	1910
Salisbury	1906-	8.84	8/31-9/1/2002	13.69	8/1967	.18	5/1911	72.59	1948	35.21	1913
MASSACHUSETTS		18.15	8/18-19/1955	26.85	8/1955			76.49	1996	21.76	1965
Boston	1818-	8.40	8/18-19/1955	17.09	8/1955	T	6/1999	67.72	1863	23.71	1965
Nantucket	1847-	6.53	5/24-25/1967	12.92	8/1946	.01	6/1947	71.69	1972	25.31	1965
Pittsfield	1895-			12.19	7/1897	.25	3/1915	58.13	1898	30.89	1908
Worcester	1841-	8.67	8/18-19/1955	18.68	8/1955	.04	3/1915	71.66	1972	27.92	1941
MICHIGAN		9.78	9/1/1914	19.44	2/1912			71.19	1855	15.64	1936
Alpena	1872-	5.14	9/2-3/1937	13.18	10/1877	.00	8/1959	45.61	1881	17.08	1925
Detroit	1836-	4.75	7/31-8/1/1925	15.01	7/1855	.04	2/1877	71.19	1855	20.49	1963
Escanaba	1871-	5.05	9/1937	12.06	8/1875	.00	3/1895	48.49	1881	20.30	1900
Flint	1889-	6.04	9/10/1950	11.18	8/1937	.04	8/1899	45.38	1975	18.08	1963
Grand Rapids	1870-	5.48	5/10-11/1981	13.22	6/1892	T	9/1979	52.14	1883	20.92	1930
Houghton	1900-	3.23	8/3/1955	9.48	12/1983	.06	2/2001	53.43	1985	17.23	1900
Houghton Lake	1906-	5.18	7/17/1957	9.49	7/1986	.01	9/1979	41.34	1951	20.23	1989
Lansing	1863-	5.47	6/6/1905	11.35	6/1883	.00	8/1894	48.36	1883	18.50	1930
Marquette	1871-	5.14	6/20/1878	12.73	9/1881	.05	5/1986	51.59	1985	19.68	1925
Muskegon	1896-	6.60	2/26/1912	19.44	2/1912	T	3/1910	46.79	1912	16.57	1930
Sault Ste. Marie	1887-	5.92	8/3/1984	9.48	8/1974	.09	3/1915	45.84	1995	20.69	1925
MINNESOTA		10.84	7/21-22/1972	16.52	8/1900			52.36	1993	7.81	1936
Bemidji	1897-	4.15	7/22/1914	13.44	7/1949	.00	12/2001	41.03	1901	12.47	1917

CITY PRECIPITATION RECORDS

State and City		Maximum 24hr Precipitation and Date		Maximum Monthly Precipitation and Date		Minimum Monthly Precipitation and Date		Maximum Annual Precipitation and Date		Least Annual Precipitation and Date	
MINNESOTA Con't											
Duluth	1871-	5.79	8/22-23/1978	13.32	9/1913	.00	6/1995	45.28	1879	16.66	1976
Int'l Falls	1892-	4.87	7/2-3/1966	11.26	8/1942	T	11/1924	34.35	1941	17.23	1952
Minn/St. Paul	1836-	9.90	7/23-24/1987	17.90	7/1987	T	12/1943	49.69	1849	10.21	1910
Rochester	1886-	7.47	7/11/1981	12.64	6/1899	T	12/1943	43.94	1990	11.65	1910
St. Cloud	1893-	5.37	9/2/1985	12.81	7/1897	.00	2/1921	41.01	1897	14.64	1910
MISSISSIPPI		15.68	7/9/1968	30.75	7/1916			104.36	1991	25.97	1936
Jackson	1873-	9.31	4/6-7/2003	23.80	4/1874	.00	10/1963	92.75	1979	31.66	1952
Meridian	1889-	9.23	4/16-17/1900	20.06	6/1900	.00	10/1963	79.03	1973	34.92	1914
Vicksburg	1840-	8.75	4/29/1953	22.24	4/1874	.00	10/1847	84.22	1880	31.20	1924
MISSOURI		18.18	7/20/1965	25.54	5/1943			92.77	1957	16.14	1957
Columbia	1900-	6.61	9/2/1918	14.86	6/1928	.05	1/1986	62.49	1993	21.35	1901
Kansas City		8.82	9/12/1977	16.17	9/1914	T	11/1989	60.25	1961	19.04	1886
St. Joseph	1869-	4.94	9/10/1986	12.92	7/1915	.00	1/2000	57.13	1877	20.53	1901
St. Louis	1830-	8.78	8/15-16/1946	20.45	8/1946	.00	11/1865	68.83	1858	20.59	1953
Springfield	1877-	6.85	7/6-7/1958	18.75	7/1958	.05	9/1928	65.31	1877	25.21	1953
MONTANA		11.50	6/19-20/1921	16.79	6/1921			55.51	1953	2.97	1960
Billings	1894-	3.19	4/27-28/1978	7.71	5/1981	.00	10/1896	26.80	1978	7.90	1948
Butte	1894-	3.00	6/8/1913	8.86	6/1913	.00	many	20.55	1909	6.89	1935
Glasgow	1894-	4.99	8/2-3/1985	10.29	6/1923	T	many	20.85	1938	6.74	1984
Great Falls	1891-	3.42	5/24-25/1980	8.13	5/1953	.00	11/1917	25.24	1975	6.68	1904
Havre	1880-	3.71	6/15-16/1887	9.67	7/1884	T	1/1973	25.67	1884	6.25	1904
Helena	1880-	3.67	6/4-5/1908	6.67	5/1927	T	8/1940	20.94	1975	6.26	1973
Kalispell	1896-	2.71	6/19/1982	6.02	7/1993	T	8/1955	25.23	1996	8.79	1952
Miles City	1877-	3.74	5/19-20/1908	9.78	6/1944	T	8/1967	22.75	1879	5.27	1988
Missoula	1870-	2.23	11/5/1927	8.58	5/1874	.00	8/1917	22.43	1891	8.62	1952
NEBRASKA		13.15	7/8/1950	25.29	7/1993			64.52	1869	6.30	1931
Grand Isle	1889-	5.88	9/1-2/1977	13.96	6/1967	.00	10/1958	45.85	1905	11.91	1940
Lincoln	1878-	8.38	8/28-29/1910	14.21	8/1910	.00	12/1889	45.15	1965	14.09	1936
Norfolk	1891-	7.78	6/4/1940	16.10	8/1875	T	10/1958	37.68	1905	13.80	1894
North Platte	1874-	6.32	9/1-2/1942	10.47	6/1951	.00	12/1894	33.44	1951	10.01	1931
Omaha	1857-	10.48	8/6-7/1999	17.01	7/1869	T	12/2002	64.52	1869	14.90	1934
Scottsbluff	1884-	3.74	6/6-7/1953	8.33	6/1947	.00	11/1939	27.48	1915	7.70	1964
Valentine	1889-	4.21	5/25-26/1920	9.35	5/1888	.00	10/1895	32.68	1977	10.14	1894
NEVADA		8.06	5/29/1896	33.03	12/1964			59.03	1969	trace	1898
Austin	1877-	2.04	10/1/1946	7.75	5/1998	.00	many	22.37	1983	5.91	1959
Elko	1870-	4.13	8/27/1970	6.00	1/1903	.00	many	18.43	1983	.94	1872
Ely	1888-	2.87	9/26-27/1982	5.52	4/1900	.00	many	18.01	1891	4.22	1974
Las Vegas	1895-	2.59	8/20-21/1957	4.80	3/1992	.00	many	10.72	1941	.56	1953
Reno	1870-	2.71	1/27/1903	6.76	1/1916	.00	many	13.73	1890	1.55	1947
Tonopah	1906-	1.62	8/17/1977	3.26	4/1915	.00	many	10.82	1998	2.19	1942
Winnemucca	1870-	1.79	6/10-11/1958	5.23	3/1884	.00	many	18.38	1884	3.13	1954
NEW HAMPSHIRE		11.07	10/21-22/1996	25.56	2/1969			130.14	1969	22.31	1930
Concord	1853-	5.97	9/16-17/1932	11.65	10/1869	T	3/1915	54.33	1888	24.17	1965
Mt. Washington	1871-	11.07	10/21-22/1996	25.56	2/1969	.07	4/1872	130.14	1969	53.30	1941
NEW JERSEY		14.81	8/19/1939	25.98	9/1882			85.99	1882	19.85	1965
Atlantic City	1874-	13.52	8/20-21/1997	16.12	8/1997	.01	9/1941	67.17	1958	25.27	1965
Newark	1843-	7.84	8/27-28/1971	22.48	8/1843	.07	6/1949	69.10	1903	26.09	1965
Trenton	1865-	7.55	8/27-28/1971	15.22	7/1880	.05	10/1963	67.23	1889	28.79	1957
NEW MEXICO		11.28	5/19/1955	16.21	5/1941			62.45	1941	1.00	1910
Albuquerque	1850-	2.26	9/27-28/1893	8.15	6/1852	.00	many	16.30	1858	3.29	1917
Clayton	1896-	6.20	4/30-5/1/1914	10.51	7/1899	.00	many	37.65	1941	5.54	1936
Gallup	1921-	2.36	4/16/1988	6.09	8/1999	.00	many	15.83	1997	5.44	1989

CITY PRECIPITATION RECORDS

State and City		Maximum 24hr Precipitation and Date		Maximum Monthly Precipitation and Date		Minimum Monthly Precipitation and Date		Maximum Annual Precipitation and Date		Least Annual Precipitation and Date	
NEW MEXICO Con't											
Roswell	1894-	5.65	11/1/1901	9.56	8/1916	.00	many	32.92	1941	4.35	1956
Santa Fe	1852-	3.61	7/26/1968	7.89	8/1855	.00	many	24.80	1854	3.12	1956
Silver City	1911-	3.88	10/9/1945	9.62	7/1881	.00	many	31.08	1905	5.79	1871
NEW YORK		13.70	9/16/1999	25.27	10/1955			90.97	1996	17.04	1834
Albany	1826-	5.98	9/16/1999	13.48	10/1869	.08	1/1860	56.78	1871	21.55	1964
Binghampton	1890-	4.55	9/29-30/1924	9.66	9/1977	.16	12/1928	49.00	1996	22.42	1900
Buffalo	1858-	5.01	6/22/1987	10.67	8/1977	.05	8/1876	60.24	1878	22.16	1941
New York City	1826-	11.17	10/8-9/1903	16.85	9/1882	.02	9/1949	67.03	1972	22.17	1965
Rochester	1829-	4.19	8/28-29/1893	9.70	7/1947	.08	10/1924	49.89	1873	17.04	1834
Syracuse	1902-	4.79	6/11/1922	15.92	6/1922	.19	5/1920	58.17	1976	26.96	1908
NORTH CAROLINA		22.22	7/15-16/1916	37.40	7/1916			129.60	1964	22.69	1930
Asheville	1902-	7.92	10/24-25/1918	13.75	8/1940	.00	10/2000	64.91	1973	22.79	1925
Cape Hatteras	1874-	14.73	6/30-7/1/1949	20.95	6/1949	.00	11/1890	102.04	1877	29.64	1911
Charlotte	1878-	6.88	7/23/1997	16.55	7/1916	T	10/1953	68.44	1884	26.33	2001
Greensboro	1892-	7.49	9/24-25/1947	13.26	9/1947	T	6/1990	62.34	2003	27.74	1941
Raleigh	1886-	8.80	9/5-6/1996	21.79	9/1999	.00	10/2000	64.22	1936	29.93	1933
Wilmington	1871	18.91	9/14-15/1999	23.45	9/1999	.02	10/1943	83.65	1877	27.68	1909
NORTH DAKOTA		8.10	6/29/1975	14.01	6/1944			37.98	1944	4.02	1934
Bismarck	1875-	5.27	7/14-15/1993	13.75	7/1993	T	11/1963	30.92	1876	5.97	1936
Devils Lake	1870-	4.82	NA	10.30	7/1993	.00	4/1980	29.87	1991	10.04	1967
Dickinson	1892-	4.03	7/28/1914	10.08	6/1941	.00	many	31.58	1941	6.72	1936
Fargo	1891-	5.17	7/3-4/1886	11.72	6/2000	T	11/1901	34.76	2000	8.84	1976
Grand Forks	1891-	4.44	7/16/1995	12.16	8/1944	.00	5/1997	29.11	1944	9.38	1910
Williston	1879-	5.03	7/9-10/1963	8.84	6/1901	T	10/1965	23.25	1880	6.13	1934
OHIO		10.75	8/7-8/1995	17.33	6/1887			70.82	1870	15.72	1901
Akron	1887-	6.78	6/30/1989	12.55	7/2003	.13	2/1968	65.70	1990	23.79	1963
Cincinnati	1835-	5.22	3/12-13/1907	13.68	1/1937	.10	10/1963	65.18	1847	15.72	1901
Cleveland	1855-	5.24	9/6-7/1996	11.05	9/1996	.13	11/1904	53.83	1990	18.63	1963
Columbus	1878-	5.16	7/12-13/1992	12.36	7/1992	.10	10/1924	53.16	1990	21.60	1930
Dayton	1882-	4.75	5/12/1886	12.41	1/1937	.03	8/1996	59.75	1990	20.81	1914
Toledo	1861-	5.98	9/4-5/1918	10.19	9/1865	.04	11/1904	47.84	1950	21.34	1894
Youngstown	1885-	4.65	7/21/2003	10.66	6/1986	.16	10/1924	50.71	1911	23.39	1921
OKLAHOMA		15.68	10/10-11/1973	23.95	5/1943			84.47	1957	6.53	1956
Oklahoma City	1890-	8.95	10/19-20/1983	14.66	6/1989	.00	8/2000	52.03	1908	15.74	1901
Tulsa	1888-	9.27	5/26-27/1984	18.81	9/1971	.00	1/1993	69.88	1973	23.24	1956
OREGON		11.65	11/19/1996	52.78	12/1917			204.12	1996	3.33	1939
Astoria	1853-	6.98	1/22/1919	36.07	12/1933	.01	7/1960	101.40	1899	41.58	1985
Bandon	1878-	6.25	11/18/1996	23.35	11/1973	.00	many	103.97	1894	32.22	1976
Bend	1902-	2.72	10/29/1950	8.74	12/1964	.00	many	25.75	1907	5.40	1994
Burns	1892-	2.15	12/22-23/1964	5.73	1/1970	.00	many	18.24	1983	5.12	1949
Eugene	1891-	5.15	12/5-6/1981	20.99	12/1964	.00	many	77.17	1996	23.26	1944
Lakeview	1884-	2.39	12/10/1937	8.96	12/1964	.00	many	27.16	1907	7.53	1924
Medford	1911-	3.75	12/21-22/1964	12.72	12/1964	.00	many	36.40	1996	10.42	1959
Pendleton	1890-	2.19	8/15-16/1993	4.68	12/1973	.00	many	21.80	1927	6.72	1967
Portland	1850-	7.66	12/12-13/1882	20.14	12/1882	.00	many	72.71	1996	22.48	1985
Roseburg	1877-	4.35	11/19/1996	15.91	11/3/1973	.00	many	60.19	1996	19.10	1976
Salem	1870-	4.30	12/6/1933	17.54	12/1933	.00	many	66.96	1996	21.97	2000-2001
PENNSYLVANIA		34.50	7/17/1942	23.66	8/1955			81.64	1952	15.71	1965
Allentown	1912-	7.85	9/26-27/1985	12.10	8/1955	.00	10/1924	67.69	1952	28.76	1941
Erie	1874-	10.42	7/22-23/1947	13.27	7/1947	.02	10/1924	61.70	1977	23.84	1934
Harrisburg	1889-	12.53	6/21-22/1972	18.55	6/1972	.02	10/1924	59.67	1863	25.52	1941
Philadelphia	1819-	6.77	9/15-16/1999	15.82	8/1867	.00	10/1963	61.20	1867	29.31	1922
Pittsburgh	1836-	4.41	7/28/1999	11.05	11/1985	.06	10/1874	52.24	1990	22.65	1930

CITY PRECIPITATION RECORDS

State and City		Maximum 24hr Precipitation and Date		Maximum Monthly Precipitation and Date		Minimum Monthly Precipitation and Date		Maximum Annual Precipitation and Date		Least Annual Precipitation and Date	
PENNSYLVANIA Con't											
WB-Scranton	1895-	6.52	9/26-27/1985	11.76	8/1955	.03	10/1963	53.72	1945	26.12	1930
Williamsport	1895-	8.66	6/22/1972	16.80	6/1972	.16	9/1943	61.27	1972	25.98	1895
RHODE ISLAND		12.13	9/16-17/1932	15.00	8/1955			70.21	1983	24.08	1965
Block Island	1880-	8.52	9/12/1960	12.93	6/1881	T	6/1957	63.15	1884	24.08	1965
Providence	1832-	6.71	8/3-4/1979	12.74	4/1983	.04	6/1949	67.52	1983	25.44	1965
SOUTH CAROLINA		17.00	8/27/1995	31.13	7/1916			110.79	1994	20.73	1954
Charleston	1738-	10.57	5/6/1933	27.24	6/1973	.00	10/2000	78.42	1876	23.69	1850
Columbia	1887-	7.66	8/16-17/1949	17.46	7/1991	T	10/1963	70.53	1964	27.11	1933
Florence	1895-	6.08	10/15/1954	18.05	7/1916	.00	10/1954	64.71	1959	27.50	1954
Greenville	1889-	12.32	8/26-27/1995	17.37	8/1995	.00	10/2000	77.83	1901	31.64	1938
SOUTH DAKOTA		15.00	6/9/1972	18.61	5/1946			48.42	1948	2.89	1948
Aberdeen	1890-	5.20	6/29-30/1978	12.39	5/1906	.00	10/1952	38.39	1896	7.88	1976
Huron	1881-	5.48	6/18-19/1967	11.56	6/1914	T	many	31.71	1962	9.72	1952
Pierre	1891-	3.72	8/?	7.25	6/1954	.00	1/1913	23.57	1915	7.82	1894
Rapid City	1888-	5.57	5/27/1926	9.66	7/1905	T	many	27.70	1946	7.51	1936
Sioux Falls	1890-	4.59	8/1/1975	9.26	9/1986	.00	11/1914	36.11	1993	10.44	1894
TENNESSEE		11.00	3/28/1902	23.90	1/1937			114.88	1957	25.23	1941
Bristol	1894-	3.65	10/16/1964	11.34	8/2003	T	10/1904	65.65	2003	30.06	1939
Chattanooga	1879-	7.61	3/29-30/1886	16.32	3/1980	.04	9/1911	73.70	1994	32.68	1904
Knoxville	1869-	6.14	7/16/1917	17.32	4/1874	T	10/2000	73.87	1875	32.48	1986
Memphis	1867-	10.48	11/20-21/1934	18.16	6/1877	.00	9/1897	76.85	1957	30.54	1941
Nashville	1871-	6.68	9/13-14/1979	14.75	1/1937	T	10/1963	70.12	1979	30.23	1987
TEXAS		43.00	7/25-26/1979	45.00	7/1979			109.38	1873	1.64	1956
Abilene	1885-	6.78	5/22-23/1908	15.70	8/1914	.00	10/1952	46.43	1932	9.78	1956
Amarillo	1892-	6.75	5/15-16/1951	10.73	6/1965	.00	11/1989	39.75	1923	9.56	1969
Austin	1856-	19.03	9/9-10/1921	20.78	9/1921	.00	7/1962	64.68	1919	11.42	1954
Brownsville	1871-	12.19	9/19-20/1967	30.57	9/1886	.00	8/1920	60.06	1886	11.59	1953
Corpus Cristi	1887-	8.92	8/9-10/1980	20.33	9/1967	.00	7/1957	48.16	1888	5.38	1917
Dallas	1874-	9.18	8/26-27/1947	15.40	4/1966	.00	8/2000	55.14	1957	17.52	1963
Del Rio	1906-	17.03	8/23/1998	20.93	8/1998	.00	8/1952	37.75	1914	4.34	1956
El Paso	1878-	6.50	7/9/1881	10.85	7/1881	.00	many	19.17	1881	2.22	1891
Fort Worth	1889-	9.57	9/4-5/1932	17.64	4/1922	.00	11/1903	53.54	1991	17.91	1921
Galveston	1871-	14.35	7/13-14/1900	26.01	9/1985	.00	8/1902	78.39	1900	21.40	1948
Great Bend NP	1948-	4.29	10/5/1966	10.71	8/1980	.00	many	33.69	1986		
Houston	1882-	15.65	8/27-28/1945	22.31	10/1949	T	10/1952	83.02	1979	17.66	1917
Laredo	1871-	8.75	8/2/1971	12.59	8/1879	.00	many	48.04	1919	4.31	1901
Lubbock	1911-	5.82	10/18-19/1983	13.95	9/1936	.00	many	40.53	1941	8.73	1917
Midland	1886-	5.99	7/21-22/1961	13.03	8/1920	.00	many	32.13	1986	4.25	1951
Port Arthur	1911-	17.76	7/27-28/1943	24.25	7/1943	.00	10/1963	82.85	1946	30.52	1924
San Angelo	1872-	11.75	9/15/1936	27.65	9/1936	.00	10/1952	42.12	1882	7.41	1956
San Antonio	1871-	13.35	10/17-18/1998	18.07	10/1998	.00	8/1952	52.28	1973	10.11	1917
Victoria	1892-	9.87	4/5-6/1991	19.05	9/1978	.00	11/1945	67.11	1997	11.15	1917
Waco	1867-	7.98	12/20/1997	15.00	5/1965	.00	9/1956	60.20	1905	13.39	1917
Wichita Falls	1891-	6.22	9/28/1980	13.22	5/1982	.00	many	47.15	1919	12.38	1896
UTAH		8.40	9/7-8/1991	25.45	12/1983			108.54	1983	1.34	1956
Blanding	1904-	4.48	8/1/1968	7.01	10/1916	.00	many	24.61	1909	4.93	1956
Milford	1906-	1.92	10/6/1916	3.75	8/1984	.00	many	13.91	1984	4.48	1950
Saint George	1890-	2.40	8/31/1909	10.25	7/1907	.00	many	18.71	1907	3.55	1894
Salt Lake City	1874-	2.41	4/22-23/1957	7.04	9/1982	.00	10/1952	24.26	1983	8.70	1979
Wendover	1911-	1.95	5/19/1987	4.19	5/1987	.00	many	10.40	1987	1.62	1992
VERMONT		9.92	9/17/1999	16.99	10/1955			100.96	1996	20.99	1881
Burlington	1837-	4.49	11/3-4/1927	11.54	8/1955	T	5/1903	50.42	1998	20.99	1881

CITY PRECIPITATION RECORDS

State and City		Maximum 24hr Precipitation and Date		Maximum Monthly Precipitation and Date		Minimum Monthly Precipitation and Date		Maximum Annual Precipitation and Date		Least Annual Precipitation and Date	
VIRGINIA		27.35	8/20/1969	24.98	6/1995			83.70	1996	12.52	1941
Lynchburg	1871-	7.59	8/10-11/1928	14.87	8/1928	.01	10/2000	60.58	1889	19.83	1930
Norfolk	1871-	11.40	8/31-9/1/1964	15.61	8/1942	.01	10/2000	70.72	1889	26.48	1986
Richmond	1871-	8.79	8/12/1955	18.87	7/1945	.01	10/2000	72.02	1889	22.91	1941
Roanoke	1901-	6.63	11/4/1985	16.71	8/1940	.02	10/2000	58.87	1948	23.17	1930
WASHINGTON		14.26	11/23-24/1986	57.04	12/1933			184.56	1931	2.61	1930
Bellingham	1891-	3.30	1/12/1976	13.32	12/1917	.00	8/1986	47.18	1971	19.94	1929
Olympia	1877-	5.90	11/24-25/1990	27.12	12/1933	.00	8/1946	73.48	1879	19.83	1930
Omak	1909-	2.21	6/2/1936	4.65	7/1992	.00	many	20.85	1941	4.79	1929
Seattle	1878-	5.02	10/20/2003	15.90	12/1888	.00	7/1922	57.57	1879	19.52	1952
Spokane	1881-	2.22	6/7/1888	5.85	11/1897	.00	7/1883	26.07	1948	7.54	1929
Tatoosh Is.	1869-1966	5.91	10/17-18/1930	27.48	11/1899	.01	7/1889	121.85	1879	49.56	1929
Walla Walla	1873-	3.64	1/23/1992	6.63	5/1991	.00	many	26.68	1981	11.15	1922
Yakima	1909-	1.74	8/20-21/1990	5.59	12/1996	.00	many	14.78	1996	3.90	1930
WEST VIRGINIA		19.00	7/18/1889	26.66	7/1889			94.01	1948	9.50	1930
Charleston	1885-	5.60	7/19/1961	13.54	7/1961	.07	10/1897	61.01	2003	26.13	1930
Elkins	1899-	5.45	7/7-8/1935	16.75	5/1996	.26	10/1924	72.88	1996	28.38	1930
Huntington	1897-	4.27	7/1-2/1962	12.07	1/1937	T	10/1963	59.98	1989	21.77	1930
Parkersburg	1885-	4.81	7/27-28/1947	13.20	6/1998	.07	10/1897	62.67	1890	19.70	1930
WISCONSIN		11.72	6/24/1946	18.33	6/1996			62.07	1884	12.00	1937
Bayfield	1891-	8.52	7/17/1942	11.53	7/1942	.00	12/1912	46.01	1991	14.85	1898
Eau Claire	1891-	5.98	9/10/2000	11.64	8/1980	T	12/1913	45.42	1975	16.71	1910
Green Bay	1886-	5.32	6/6/1910	10.29	6/1990	T	10/1952	38.36	1985	16.31	1930
La Crosse	1872-	7.23	10/27-28/1900	12.09	10/1900	T	11/1976	44.74	1881	16.77	1910
Madison	1869-	5.31	9/7-8/1941	10.93	7/1950	T	10/1889	52.93	1881	13.12	1895
Milwaukee	1871-	6.84	8/6/1986	10.17	6/1968	.01	2/1969	50.36	1876	18.50	1963
Wausau	1895-	4.46	6/5/1980	13.87	8/1995	.00	12/1943	48.28	1900	21.43	1976
WYOMING		6.06	8/1/1905	12.78	5/1962			55.46	1945	1.28	1960
Casper	1910-	3.09	4/13-14/1941	6.59	5/1978	.00	1/1919	20.48	1982	6.56	1988
Cheyenne	1871-	6.06	8/1/1985	7.66	4/1900	.00	9/1879	23.69	1942	5.04	1876
Jackson	1905-	1.80	3/3/1995	6.02	5/1980	.00	many	25.29	1995	8.26	1979
Lander	1891-	3.66	5/29-30/1924	7.19	4/1900	.00	1/1919	21.89	1957	6.41	1954
Rawlins	1898-	2.06	9/16/1965	4.97	5/1908	.00	many	17.00	1912	3.82	1907
Rock Springs	1889-	2.06	6/9/1965	3.67	9/1973	.00	many	14.54	1965	3.79	1953
Sheridan	1893-	4.41	7/22-23/1923	9.54	6/1944	T	8/1970	29.79	1923	8.23	1960
W. Yellowstone	1924-	2.70	6/17/1925	6.62	12/1994	.00	many	29.32	1955	11.52	1929
UNTIED STATES		43.00	7/25-26/1979 Alvin, TX	107.00	3/1942 Kukui, HI			704.83	1982 Kukui, HI	.00	1929 Greenland Ranch, CA

*all figures in Inches
**T= trace amount

CITY SNOWFALL RECORDS

State and City	Maximum 24hr Snowfall and Date		Largest Single Snowstorm and Date		Maximum Monthly Snowfall and Date		Maximum Seasonal Snowfall and Date		Maximum Depth of Snowfall and Date	
ALABAMA	19.2	12/31-1/1/1964	19.5	12/31-1/1/1964	24.0	1/1940	27.9	1963-1964		
Birmingham	13.0	3/8-9/1993	13.0	3/8-9/1993	13.0	3/1993	13.0	1992-1993	13.0	3/9/1993
Huntsville	9.5	1/30/1966			13.5	1/1964	24.1	1963-1964		
Mobile	6.0	2/14-15/1895	6.0	2/14-15/1895	6.0	2/1895	6.0	1894-1895	6.0	2/15/1895
Montgomery	11.0	12/5-6/1886	11.0	12/5-6/1886	11.0	12/1886	12.7	1886-1887	11.0	12/6/1886
ALASKA	62.0	12/29/1955	175.4	12/26-31/1955	346.1	2/1964	974.4	1952-1953		
Anchorage	28.7	3/16-17/2002	35.7	12/26-30/1955	52.1	2/1996	132.6	1954-1955	47.0	12/30/1955
Barrow	15.0	10/26/1926	17.8	10/26-30/1926	26.5	4/1916	67.3	1925-1926	29.0	4/25/1948
Fairbanks	20.1	2/11-12/1966	38.1	1/18-30/1937	65.6	1/1937	147.3	1990-1991	62.0	1/20/1937
Juneau	31.0	3/20-21/1948	45.9	4/1-4/1963	86.3	2/1965	246.3	1917-1918	43.8	3/13/1918
Nome	9.0	2/3-4/1975	12.8	12/28-31/1951	39.7	11/1915	110.0	1994-1995	60.0	4/1/1920
St. Paul Island	13.8	1/30/1964			55.8	2/1964	100.6	1972-1973		
ARIZONA	38.0	12/14/1967	82.0	2/23-26/1987	126.0	1/1895	400.9	1972-1973		
Flagstaff	27.3	12/13-14/1967	50.6	1/19-25/1949	126.0	1/1895	210.0	1972-1973	83.0	12/20/1967
Lake Havasu City*	3.0	12/13/1932	3.0	12/13/1932	3.0	12/1932	3.0	1931-1932	3.0	12/13/1932
Phoenix	1.0	1/21/1937	1.0	1/21/1937	1.0	1/1937	1.0	1936-1937	1.0	1/21/1937
Tucson	6.8	12/8/1971	6.8	12/8/1971	6.8	12/1971	8.0	1971-1972	5.5	12/8/1971
Winslow	17.0	12/13-14/1967	39.6	12/13-19/1967	39.6	12/1967	43.1	1967-1968	29.0	12/19/1967
Yuma	1.5	12/13/1932	1.5	12/13/1932	1.5	12/1932	1.5	1931-1932	1.0	12/13/1932
*early record from Parker										
ARKANSAS	25.0	1/22/1918	25.0	1/22/1918	48.0	1/1918	61.0	1917-1918		
Fort Smith	17.5	2/18-19/1921	18.3	2/18-19/1921	18.3	2/1921	22.6	1920-1921	16.0	2/19/1921
Little Rock	13.0	1/17-18/1893	13.0	1/17-18/1893	19.4	1/1918	26.6	1959-1960	13.0	1/18/1893
Texarkana					7.5	1/1962				
CALIFORNIA	68.0	1/1/1997	194.0	4/20-23/1880	390.0	1/1911	884.0	1906-1907		
Bakersfield	4.0	12/11/1932	4.0	12/11/1932	4.0	12/1932	4.0	1931-1932	3.5	12/11/1932
Bishop	14.2	1/22-23/1969	18.0	1/22-25/1969	59.0	1/1916	59.8	1915-1916	12.0	1/6/1969
Blue Canyon	60.1	1/13-14/1980			230.5	1/1916	526.2	1951-1952	155.0	NA
Burbank	4.7	1/10-11/1949	4.7	1/10-11/1949	4.7	1/1949	4.7	1948-1949	3.0	1/11/1949
Death Valley	1.1	12/26/1994	1.1	12/26/1994	1.1	12/1994	1.1	1994-1995	T	12/26/1994
Eureka	3.4	1/13/1907	3.4	1/13/1907	6.9	1/1907	6.9	1906-1907	3.4	1/13/1907
Fort Bragg										
Fresno	2.5	1/12/1930	2.5	1/12/1930	2.5	1/1912	2.5	1929-1930	2.5	1/12/1930
Los Angeles	3.0	1/19-20/1949	3.0	1/19-20/1949	3.0	1/1949	3.0	1948-1949	2.0	1/15/1932
Monterey	1.5	1/20/1962	1.5	1/20/1962	1.5	1/1962	1.5	1961-1962	1.0	1/20/1962
Mt. Shasta City*	37.4	12/6/1945	70.5	12/4-9/1952	150.0	1/1916	270.5	1889-1890	80.0	2/4/1937
Palm Springs	2.0	1/10/1930	2.0	1/10/1930	2.0	1/1930	2.0	1929-1930	2.0	1/10/1930
Red Bluff	11.4	12/20-21/1924	15.0	1/10-11/1937	15.0	1/1937	22.5	1931-1932	15.0	1/11/1937
Sacramento	3.5	1/1/1916	4.0	1/5-6/1888	4.0	1/1888	4.0	1887-1888	3.0	1/1/1916
San Diego	T	1/10/1949	T	1/10/1949	T	1/1949	T	1948-1949	0	never
San Francisco	3.7	2/5/1887	3.7	2/5/1887	3.7	2/1887	3.7	1886-1887	3.7	2/5/1887
San Jose	.5	2/3/1976	.5	2/3/1976	.5	2/1976	.5	1975-1976	T	2/3/1976
Santa Barbara	T	2/9/1939	T	2/9/1939	T	2/1939	T	1938-1939	0	never
Santa Maria	T	11/28/1975	T	11/28/1975	T	11/1975	T	1975-1976	0	never
*formerly Sisson										
COLORADO	75.8	4/15-16/1921	141.0	3/23-30/1899	249.0	3/1899	838.0	1978-1979		
Alamosa	15.8	12/13/1967			29.2	3/1973	97.5	1972-1973	18.0	12/4/1964
Aspen	25.0	NA			76.5	3/1965	279.0	1983-1984	60.0	NA
Colorado Springs	22.0	1/8-9/1987	34.1	9/28-10/3/1959	42.7	4/1957	89.4	1956-1957	20.0	10/3/1959
Denver	23.6	12/24/1982	45.7	12/1-6/1913	57.4	12/1913	118.7	1908-1909	32.6	12/6/1913
Durango	21.0	2/13/1915			100.2	2/1915	133.4	1978-1979	37.0	NA
Grand Junction	17.0	11/26-27/1919	23.6	11/26-28/1919	33.7	1/1957	65.0	1919-1920	17.5	11/28/1919
Lamar	22.0	11/25/1946			38.1	11/1946	61.6	1979-1980		
Pueblo	16.8	4/29-30/1990	21.5	11/2-5/1946	29.3	11/1946	69.6	1989-1990	16.8	11/5/1946
CONNECTICUT	28.0	3/12/1888	50.0	3/11-14/1888	73.6	3/1956	177.4	1955-1956		
Hartford	21.0	2/11-12/1983	21.0	2/11-12/1983	45.3	12/1945	115.2	1995-1996	32.8	2/5/1948
New Haven	28.0	3/12/1888	44.7	3/11-14/1888	46.3	2/1934	76.0	1915-1916	40.0	3/14/1888

CITY SNOWFALL RECORDS

State and City	Maximum 24hr Snowfall and Date		Largest Single Snowstorm and Date		Maximum Monthly Snowfall and Date		Maximum Seasonal Snowfall and Date		Maximum Depth of Snowfall and Date	
DELAWARE	25.0	2/18-19/1979	27.0	3/20-21/1958	36.5	2/1979	59.4	1995-1996		
Wilmington	22.0	12/25-26/1909	22.2	2/15-18/2003	31.6	2/2003	59.4	1995-1996	22.0	1/26/1909
DC/Washington	25.0	1/27-28/1922	28.0	1/27-28/1922	35.2	2/1899	61.9	1995-1996	34.2	2/14/1899
Dulles										
FLORIDA	4.0	3/6/1954	4.0	3/6/1954	4.0	3/1954	4.0	1953-1954		
Apalachicola	1.2	2/12/1958	1.2	2/12/1958	1.2	2/1958	1.2	1957-1958	1.2	2/12/1958
Daytona Beach	T	1/1/1977	T	1/19/1977	T	1/1977	T	1976-1977	0	never
Everglades City	0		0		0		0		0	never
Fort Meyers	T	2/13/1899	T	2/13/1899	T	2/1899	T	1898-1899	0	never
Gainesville	1.0	2/13/1899	1.0	2/13/1899	1.0	2/1899	1.0	1898-1899	1.0	2/13/1899
Jacksonville	1.9	2/12-13/1899	1.9	2/12-13/1899	1.9	2/1899	1.9	1898-1899	1.9	2/13/1899
Key West	0	0	0	0	0	0	0	0	0	never
Lakeland	1.0	1/19/1977	1.0	1/19/1977	1.0	1/1977	1.0	1976-1977	1.0	1/1/1970
Melbourne	0		0	0	0		0		0	never
Miami	T	1/19/1977	T	1/19/1977	T	1/1977	T	1976-1977	0	never
Orlando	T	1/19/1977	T	1/19/1977	T	1/1977	T	1976-1977	0	never
Pensacola	2.5	1/19/1977	3.0	2/15/1895	3.0	2/1895	3.0	1894-1895	3.0	2/15/1895
Tallahassee	2.8	2/12/1958	2.8	2/12/1958	2.8	2/1958	2.8	1957-1958	2.8	2/12/1958
Tampa	0.2	1/18-19/1977	0.2	1/18-19/1977	0.2	1/1977	0.2	1976-1977	T	1/19/1977
West Palm Beach	T	1/9/1977	T	1/19/1977	T	1/1977	T	1976-1977	0	never
GEORGIA	24.0	3/12-13/1993	24.0	3/12-13/1993	26.5	2/1895	39.0	1894-1895		
Albany	3.0	2/10/1973	3.0	2/10/1973	3.0	2/1973	3.0	1972-1973	3.0	2/10/1973
Athens	10.0	3/3/1942	10.0	3/3/1942	10.0	3/1942	14.5	1981-1982	10.0	3/3/1942
Atlanta	10.3	1/23/1940	10.3	1/23/1940	19.5	1/1940	19.5	1939-1940	10.3	1/23/1940
Augusta	14.0	2/9-10/1973	14.0	2/9-10/1973	14.0	2/1973	14.4	1972-1973	14.0	2/10/1973
Columbus	14.0	2/9-10/1973	14.0	2/9-10/1973	14.0	2/1973	14.0	1972-1973	14.0	2/10/1973
Macon	16.5	2/9-10/1973	16.5	2/9-10/1973	16.5	2/1973	16.5	1972-1973	13.0	2/10/1973
Rome	18.5	12/5-6/1886	24.0	12/3-6/1886	24.0	12/1886	24.0	1886-1887	24.0	12/6/1886
Savannah	3.6	12/8/1989	3.6	12/8/1989	3.6	12/1989	3.6	1989-1990	3.6	12/8/1989
HAWAII			12.0	2/9/1922						
Honolulu	0	0	0	0	0	0	0	0	0	0
Hilo	0	0	0	0	0	0	0	0	0	0
Kahului	0	0	0	0	0	0	0	0	0	0
Lihue	0	0	0	0	0	0	0	0	0	0
IDAHO	38.0	2/11/1959	60.0	12/25-27/1937	143.8	1/1954	441.8	1949-1950		
Boise	17.0	12/16-17/1884	23.6	12/15-17/1884	36.6	12/1884	50.0	1916-1917	22.0	12/17/1884
Lewiston	12.8	1/2-3/1966	14.4	11/23-25/1961	27.7	2/1916	55.0	1915-1916	14.0	1/16/1950
Pocatello	14.6	3/23/1916	14.6	3/23/1916	33.7	12/1983	93.3	1992-1993	21.0	1/31/1949
Salmon	11.0	NA			23.2	12/1983	56.2	1983-1984	17.0	NA
Sandpoint	19.0	NA			68.8	1/1969	181.5	1915-1916	46.0	1/31/1969
ILLINOIS	36.0	2/27-28/1900	37.8	2/27-28/1900	52.0	1/1979	103.3	1978-1979		
Cairo	17.0	12/7-8/1917	17.0	12/7-8/1917	24.2	1/1918	47.7	1917-1918	19.0	1/15/1918
Chicago	19.8	1/26-27/1967	23.0	1/26-27/1967	42.5	1/1918	93.3	1978-1979	29.0	1/14/1979
Moline	16.4	1/3/1971			32.9	12/2000	73.6	1974-1975	28.0	1/19/1979
Peoria	18.0	2/27-28/1900	18.0	2/27-28/1900	27.4	1/1979	54.3	1978-1979	20.0	1/14/1979
Quincy	10.9	2/25/1993			24.4	3/1960	45.2	1977-1978		
Rockford	16.3	1/6-7/1918			36.1	1/1918	74.5	1978-1979	27.0	1/16/1979
Springfield	17.0	12/12/1973	17.5	2/27-28/1900	24.4	2/1900	52.1	1977-1978	16.3	1/11/1918
Urbana	14.0	3/20/1906			32.0	3/1906	67.2	1977-1978	18.0	2/10/1979
INDIANA	20.0	1/14-15/1918	37.0	2/14-19/1958	86.1	1/1978	172.0	1977-1978		
Evansville	20.0	1/14-15/1918	22.3	1/13-15/1918	41.0	1/1918	67.7	1917-1918	28.0	1/15/1918
Fort Wayne	13.6	3/9-10/1964	13.9	2/20-21/1912	29.5	1/1982	81.2	1981-1982	17.0	2/12/1978
Indianapolis	12.5	2/24-25/1965	16.1	2/16-18/1910	30.6	1/1978	58.2	1981-1982	20.0	1/20/1978
South Bend	17.5	11/25-26/1977	17.9	1/26-27/1967	86.1	1/1978	172.0	1977-1978	41.0	1/30/1978

CITY SNOWFALL RECORDS

State and City	Maximum 24hr Snowfall and Date		Largest Single Snowstorm and Date		Maximum Monthly Snowfall and Date		Maximum Seasonal Snowfall and Date		Maximum Depth of Snowfall and Date	
INDIANA Con't										
Terre Haute	14.0	12/31/1973			25.2	12/1973	29.3	1995-1996		
IOWA	21.0	2/18/1962	30.8	2/17-21/1912	43.7	12/1968	90.4	1908-1909		
Burlington	13.0	3/3/1912	16.0	3/10-14/1951	28.3	3/1912	68.1	1911-1912	19.0	1/18/1936
Des Moines	19.8	12/31-1/1/1942	21.5	12/29-1/1/1942	37.0	1/1886	72.0	1911-1912	21.8	1/1/1942
Dubuque	15.5	3/4-5/1959	19.3	4/9-10/1973	37.6	12/2000	77.4	1974-1975	27.0	2/9/1893
Mason City	16.0	1/6/1918			38.2	2/1924	86.0	1928-1929		
Sioux City	20.0	4/10/1913	22.9	4/9-10/1913	29.1	1/1982	72.8	1916-1917	28.0	3/13/1962
Waterloo	14.8	1/3-4/1971	14.8	1/3-4/1971	34.0	12/2000	59.4	1961-1962	20.0	3/5/1962
KANSAS	26.0	12/28-29/1954	37.0	3/23-24/1912	55.9	3/1912	103.6	1983-1984		
Concordia	17.2	3/16-17/1924	17.2	3/16-17/1924	25.0	3/1891	59.1	1959-1960	25.0	3/16/1960
Dodge City	17.5	3/9-10/1922	20.5	2/24-28/1903	27.7	2/1903	58.8	1992-1993	18.0	3/10/1922
Goodland	19.3	10/26/1997	21.7	11/26-28/1983	32.0	3/1912	102.0	1979-1980	19.0	10/26/1997
Topeka	18.7	2/27-28/1900	18.7	2/27-28/1900	27.1	2/1900	47.9	1911-1912	19.0	3/15/1960
Wichita	13.5	3/15-16/1970	15.0	1/17-18//1962	20.5	2/1913	39.7	1911-1912	17.0	1/19/1962
KENTUCKY	25.0	3/14/1993	28.3	1/5-8/1996	46.5	3/1960	108.2	1959-1960		
Bowling Green	18.0	3/9/1960	27.0	3/7-11/1960	32.0	3/1960	48.0	1959-1960		
Lexington	13.4	1/26/1943	16.1	1/13-15/1917	26.4	1/1978	53.0	1916-1917	14.0	1/22/1978
Louisville	20.0	2/5-6/1998	22.4	2/4-6/1998	28.4	1/1978	50.2	1917-1918	22.0	2/6/1998
LOUISIANA	24.0	2/14-15/1895	24.0	2/14-15/1895	24.0	2/1895	24.0	1894-1895		
Baton Rouge	12.5	2/14-15/1895	12.5	2/14-15/1895	12.5	2/1895	4.5	1851-1852	4.5	1/13/1852
Lake Charles	22.0	2/14-15/1895	22.0	2/14-15/1895	22.0	2/1895	22.0	1894-1895	22.0	2/15/1895
New Orleans	8.2	2/14-15/1895	8.2	2/14-15/1895	8.2	2/1895	8.2	1894-1895	8.0	2/15/1895
Shreveport	11.0	12/21-22/1929	11.0	12/11-12/1929	12.4	1/1948	13.0	1929-1930	11.0	12/21/1929
MAINE	40.0	12/31/1962	57.0	2/25-27/1969	89.0	2/1969	238.5	1968-1969		
Bangor										
Caribou	28.6	3/14/1984	30.0	11/26-27/1974	59.9	12/1972	181.1	1954-1955	62.0	2/1/1977
Eastport	18.5	4/9-10/1946	18.5	4/9-10/1946	53.7	1/1952	187.5	1906-1907	42.0	3/14/1923
Portland	24.4	1/17-18/1979	27.1	1/17-18/1979	62.4	1/1979	141.5	1970-1971	55.0	1/17/1923
MARYLAND	31.0	3/29/1942	49.0	2/15-17/2003	81.5	2/2003	174.9	1901-1902		
Baltimore	24.5	1/28/1922	28.2	2/15-18/2003	40.5	2/2003	62.5	1995-1996	30.0	2/13/1899
Frederick	29.5	1/7/1996	32.0	1/6-8/1996	35.0	1/1918	60.4	1960-1961	32.0	1/8/1996
Salisbury					25.5	2/1967				
MASSACHUSETTS	33.0	3/31-4/1/1997	47.0	3/2-3/1947	78.0	2/1893	162.0	1892-1893		
Boston	25.4	3/31-4/1/1997	27.6	2/17-18/2003	41.6	2/2003	107.6	1995-1996	32.0	1/10/1996
Nantucket	20.1	2/27-28/1952	31.3	3/3-5/1960	40.2	3/1960	82.0	1903-1904	23.0	2/28/1952
Pittsfield	24.0	3/22/1977	36.0	3/11-14/1888	51.7	12/1969	121.9	1955-1956	40.2	3/20/1956
Worcester	33.0	3/31-4/1/1997	33.0	3/31/1997	46.8	1/1987	132.9	1995-1996	42.0	2/5/1961
MICHIGAN	33.5	11/30/1960	61.7	12/8-12/1995	129.5	1/1982	391.9	1978-1979		
Alpena	19.3	3/15-16/1997	21.3	12/27-29/1962	55.8	1/1997	180.0	1996-1997	33.0	2/29/1904
Detroit	24.5	4/6/1886	24.5	4/6/1886	38.4	2/1908	78.0	1925-1926	26.0	3/5/1900
Escanaba	14.5	1/1/1915	18.8	4/29-30/1909	40.4	1/1888	93.1	1922-1923	37.0	2/26/1962
Flint	19.8	12/26-27/1967			35.3	12/2000	82.9	1974-1975		
Grand Rapids	16.1	1/26/1978	28.2	12/25-27/2001	59.2	12/2000	144.1	1951-1952	27.0	1/27/1978
Houghton	26.5	1/18/1996			119.0	12/1978	376.1	1978-1979		
Houghton Lake	24.0	3/4-5/1985			43.5	3/1923	124.1	1970-1971	24.0	1/26/1979
Lansing	20.4	1/6-27/1967	24.0	1/26-27/1967	34.6	1/1999	88.8	1951-1952	24.0	1/28/1978
Marquette	28.0	3/13-14/1997	43.0	3/11-15/1988	91.9	2/2002	319.8	2001-2002	80.0	2/16/1971
Muskegon	22.0	1/10/1982	37.4	12/16-19/1963	102.4	1/1982	173.9	1981-1982	33.0	1/16/1979
Sault Ste. Marie	26.6	12/10/1995	61.7	12/8-12/1995	98.8	12/1995	222.1	1995-1996	50.0	12/11/1995
MINNESOTA	36.0	1/6-7/1994	46.5	1/6-8/1994	66.4	3/1965	147.5	1936-1937		
Bemidji	14.0	1/18/1996			29.0	3/1951	82.7	1950-1951		

CITY SNOWFALL RECORDS

State and City	Maximum 24hr Snowfall and Date		Largest Single Snowstorm and Date		Maximum Monthly Snowfall and Date		Maximum Seasonal Snowfall and Date		Maximum Depth of Snowfall and Date	
MINNESOTA Con't										
Duluth	25.4	12/5-6/1950	35.2	12/5-8/1950	58.5	1/1969	135.4	1995-1996	48.0	3/18/1965
International Falls	17.7	1/10/1975	21.2	3/2-5/1966	43.9	12/1992	116.0	1995-1996	38.0	3/7/1966
Minneapolis/St. Paul	21.0	10/31-11/1/1991	28.4	10/31-11/1/1991	46.9	11/1991	98.6	1983-1984	38.0	1/20/1982
Rochester	15.4	1/22/1982	16.0	3/29-31/1934	35.3	12/2000	89.0	1996-1997	30.0	2/15/1917
St. Cloud	14.5	3/3-4/1985			51.7	3/1965	87.9	1964-1965		
MISSISSIPPI	18.0	12/23/1963	18.0	12/23/1963	23.0	1/1966	25.2	1967-1968		
Jackson	11.7	1/28/1904	11.7	1/28/1904	11.7	1/1904	11.7	1903-1904	11.7	1/28/1904
Meridian	15.0	12/31/1963	15.0	12/31/1963	17.6	12/1963	18.6	1963-1964	15.0	1/1/1964
Vicksburg	10.1	1/2/1919	10.1	1/2/1919	10.1	1/1919	10.1	1918-1919	10.1	1/2/191999
MISSOURI	27.6	3/16-17/1970	35.0	1/22-25/1898	47.5	1/1918	74.0	1888-1889		
Columbia	19.7	1/18-19/1995	19.7	1/18-19/1995	24.5	3/1960	58.5	1977-1978	19.0	1/19/1995
Kansas City	25.0	3/23-24/1912	25.0	3/23-24/1912	40.2	3/1912	67.0	1911-1912	25.0	3/24/1912
St. Joseph	12.4	2/23-24/1942	13.6	1/20-21/1958	24.0	1/1936	43.4	1977-1978	18.0	3/17/1960
St. Louis	20.4	3/30-31/1890	20.4	3/30-31/1890	28.8	3/1912	67.6	1911-1912	20.4	3/31/1890
Springfield	20.0	2/20-21/1912	20.0	2/20-21/1912	24.1	2/1912	54.5	1911-1912	20.0	2/21/1912
MONTANA	48.0	5/28-29/1982	77.5	1/17-20/1972	131.1	1/1972	418.1	1977-1978		
Billings	23.7	4/4/1955	42.3	4/2-5/1955	42.3	4/1955	98.7	1996-1997	35.0	4/5/1955
Butte	40.5	5/13/1927			41.5	5/1927	115.8	1926-1927		
Glasgow	12.2	4/8-9/1995			32.9	1/2004	70.2	2003-2004	29.0	2/11/2004
Great Falls	16.8	4/20/1973	25.0	4/3-9/1975	35.4	4/1967	117.5	1988-1989	24.0	4/9/1975
Havre	26.9	3/28-29/1977	33.8	5/2-4/1899	41.4	1/1971	74.1	1978-1979	32.2	5/2/1899
Helena	21.5	11/11-12/1959	28.5	12/5-14/1917	46.4	12/1880	112.8	1880-1881	24.0	12/13/1917
Kalispell	20.1	11/19/1996	23.0	11/18-19/1996	52.1	12/1990	146.3	1996-1997	30.0	12/23/1951
Miles City	28.0	3/1894			35.3	3/1894	74.1	1911-1912		
Missoula	22.8	2/12/1936	28.3	2/11-13/1936	54.1	12/1996	111.2	1996-1997	29.3	2/19/1936
NEBRASKA	24.0	2/11/1965	41.0	1/2-4/1949	59.6	1/1949	112.0	1972-1973		
Grand Isle	15.0	2/17-18/1984	19.0	3/3-6/1915	29.7	2/1915	86.7	1914-1915	28.4	2/11/1905
Lincoln	19.0	2/11/1965	19.0	2/11/1965	26.1	2/1965	59.4	1914-1915	21.0	2/11/1965
Norfolk	22.5	2/18-19/1984	25.0	2/18-19/1984	25.0	2/1984	86.1	1983-1984	25.0	2/19/1984
North Platte	15.1	3/27-28/1980	17.7	3/29-30/1949	27.8	3/1912	66.3	1979-1980	18.0	1/4/1949
Omaha	18.3	2/11/1965	18.9	3/14-15/1923	29.2	3/1912	67.5	1911-1912	27.0	3/16/1960
Scottsbluff	17.6	4/25/1935	28.7	4/11-16/1920	29.7	4/1927	78.5	1979-1980	23.0	4/14/1927
Valentine	24.0	3/11-12/1977	28.1	4/17-19/1920	51.0	3/1977	90.3	1919-1920	29.2	2/17/1936
NEVADA	38.0	2/7-8/1985	75.0	1/18-22/1969	139.0	1/1969	412.0	1968-1969		
Austin	24.0	NA			59.8	3/1946	144.0	1974-1975	27.0	NA
Elko	18.3	1/1996			48.5	1/1916	79.1	1931-1932	20.0	1/27/1996
Ely	13.1	1/1943			24.8	1/1967	79.0	1963-1964		
Las Vegas	12.0	12/21//1909	12.0	12/21/1909	16.7	1/1949	16.7	1948-1949	12.0	12/21/1909
Reno	22.5	1/17/1916	28.5	1/17-18/1916	65.7	1/1916	72.3	1915-1916	30.0	1/18/1890
Tonopah	16.0	11/9/1946			37.0	11/1946	59.1	1946-1947	22.0	11/10/1946
Winnemucca	18.2	1/3-4/1890	18.2	1/3-4/1890	33.0	1/1890	75.0	1889-1890	32.6	1/13/1890
NEW HAMPSHIRE	56.0	11/23-24/1943	97.8	2/24-28/1969	172.8	2/1969	566.4	1968-1969		
Concord	19.0	1/6-7/1944	27.5	3/11-13/1888	59.0	2/1893	112.3	1995-1996	37.0	2/13/1923
Mt. Washington	49.3	2/24-25/1969	97.8	2/24-28/1969	172.8	2/1969	566.4	1968-1969		
NEW JERSEY	35.0	1/6-7/1996	35.0	1/6-7/1996	50.1	12/1880	122.0	1995-1996		
Atlantic City	18.0	2/17/1902	21.6	2/15-18/1902	35.2	2/1967	46.9	1966-1967	27.9	2/14/1899
Newark	26.0	12/26-27/1947	27.8	1/7-8/1996	33.4	1/1994	78.4	1995-1996	26.0	12/27/1947
Trenton	22.0	2/12-13/1899	23.2	3/13-14/1993	34.0	2/1899	61.0	1898-1899	30.0	2/14/1899
NEW MEXICO	36.0	11/22/1931	51.0	2/4-6/1989	144.0	3/1912	483.0	1911-1912		
Albuquerque	14.2	12/28-29/1958	14.2	12/28-29/1958	14.7	12/1958	37.4	1972-1973	14.2	12/29/1958
Clayton	12.7	11/14/1961			21.0	12/1918	55.3	1923-1924		
Gallup	11.5	12/4/1992			29.1	12/1992	65.1	1990-1991	13.0	12/22/1992

CITY SNOWFALL RECORDS

State and City	Maximum 24hr Snowfall and Date		Largest Single Snowstorm and Date		Maximum Monthly Snowfall and Date		Maximum Seasonal Snowfall and Date		Maximum Depth of Snowfall and Date	
NEW MEXICO Con't										
Roswell	16.5	2/4-5/1988	17.5	12/7-10/1960	23.3	2/1905	32.6	1987-1988	16.5	2/5/1988
Santa Fe	16.0	11/23/1986			31.5	12/1967	66.0	1963-1964	16.0	11/24/1986
Silver City	13.0	NA			24.4	3/1958	37.0	1912-1913		
NEW YORK	68.0	1/9/1976	102.0	1/27-31/1966	192.0	1/1978	466.9	1976-1977		
Albany	30.4	3/12/1888	46.7	3/11-14/1888	57.5	12/1969	112.5	1970-1971	46.0	3/14/1888
Binghampton	23.0	2/3-4/1961	23.1	2/2-3/1961	59.6	12/1969	138.6	1995-1996	33.0	12/28/1969
Buffalo	37.9	12/9-10/1995	81.5	12/24-28/2001	82.7	12/2001	199.4	1976-1977	45.0	12/30/2001
New York City	26.4	12/26-27/1947	26.4	12/26-27/1947	37.9	2/1894	87.2	1867-1868	26.4	12/27/1947
Rochester	29.8	2/28-3/1/1900	43.5	2/28-3/2/1900	64.8	2/1958	161.7	1959-1960	40.0	12/1959
Syracuse	35.6	3/13-14/1993	43.4	3/12-15/1993	72.6	2/1958	192.1	1992-1993	48.0	2/19/1958
NORTH CAROLINA	31.0	3/2/1927	60.0	5/7-8/1992	60.0	5/1992	103.4	1959-1960		
Asheville	16.5	3/12-13/1993	33.0	12/3-6/1886	33.0	12/1886	48.2	1968-1969	33.0	12/6/1886
Cape Hatteras	13.3	1/24-25/1989	13.3	1/24-25/1989	13.5	12/1989	14.4	1917-1918	13.0	1/25/1989
Charlotte	16.5	2/14-15/1902	17.4	2/14-17/1902	19.3	3/1960	22.6	1959-1960	16.0	2/17/1902
Greensboro	16.5	2/27/1987			25.4	2/1987	39.8	1986-1987		
Raleigh	20.3	1/24-25/2000	20.3	1/24-25/2000	25.8	1/2000	31.6	1892-1893	20.0	1/25/2000
Wilmington	15.0	12/23-24/1989	15.0	12/23-24/1989	15.3	12/1989	18.4	1911-1912	15.0	12/24/1989
NORTH DAKOTA	24.0	2/25/1930	40.4	3/2-5/1966	46.4	12/1996	117.0	1996-1997		
Bismarck	15.9	10/28-29/1991	28.3	11/22-27/1993	31.1	3/1975	101.6	1996-1997	18.0	3/5/1966
Devils Lake	17.5	3/4/1966	30.5	3/3-5/1966	30.7	3/1966	90.2	1949-1950	35.0	3/5/1966
Dickinson	23.8	4/28/1984			28.5	4/1984	75.9	1981-1982		
Fargo	19.4	1/6-7/1989	24.4	1/6-8/1989	31.5	1/1989	117.0	1996-1997	32.0	3/5/1997
Grand Forks	18.0	3/4/1966	40.4	3/2-5/1966	46.4	12/1996	98.2	1995-1996	38.0	1/11/1997
Williston	15.0	4/13-14/1986			30.9	3/1975	72.2	1993-1994	27.1	3/15/1897
OHIO	22.0	1/31/1878	68.9	11/9-14/1996	94.8	11/1996	161.5	1959-1960		
Akron	20.6	4/3-4/1987	24.6	4/19-20/1901	37.5	1/1978	82.0	1977-1978	20.0	12/12/1974
Cincinnati	20.0	1/14-15/1863	20.0	1/14-15/1863	31.5	1/1978	53.9	1977-1978	18.5	2/6/1998
Cleveland	22.0	1/31/1878	22.2	11/9-11/1913	48.2	1/1978	101.1	1995-1996	21.0	2/2/1978
Columbus	12.3	4/3-4/1987	15.3	2/16-18/1910	36.2	1/1978	67.8	1909-1910	17.0	1/23/1978
Dayton	13.4	12/22/1981			40.2	1/1978	62.7	1977-1978	22.0	2/21/1978
Toledo	19.0	2/28/1900	20.2	2/28-3/1/1900	32.6	1/1978	74.9	1977-1978	21.0	3/1/1900
Youngstown	20.7	11/24-25/1950	28.7	11/24-27/1950	36.4	1/1999	85.3	1950-1951	26.0	11/26/1950
OKLAHOMA	23.0	2/21/1971	36.0	2/21-22/1971	46.0	2/1903	87.3	1911-1912		
Oklahoma City	11.7	3/19/1924	11.7	3/19-20/1924	20.7	3/1924	25.2	1947-1948	12.0	1/8/1988
Tulsa	12.9	3/8-9/1994	14.1	3/8-9/1994	19.7	3/1924	28.7	1923-1924	14.0	3/9/1994
OREGON	39.0	1/9/1980	119.0	3/13-16/1975	313.0	1/1950	903.0	1949-1950		
Astoria	20.5	1/24-25/1972			58.7	1/1972	58.7	1971-1972	18.0	1/27/1969
Bandon	6.0	1/27/1969			18.0	1/1969	18.0	1968-1969	6.0	1/27/1969
Bend	28.0	12/10/1919	47.0	12/9-10/1919	56.5	1/1950	90.6	1973-1974	55.0	12/10/1919
Burns	11.0	3/17/1967			36.6	1/1956	96.7	1955-1956	26.0	NA
Eugene	22.9	1/25-26/1969	29.0	1/25-26/1969	55.0	2/1937	57.5	1936-1937	34.0	1/27/1969
Lakeview	20.0	2/7/1999			50.0	2/1999	147.0	1998-1999		
Medford	11.0	12/11-12/1919	12.4	2/19-22/1917	22.6	1/1930	31.6	1955-1956	10.3	1/13/1930
Pendleton	16.1	2/24-25/1994	32.0	11/19-21/1921	41.6	1/1950	80.0	1921-1922	32.0	11/21/1921
Portland	16.0	1/31-2/1/1937	27.5	12/21-24/1892	41.4	1/1950	60.9	1892-1893	19.0	2/6/1893
Roseburg	13.5	1/26/1969	39.5	1/26-30/1969	44.2	1/1969	46.3	1968-1969	27.0	1/27/1969
Salem	25.0	2/1/1937	25.2	1/31-2/1/1937	32.8	1/1950	34.9	1968-1969	25.2	2/1/1937
PENNSYLVANIA	40.0	3/13-14/1993	50.0	3/19-21/1958	86.0	12/1890	225.0	1890-1891		
Allentown	25.2	1/6-7/1996	25.6	1/6-8/1996	43.2	1/1925	75.2	1993-1994	28.0	2/12/1983
Erie	26.5	12/11-12/1944	30.2	12/11-14/1944	66.9	12/1989	149.1	2000-2001	39.0	12/20/1989
Harrisburg	25.0	2/11-12/1983	32.0	1/6-8/1996	38.9	1/1996	81.3	1960-1961	32.0	1/13/1996
Philadelphia	27.6	1/6-7/1996	30.7	1/6-7/1996	33.8	1/1996	65.5	1995-1996	28.0	1/8/1996
Pittsburgh	23.8	3/13-14/1993	27.7	11/24-29/1950	41.3	12/1890	82.0	1950-1951	26.0	1/22/1978

CITY SNOWFALL RECORDS

State and City	Maximum 24hr Snowfall and Date		Largest Single Snowstorm and Date		Maximum Monthly Snowfall and Date		Maximum Seasonal Snowfall and Date		Maximum Depth of Snowfall and Date	
PENNSYLVANIA Con't										
WB-Scranton	20.6	1/7-8/1996	21.5	1/6-8/1996	42.3	1/1994	98.3	1995-1996	29.0	1/13/1996
Williamsport	23.1	1/12-13/1964	24.1	1/12-13/1964	40.1	1/1987	87.7	1995-1996	29.0	2/23/1902
RHODE ISLAND	34.0	2/8-9/1945	38.0	2/5-7/1978	62.0	3/1956	129.1	1995-1996		
Block Island	21.7	2/6-7/1978			30.0	1/1978	51.7	1977-1978		
Providence	27.6	2/6-7/1978	28.6	2/6-7/1978	37.4	1/1996	106.1	1995-1996	30.0	2/5/1961
SOUTH CAROLINA	24.0	2/9-10/1973	28.9	2/15-16/1929	33.9	2/1969	60.3	1968-1969		
Charleston	6.6	12/22-23/1989	7.1	2/9-10/1973	8.0	12/1989	8.0	1989-1990	8.0	12/25/1989
Columbia	15.7	2/9-10/1973	16.0	2/9-10/1973	16.0	2/1973	18.2	1972-1973	14.0	2/11/1973
Florence	17.0	2/9-10/1973	17.0	2/9-10/1973	17.0	2/1973	17.0	1972-1973	17.0	2/11/1973
Greenville	14.4	12/16-17/1930	14.4	12/16-17/1930	18.8	12/1945	20.4	1935-1936	14.0	12/17/1930
SOUTH DAKOTA	38.9	10/26/1996	114.6	2/25-3/2/1998	94.0	3/1950	324.0	1996-1997		
Aberdeen	15.0	4/18-19/1970	25.3	11/22-23/1993	30.1	11/1993	110.8	1936-1937	30.0	1/18/1997
Huron	19.5	1/29-30/2001	25.0	3/3-4/1985	39.9	2/1962	85.1	2000-2001	36.0	2/25/2001
Pierre	15.5	2/7-8/2001								
Rapid City	22.0	4/17-18/1970	32.2	4/11-16/1927	38.5	4/1927	80.9	1985-1986	22.0	4/18/1970
Sioux Falls	26.0	2/17-18/1962	32.2	2/16-18/1962	48.4	2/1962	94.7	1968-1969	34.0	2/28/1969
TENNESSEE	22.0	3/9/1960	56.0	3/12-14/1993	56.0	3/1993	75.5	1959-1960		
Bristol	16.2	11/21/1952			29.9	3/1960	51.0	1959-1960		
Chattanooga	20.3	3/13-14/1993	20.3	3/13-14/1993	20.3	3/1993	47.0	1947-1948	20.0	3/14/1993
Knoxville	18.2	11/21-22/1952	22.5	12/4-6/1886	25.7	2/1895	56.7	1959-1960	22.5	12/6/1886
Memphis	18.0	3/16-17/1892	18.5	3/15-17/1892	18.5	3/1892	25.1	1917-1918	18.0	3/17/1892
Nashville	17.0	3/17/1892	21.5	3/15-18/1892	21.5	3/1892	38.5	1959-1960	17.0	3/17/1892
TEXAS	24.0	2/3-4/1956	61.0	2/1-8/1956	61.0	2/1956	65.0	1923-1924		
Abilene	9.3	4/5/1996	9.4	2/12-13/1890	13.5	1/1973	18.4	1918-1919	8.0	1/11/1973
Amarillo	20.6	3/25-26/1934	20.6	3/25-26/1934	28.7	2/1903	48.7	1918-1919	16.5	2/26/1903
Austin	9.7	11/22-23/1937	9.7	11/22-23/1937	9.7	11/1973	9.7	1937-1938	9.7	11/23/1937
Brownsville	6.0	2/14-15/1895	6.0	2/14-15/1895	6.0	2/1895	6.0	1894-1895	6.0	2/15/1895
Corpus Cristi	5.0	1/28 29/1897	5.0	1/28 29/1897	6.0	1/1897	6.0	1896 1897	5.0	1/29/1897
Dallas	7.4	1/15-16/1964	7.4	1/15-16/1964	9.0	1/1918	9.9	1963-1964	7.4	1/16/1964
Del Rio	8.6	1/12/1985	8.7	1/12-13/1985	9.8	1/1985	9.8	1984-1985	8.0	1/12/1985
El Paso	16.8	12/13-14/1987	22.4	12/13-15/1987	25.9	12/1987	35.0	1982-1983	16.0	12/15/1987
Fort Worth	12.1	1/15-16/1964	12.1	1/15-16/1964	13.5	2/1978	17.6	1977-1978	12.1	1/16/1964
Galveston	15.4	2/14-15/1895	15.4	2/14-15/1895	15.4	2/1895	15.4	1894-1895	15.4	2/15/1895
Great Bend NP					26.0	NA				
Houston	20.0	2/14-15/1895	20.0	2/14-15/1895	20.0	2/1895	20.0	1894-1895	20.0	2/15/1895
Laredo	4.5	3/10/1932	4.5	3/10/1932	4.5	3/1932	4.5	1931-1932	4.5	3/10/1932
Lubbock	16.3	1/20-21/1983	16.9	1/20-21/1983	25.3	1/1983	41.2	1982-1983	17.0	1/21/1983
Midland	9.8	12/10-11/1998	9.8	12/10-11/1998	9.8	12/1998	12.3	1982-1983	9.5	12/11/1998
Port Arthur	12.0+	2/14-15/1895	12.0+	2/14-15/1895	12.0+	2/1895	12.0+	1894-1895	12.0+	2/15/1895
San Angelo	11.5	1/14/1917	13.0	2/25-26/1924	13.0	1/1926	17.3	1918-1919	10.0	1/16/1919
San Antonio	13.2	1/12/1985	13.5	1/12-13/1985	15.9	1/1985	15.9	1984-1985	9.0	1/13/1985
Victoria	4.0	2/12/1960	5.0	1/23-24/1940	5.0	1/1940	5.0	1939-1940	5.0	1/24/1940
Waco	13.0	2/26/1924	13.0	2/26/1924	13.0	12/1929	15.0	1929-1930	13.0	2/26/1924
Wichita Falls	9.7	3/4-5/1989			11.9	1/1966	14.3	1957-1958		
UTAH	55.5	1/5/1994	105.0	1/23-26/1965	244.5	12/1983	846.8	1982-1983		
Blanding	16.6	4/3/1911			55.0	12/1909	117.5	1978-1979		
Milford	16.3	12/19-20/1919	18.6	12/16-20/1949	30.6	12/1972	101.8	1972-1973	24.0	12/20/1949
Saint George	10.0	1/5/1974			17.0	1/1898	22.5	1897-1898		
Salt Lake City	18.4	10/17-18/1984	23.6	1/6-10/1993	50.3	1/1993	117.3	1951-1952	26.0	1/11/1993
Wendover	10.0	1/23/1934			15.2	12/1921	22.4	1961-1962		
VERMONT	37.0	3/14/1984	50.0	3/2-6/1947	90.3	12/2000	318.6	1970-1971		
Burlington	24.2	1/13-14/1934	29.8	12/25-28/1969	56.7	12/1970	145.4	1970-1971	40.0	1/1964

CITY SNOWFALL RECORDS

State and City	Maximum 24hr Snowfall and Date		Largest Single Snowstorm and Date		Maximum Monthly Snowfall and Date		Maximum Seasonal Snowfall and Date		Maximum Depth of Snowfall and Date	
VIRGINIA	33.5	3/2-3/1994	48.0	1/6-7/1996	54.0	2/1899	124.2	1995-1996		
Lynchburg	21.0	1/7/1996	21.4	1/6-8/1996	31.8	1/1966	58.4	1986-1987	25.0	1/30/1966
Norfolk	17.7	12/27-28/1892	18.6	12/26-28/1892	24.4	2/1989	41.9	1979-1980	18.6	12/28/1892
Richmond	21.6	1/23-24/1940	21.6	1/23-24/1940	28.5	1/1940	38.9	1961-1962	21.6	1/24/1940
Roanoke	22.2	1/6-7/1996	24.9	1/6-8/1996	41.2	1/1966	72.9	1986-1987	23.0	1/1987
WASHINGTON	65.0	1/23-24/1994	129.0	2/24-26/1910	363.0	1/1925	1140.0	1998-1999		
Bellingham	10.0	12/31/1968			34.5	3/1951	44.9	1970-1971	22.5	3/8/1951
Olympia	20.5	1/1972	21.0	1/1972	58.7	1/1969	81.5	1968-1969	24.4	1/31/1969
Omak	14.0	NA			36.3	12/1931	56.0	1937-1938	40.0	NA
Seattle	21.5	2/2/1916	32.5	1/31-2/2/1916	57.2	1/1950	67.5	1968-1969	29.0	2/2/1916
Spokane	13.0	1/6-7/1950	20.9	12/18-21/1965	56.9	1/1950	93.5	1949-1950	42.0	2/1/1969
Tatoosh Island	13.8	2/3/1893	24.7	3/5-10/1951	32.3	1/1950	36.8	1955-1956	16.0	3/7/1951
Walla Walla	14.0	2/1-2/1916	27.7	1/31-2/2/1916	33.4	2/1916	76.9	1915-1916	37.5	2/4/1916
Yakima	16.0	2/1-2/1916	21.4	12/18-21/1964	37.5	12/1964	74.4	1955-1956	27.0	12/30/1996
WEST VIRGINIA	35.0	1/27-28/1998	63.2	11/23-30/1950	104.0	1/1977	301.0	1959-1960		
Charleston	17.1	3/13-14/1993	25.6	11/24-25/1950	39.5	1/1978	110.2	1995-1996	23.0	1/1978
Elkins	20.7	11/24-25/1950	29.5	11/24-25/1950	54.1	1/1985	136.6	1995-1996	23.0	1/8/1996
Huntington	21.1	3/13-14/1993	21.7	3/12-14/1993	30.3	1/1978	60.9	1935-1936	22.0	3/15/1993
Parkersburg	18.3	12/17/1890	34.4	11/24-29/1950	39.6	1/1994	55.1	1935-1936	28.3	11/28/1950
WISCONSIN	26.0	12/27/1904	39.0	11/1-3/1989	103.5	1/1997	277.0	1996-1997		
Bayfield	18.2	12/28/1982			58.3	12/1996	150.0	1996-1997		
Eau Claire	14.0	3/28/1965			32.1	1/1999	89.3	1996-1997		
Green Bay	22.0	1/1889			32.5	1/1889	79.6	1922-1923	23.8	2/10/1893
La Crosse	16.0	1/3/1971	18.5	3/5-6/1959	39.6	1/1929	78.7	1961-1962	31.0	3/6/1959
Madison	17.3	12/3/1990			37.0	2/1994	76.1	1978-1979	33.0	12/21/2000
Milwaukee	20.3	2/4-5/1924	26.0	2/19-21/1898	52.6	1/1918	109.8	1885-1886	33.0	1/27/1979
Wausau	11.7	11/23/1991			36.8	12/1972	100.3	1995-1996		
WYOMING	41.0	4/19/1973	58.0	4/18-20/1973	188.5	1/1933	491.6	1932-1933		
Casper	31.1	12/23-24/1982	31.1	12/23-24/1982	62.8	12/1982	151.6	1982-1983	31.0	12/24/1982
Cheyenne	19.8	11/20/1979	25.2	11/20-21/1979	46.5	4/1905	121.5	1979-1980	26.0	11/21/1979
Jackson	22.0	1/25-26/1969	22.0	1/25-26/1969	56.0	1/1969	153.3	1966-1967	33.0	1/26/1969
Lander	28.6	4/24/1999	52.7	4/24-26/1999	70.4	4/1999	219.3	1972-1973	30.2	4/14/1945
Rawlins	19.0	1/29/1980			42.9	1/1980	105.2	1972-1973	27.0	1/30/1980
Rock Springs	14.0	5/3/1975	19.3	9/16-17/1965	26.7	1/1980	81.4	1949-1950	17.0	NA
Sheridan	26.7	4/3-4/1955	38.9	4/3-4/1955	43.5	12/1989	128.8	1954-1955	33.0	4/4/1955
West Yellowstone	24.0	1/19/1962			93.9	12/1994	316.6	1994-1995	76.0	NA
UNTIED STATES	75.8	4/14-15/1921 Silver Lake, CO	194.0	4/20-23/1880 Norden, CA	390.0	1/1911 Tamarack, CA	1140.0	1998-1999 Mt. Baker, WA	451.0	3/11/1911 Tamarack, CA

*all figures in Inches

BIBLIOGRAPHY & SOURCES

Abley, Mark, ed. *Stories From the Ice Storm* McClelland & Stewart Publishers, Toronto, Canada, 1999

Abramovich, Ron et al *Climates of Idaho* University of Idaho, Moscow, Idaho, 1998

Allaby, Michael *The Facts on File Weather and Climate Handbook* Checkmark Books, New York, 2002

Aldrich, John H. and Myra Southland Meadows *Weather Handbook: A Guide to the Weather and Climate of Southern California.* Brewster Publications, Los Angeles, California, 1966

Ambrose, Kevin *Blizzards and Snowstorms of Washington, D.C.* Historical Enterprises, Merrifield, Virginia, 1993

Anderson, Bette Roda *Weather in the West* American West Publishing, Palo Alto, Calfornia, 1975

Andrew! Savagery From the Sea, Aug. 24, 1992 Sun-Sentinel Newspaper souvenir book, Fort Lauderdale, Florida, 1992

Ashcroft, Frances *Life at the Extremes* Harper Collins, 2001

Bahr, Robert *The Blizzard* (of Buffalo, New York January, 1977) Prentice-Hall Inc, Englewood Cliffs, New Jersey, 1980

Barnes, Jay *North Carolina's Hurricane History* University of North Carolina Press, Chapel Hill, North Carolina, 1995

Bee, Ooi Jin and Sien, Lin Sien *The Climate of West Malaysia and Singapore* Oxford University Press, Singapore, 1974

Blazyk, Stan *A Century of Galveston Weather* Eakin Press, Austin, Texas, 2000

Blij, H.J. de *Nature on the Rampage* Smithsonian Books, Washington D.C., 1994

Bluestein, Howard B. *Tornado Alley: Monster Storms of the Great Plains* Oxford University Press, New York, Oxford, 1999

Bomer, George W. *Texas Weather* University of Texas Press, Austin, Texas, 1995

Borisov, A.A. *Climates of the U.S.S.R.* Aldine Publishing Company, Chicago, Illinois, 1965

Brazell, J. H. *London Weather* Meteorological Office, London, Great Britain, 1968

Brinkman, Waltraud A.R. *Local Windstorm Hazard in the United States: A Research Assessment* Monograph NSF-RA-E-75-019 Institute for Behavioral Science, Univesity of Colorado, Boulder, Colorado, 1975

Brown, Dave *United States Weather: Memphis-Mid-South* United States Weather Corporation, Oklahoma City, Oklahoma, 1976

Bryant, Edward *Tsunami: The Underrated Hazard* Cambridege University Press, Cambridge, U.K., 2001

Burroughs, William I. et al *The Nature Company Guide to Weather* Weldon Owen Publishing, San Francisco, California, 1996

Caplovich, Judd *Blizzard! The Great Storm of '88* VeRo Publishing, Vernon, Connecticut, 1987

Carter, E.A., and Seaquist, V.G. *Extreme Weather History and Climate Atlas for Alabama* The Strode Publishers, Huntsville, Alabama, 1984

Chaboud, Rene *Weather: Drama of the Heavens* Harry N. Abrams, Inc., New York, NY, 1996

Changnon, Stanley A. and David *Record Winter Storms in Illinois 1977-1978* Illinois State Water Survey, Urbana, Illinois, 1978

Changnon, Stanley A. *The Great Flood of 1993* Westview Press, Boulder, Colorado, 1996

Changnon, Stanley A. *Thunderstorms Across the Nation; An Atlas of Storms, Hail, and Their Damages in the 20th Century.* Changnon Climatoligist, Mahomet, Illinois, 2001

Clark, Champ *Flood* Time-Life Books, Alexandria, Virginia, 1982

Climate Change Impacts on the United States: Potential Consequences of Climate Variability and Change Cambridge University Press, Cambridge, U.K., 2000

Climate Into the 21st Century World Meteorological Organization, edited by William Burroughs, Cambridge Univ. Press, Cambridge, U.K., 2003

Climates of the States Vols. 1-2 Third Edition Gale Research Company, Detroit, Michigan, 1985

Corliss, William R. *Tornadoes, Dark Days, Anomalous Precipitation and Related Weather Phenomena* The Sourcebook Project, Glen Arm, Maryland, 1983

Corliss, William R. *Lightning, Auroras, Nocturnal Lights, and Related Luminous Phenomena* The Sourcebook Project, Glen Arm, Maryland, 1983

Cox, Henry J. and Armington, John H. *The Weather and Climate of Chicago.* The Univ. of Chicago Press, Chicago, 1914

Davis, Lee *Natural Disasters* Checkmark Books, New York, 2002

Dirks, R.A. and Martner, B.E. *The Climate of Yellowstone and Grand Teton National Parks* National Parks Service occasional paper 6, U.S. Dept. of Interior, 1982

Doswell, Charles A. editor *Severe Convective Storms* American Meteorological Society Monograph Number 50, Boston, Massachusetts, 2001

Douglas, Paul *The Minnesota Weather Book* Voyageur Press, Stillwater, Minnesota, 1990

Dunn, Gordon E. and Miller I. Banner *Atlantic Hurricanes* Louisiana State University Press, 1964

Eagleman, Joe R. *Severe and Unusual Weather* Trimedia Publishing Company, Lenexa, Kansas, 1990

Eden, Philip *The Daily Telegraph Book of the Weather* Continuum Books, London, England, 2003

Eichenlaub, Val. L. *Weather and Climate of the Great Lakes Region* University of Notre Dame Press, Notre Dame, Indiana, 1979

Eichenlaub, Val L. et al *Climatic Atlas of Michigan* Univ. of Notre Dame Press, Notre Dame, Indiana, 1990

England, Gary *Oklahoma Weather* England and May, Oklahoma City, Oklahoma, 1975

Eubank, Mark *Utah Weather* Horizon Publishers, Bountiful, Utah, 1979

Faidley, Warren *Storm Chasers; In Pursuit of Untamed Skies* The Weather Channel Enterprises, Atlanta, Georgia, 1996

Felton, Ernest L. *California's Many Climates* Pacific Books, Palo Alto, California, 1965

Finkin, Arnold I. *Weather and Climate of the Selway-Bitterroot Wilderness* University Press of Idaho, Moscow, Idaho, 1983

Fitzpatrick, Patrick J. *Natural Disasters: Hurricanes* Contemporary World Issues, ABC-CLIO, Santa Barbara, California, 1999

Flora, Snowden D. *Hailstorms of the United States* University of Oklahoma Press, Norman, Oklahoma, 1956

Flora, Snowden D. *Tornadoes of the United States* University of Oklahoma Press, Norman, Oklahoma, 1953

Floyd, Candace *America's Great Disasters* Mallard Press, New York, New York, 1990

Fukui, E. ed. *The Climate of Japan* Kodansha, Tokyo, Japan, 1977

Garnier, B.J. *The Climate of New Zealand: A Geographical Survey* Edward Arnold Publishes, London, U.K., 1958

Gelber, Ben *The Pennsylvania Weather Book* Rutgers University Press, New Brunswick, New Jersey, 2002

Gelber, Ben *Pocono Weather* Uriel Publishing, Stroudsberg, Pennsylvania, 1992

Goddard, Dick *Dick Goddard's Weather Guide and Almanac for Northeast Ohio* Gray & Company Publishers, Cleveland, Ohio, 1998

Grazulis, Thomas P. *Tornado: Nature's Ultimate Windstorm* University of Oklahoma Press, Norman, Oklahoma, 2001

Grazulis, Thomas P. *Significant Tornadoes 1680-1991, Update 1992-1995.* Environmental Films, St. Johnsbury, Vermont, 1993

Hall, Donald et al *New England's Disastrous Weather* Yankee Books, 1990

Harding, Maria *Weather to Travel: A Guide to the World's Weather and What to Wear* Tomorrow Guides, London, U.K., 1996

Henry, James A. et al *The Climate and Weather of Florida* Pineapple Press, Sarasota, Florida, 1994

Henso, Robert *The Rough Guide to Weather* Rough Guides Ltd., London, U.K., 2002

Hidore, John J. *Workbook of Weather Maps* Wm. C. Brown Company Publishers, Dubuque, Iowa, 1971

Holford, Ingrid *British Weather Disasters* David & Charles, London, U.K., 1976

Holford, Ingrid *The Guinness Book of Weather Facts and Feats* Guinness Superlatives Ltd., Enfield, U.K., 1977

Houghton, Sakamoto, and Richard Gifford *Nevada's Weather and Climate.* Nevada Bureau of Mines and Geology, Special Publication 2, University of Nevada, Reno, 1975

Hoyt, William G. and Walter B. Langbein *Floods* Princeton University Press, Princeton, New Jersey, 1955

Hull, William H. *All Hell Broke Loose: Experiences of Young People During the Armistice Day 1940 Blizzard* Stanton Publication Services, Edina, Minnesota, 1985

Hulme, Mike and Elaine Barrow *Climates of the British Isles* Routledge Press, London, England, 1997

Hurricane Lashes Rhode Island, Aug. 31, 1954 Providence Journal Company souvenir book, Providence, Rhode Island, 1954

Jennings, Arthur H. *World's Greatest Observed Point Rainfalls* Monthly Weather Review, January 1950

Johnson, Howard L. and Claude E. Duchon *Atlas of Oklahoma Climate* University of Oklahoma Press, Norman, Oklahoma, 1995

Jolls, Tom *Western New York Weather Guide* Western New York Wares, Inc., Buffalo, New York, 1996

Keen, Richard *Skywatch East: A Weather Guide* Fulcrum Publishing, Golden, Colorado, 1992

Keen, Richard *Skywatch: The Western Weather Guide* Fulcrum Publishing, Golden, Colorado, 1987

Keen, Richard A. *Minnesota Weather* American and Geographic Publishing, Helena, Montana, 1992

Kennedy, James L. *How's The Weather? Find Your Outdoor Comfort Paradise in the U.S.* ASK Analytic Services, Inc., Bloomington, Minnesota, 2002

Kessler, Edwin (editor) *The Thunderstorm in Human Affairs* University of Oklahoma Press, Norman, Oklahoma, 1981

Kessler, Jacques and Andre Chambraud *Meteo de la France* Editions J.C. Lattes, Paris, France, 1990

Kimble, George H.T. *Our American Weather* McGraw-Hill Book Company, New York, 1955

Kocin, Paul J. and Louis W. Uccellini *Snowstorms Along the Northeastern Coast of the United States: 1955-1985* American Meteorological Society, Boston, Massachusetts, 1990

Lane, Frank W. *The Elements Rage* Chilton Books, Philadelphia, Pennsylvania, 1965

Leslie, Dr. Celia M. *A Climatic Guide to Asia* White Orchid Press, Bangkok, Thailand, 1985

Longshore, David *Encyclopedia of Hurricanes, Typhoons, and Cyclones* Checkmark Books, New York, 2000

Ludlum, David M. *The Vermont Weather Book* Vermont Historical Society, Montpelier, Vermont, 1985

Ludlum, David M. *The Country Journal New England Weather Book* Houghton Mifflin Company, Boston, Massachusetts, 1976

Ludlum, David M. *The Nantucket Weather Book* Historic Nantucket Press, Nantucket, Massachusetts, 1976

Ludlum, David M. *The New Jersey Weather Book* Rutgers University Press, New Brunswick, New Jersey, 1983

Ludlum, David M. *Early American Hurricanes 1492-1870* American Meteorological Association, 1965

Ludlum, David M. *Early American Winters 1604-1820* American Meteorological Association, 1966

Ludlum, David M. *Early American Winters 1821-1870* American Meteorological Association, 1968

Ludlum, David M. *Early American Tornadoes 1586-1870* American Meteorological Association, 1970

Ludlum, David M. *The American Weather Book* American Meteorological Society, Boston, Massachusetts, 1982

Ludlum, David M. *The Audubon Society Field Guide to North American Weather* Alfred A. Knopf, New York, 1991

Ludlum, David M. *Weather Record Book: United States and Canada* Weatherwise Inc., Princeton, New Jersey, 1971

Lynch, John *The Weather* Firefly Books, Toronto, Ontario, Canada, 2002

Lyons, Walter A. *The Handy Weather Answer Book* Accord Publishing Ltd., Boulder, Colorado, 1997,

Martner, Brooks E. *Wyoming Climate Atlas* University of Nebraska Press, Lincoln, Nebraska, 1986

Milne, Antony *Flood Shock* Alan Sutton Publishers, Gloucester, England, 1986

Monmonier, Mark *Cartographies of Danger: Mapping Hazards in America* The University of Chicago Press, Chicago, Illinois, 1997

Moore, Willis L. *Descriptive Meteorology* D. Appleton and Company, New York, New York, 1910

Moran, Joseph M. and Hopkins, Edward J. *Wisconsin's Weather and Climate* University of Wisconsin Press, Madison, Wisconsin, 2002

Nalivkin, D.V. *Hurricanes, Storms, and Tornadoes: Geographic Characterisitcs and Geological Activity* Amerind Publishing Co., New Delhi, India, 1982

Nese, Jon and Glenn Schwartz *The Philadelphia Area Weather Book Including Delaware, the Poconos, and the Jersey Shore* Temple University Press, Philadelphia, Pennsylvania, 2002

Nelson, Mike *The Colorado Weather Book* Westcliffe Publishers, Englewood, Colorado, 1999

New England Hurricane: A Factual, Pictorial Record Hale, Cushman & Flint, Boston, Massachusetts, 1938

Page, John L. *Climate of Illinois; Bulletin 532* University of Illinois, Agricultural Experimentation Center, Urbana, Illinois, 1949

Pant, G.B. and K. Rupa Kumar *Climates of South Asia* John Wiley & Sons, 1997

Paulhus, J.L.H. *Indian Ocean and Taiwan Rainfalls Set New Records* Monthly Weather Review, May 1965

Pearce, E.A., and C. G. Smith *World Weather Guide* Helicon Publishing, New York, 1998, Times Books, 1990

Phillips, David *The Day Niagara Ran Dry: Canadian Weather Facts and Trivia* Keyporter Books, Toronto, Canada, 1993

Phillips, David *The Climates of Canada* Environment Canada, Montreal, Canada, 1991

Phillips, David et al *Blame it on the Weather* Portable Press, San Diego, California, 2002

Pope, Dan and Brough, Clayton *Utah's Weather and Climate* Publisher's Press, Salt Lake City, Utah, 1996

Powers, Edward and James Witt *Travelling Weatherwise in the U.S.A.* Dodd, Mead & Company, 1973

Preston-Whyte, R.E. and P.D.Tyson *The Atmosphere and Weather of Southern Africa* Oxford University Press, U.K., 1988

Rakov, Vladimir A. and Uman, Martin A. *Lightning: Physics and Effects*, Cambridge Univ. Press, Cambridge, Massachusetts, 2003

Reifsnyder, William F. *Weathering the Wilderness: The Sierra Club Guide to Practical Meteorology* Sierra Club Books, San Francisco, California, 1980

Robinson, Andrew *Earth Shock: Hurricanes, Volcanoes, Earthquakes, Tornadoes, and other Forces of Nature* Thames & Hudson, London, U.K. 2002

Rumney, George R. *Climatology and the World's Climates* The Macmillan Company, New York, 1968

Sanderson, Marie *Prevailing Trade Winds: Weather and Climate in Hawaii* University of Hawaii Press, Honolulu, Hawaii, 1993

Savadove, Larry and Margaret Thomas Buchholz *Great Storms of the Jersey Shore* Down the Shore Publishing/The Sandpiper Inc., Harvey Cedars, New Jersey, 1993

Schmidlin, Thomas W. and Jeanne A. *Thunder in the Heartland: A Chronicle of Outstanding Weather Events in Ohio* The Kent State University Press, Kent, Ohio, 1996

Sellers, William D. and Richard H. Hill *Arizona Climate 1931-1972* The University of Arizona Press, Tucson, Arizona, 1974

Shaw, Sir Napier *The Drama of Weather* Macmillan Co., New York, 1933

Sheets, Dr. Bob and Jack Williams *Hurricane Watch* Vintage Books, New York, New York, 2001

Stevens, William K. *The Change in the Weather: People, Weather, and the Science of Climate* Random House, New York, 1999

Stombach, J.J. *Climate of Connecticut* State Geological and Natural History Survey of Connecticut Bulletin, 1965

Sturman, Andrew and Nigel Tapper *The Weather and Climate of Australia and New Zealand* Oxford University Press, Melbourne, Australia, 1996

Svarney, Thomas E. and Patricia Barnes *Skies of Fury* Simon & Schuster, New York, New York, 1999

Tannehill, Ivan Ray *Weather Around the World* Princeton University Press, Princeton, New Jersey, 1943

Tannehill, Ivan Ray *Hurricane* Princeton University Press, Princeton, New Jersey, 1950 (9 editions 1938-1950)

Taylor, George H. and Raymond Hatton *The Oregon Weather Book: A State of Extremes* Oregon State University Press, Corvallis, Oregon, 1999

Taylor, George H. and Chris Hannan *The Climate of Oregon: From Rain Forest to Desert* Oregon State University Press, Corvallis, Oregon, 1999

Thomas, Morley K. *Climatological Atlas of Canada* National Research Council, Ottawa, Canada, 1953

Tirado, Robert *Robert Torado's Weatherbook* Peregrine Press, Old Saybrook, Connecticut, 1981

Trewartha, Glenn T. *The Earth's Problem Climates* The University of Wisconsin Press, Madison, Wisconsin, 1961

Trobec, Jay *State of Extremes; Guide to the Wild Weather of South Dakota* Where's Where Publishing, Sioux Falls, South Dakota, 1995

Tufty, Barbara *1001 Questions Answered About Hurricanes, Tornadoes, and other Natural Air Disasters* Dover Publications, New York, 1987

The USA TODAY Weather Almanac Vintage Books, New York, 1995

Verkaik, Jerrine and Arjen *Under the Whirlwind: Everything You Need to Know About Tornadoes* Whirlwind Books, Elmwood, Ontario, Canada, 1997

Viemeister, Peter E. *The Lightning Book* The MIT Press, Cambridge, Massachusetts, 1972

Visher, Stephen S. *Climatic Atlas of the United States* Harvard University Press, Cambridge, 1954

Visher, Stephen S. *Climate of Indiana* Indiana University Press, Bloomington, Indiana, 1944

Wagner, Ronald L. and Bill Adler Jr. *The Weather Sourcebook* The Globe Pequot Press, Guilford, Connecticut, 1997

Watson, Benjamin A. *The Old Farmer's Almanac Book of Weather & Natural Disasters* Random House, New York, 1993

Weather America Grey House Publishers, Lakeville, Connecticut, 2001

Weather and Climate of Nebraska. Nebraskaland Magazine, Lincoln, Nebraska, 1996

Werstein, Irving *The Blizzard of '88* Thomas & Crowell Company, New York, 1960

Wheeler, Dennis and Julian Mayes *Regional Climates of the British Isles* Routledge Press, London, England, 1997

Williams, Jack *USA Today: The Weather Book: An Easy-to Understand Guide to USA's Weather* Vintage Books, New York, 1997

Williams, John M. and Iver W. Duedall *Florida Hurricanes and Tropical Storms* University Press of Florida, 1997, 2002

World Weather Records; Collected from Official Sources, Volume 79 Smithsonian Institution, Washington D.C., 1927

World Weather Survey, Vols. 1-15 ed. by H. E. Landsberg Elsevier Science Publishers B. V., Amsterdam, Holland, 1985

Yihui, Ding *Monsoons Over China* Kluwer Academic Publishers, Boston, Massachusetts, 1994

Zielinski, Gregory A. and Barry D Keim *New England Weather, New England Climate.* Univ. of New Hampshire Press, Hanover, New Hampshire, 2003

GOVERNMENT PUBLICATIONS

Climate and Man: 1941 Yearbook of Agriculture U.S. Dept. of Agriculture, Washington D.C., 1941

Climatological Data of the United States U.S. National Oceanographic and Atmospheric Administration (NOAA-NCDC), 1890–present

Climatology of the United States, 1906 U.S. Dept. of Agriculture, Weather Bureau Bulletin Q, Washington, D.C. 1906

Cold Waves and Frosts in the United States U.S. Dept. of Agriculture, Weather Bureau Bulletin F, Washington, D.C., 1906

Comparative Climate Data NOAA, Asheville, North Carolina, 1999

Flood, Katherine L. and Dr. Paul F. Krause *Weather and Climate Extremes*. U.S. Army Corps of Engineers TEC-0099, Alexandria, Virginia, 1997

Hughes, Patrick *American Weather Stories* U.S. Dept. of Commerce, Washington D.C., 1976

McAdie, Alexander *Climatology of California* U.S. Dept. of Agriculture, Weather Bureau, Bulletin L, Washington D.C., 1903

Monthly Weather Review Weather Bureau, Washington, D.C., 1872–present

Tables for Temperature, Relative Humidity, and Precipitation for the World, Parts 1-6 Meteorological Office, London, U.K., 1967

The Weather Almanac; Editions 1-11 Gale Research Company, Detroit, Michigan, 1980-2003

Thunderstorm Morphology and Dynamics Vol.2 U.S. Dept. of Commerce, Washington, D.C. 1982

Schmidli, Robert J. *Maximum 24-hour Precipitation in the Untited States*. NOAA technical memorandum NWS-WR-28, Feb., 1981 and Technical Paper, Washington. No.16, 1952 by A.H. Jennings

Shands, A.L. and Ammerman, D. *Maximum Recorded United States Point Rainfall for 5 Minutes to 24 Hours at 207 First-Order Stations* U.S. Dept. of Commerce, Weather Bureau Technical Paper No.2, Washington D.C., 1947

Storm Rainfall Department of the Army, Corps of Engineers, Washington, D.C. 1887-1955

Summaries of Climatological Data by Sections, 1912 Vols. 1-2 U.S. Dept. of Agriculture, Weather Bureau Bulletin W, Washington, D.C.

Summaries of Climatological Data by Sections, 1926 Vols. 1-3 U.S. Dept. of Agriculture, Weather Bureau Bulletin W, Washington, D.C.

World Weather Records Vols. 1–6 U.S.. Dept. of Commerce, Weather Bureau, Washington D.C., 1965

INDEX

Conversion Tables

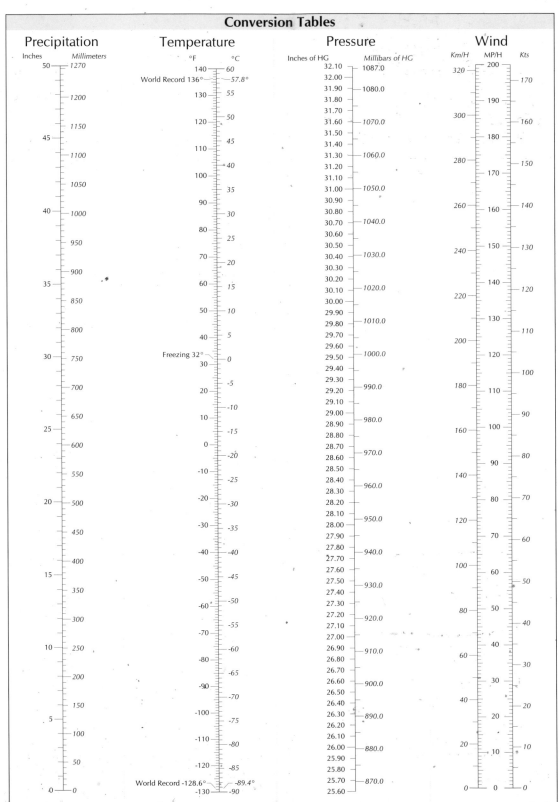

Precipitation

Inches	Millimeters
50	1270
	1200
	1150
45	1100
	1050
40	1000
	950
	900
35	850
	800
30	750
	700
	650
25	600
	550
20	500
	450
	400
15	350
	300
10	250
	200
	150
5	100
	50
0	0

Temperature

°F	°C
140	60
World Record 136°	57.8°
130	55
120	50
110	45
100	40
90	35
	30
80	25
70	20
60	15
50	10
40	5
Freezing 32° 30	0
20	-5
10	-10
0	-15
	-20
-10	-25
-20	-30
-30	-35
-40	-40
-50	-45
-60	-50
-70	-55
-80	-60
	-65
-90	-70
-100	-75
-110	-80
-120	-85
World Record -128.6°	-89.4°
-130	-90

Pressure

Inches of HG	Millibars of HG
32.10	1087.0
32.00	
31.90	1080.0
31.80	
31.70	
31.60	1070.0
31.50	
31.40	
31.30	1060.0
31.20	
31.10	
31.00	1050.0
30.90	
30.80	
30.70	1040.0
30.60	
30.50	
30.40	1030.0
30.30	
30.20	
30.10	1020.0
30.00	
29.90	
29.80	1010.0
29.70	
29.60	
29.50	1000.0
29.40	
29.30	
29.20	990.0
29.10	
29.00	
28.90	980.0
28.80	
28.70	
28.60	970.0
28.50	
28.40	960.0
28.30	
28.20	
28.10	
28.00	950.0
27.90	
27.80	940.0
27.70	
27.60	
27.50	930.0
27.40	
27.30	
27.20	920.0
27.10	
27.00	
26.90	910.0
26.80	
26.70	
26.60	900.0
26.50	
26.40	
26.30	890.0
26.20	
26.10	
26.00	880.0
25.90	
25.80	
25.70	870.0
25.60	

Wind

Km/H	MP/H	Kts
	200	
320		170
	190	
300		160
	180	
280		150
	170	
260	160	140
240	150	130
	140	120
220		
	130	110
200	120	
		100
180	110	
		90
160	100	
	90	80
140		
	80	70
120		
	70	60
100	60	
		50
80	50	
		40
	40	
60		30
	30	
40		20
	20	
20	10	10
0	0	0